Nuclear Non-proliferation and Arms Control Verification

Irmgard Niemeyer • Mona Dreicer • Gotthard Stein
Editors

Nuclear Non-proliferation and Arms Control Verification

Innovative Systems Concepts

Springer

Editors
Irmgard Niemeyer
IEK-6: Nuclear Waste Management
and Reactor Safety
Forschungszentrum Jülich GmbH
Jülich, Germany

Mona Dreicer
Center for Global Security Research
Lawrence Livermore National Laboratory
Livermore, CA, USA

Gotthard Stein
Consultant
Bonn, Germany

ISBN 978-3-030-29536-3 ISBN 978-3-030-29537-0 (eBook)
https://doi.org/10.1007/978-3-030-29537-0

This Springer imprint is published by the registered company Springer Nature Switzerland AG.
The registered company address is: Gewerbestrasse 11, 6330 Cham, Switzerland

Foreword

There is perhaps no phrase in nuclear policy more famous than "Trust, but verify." President Reagan's utterance is an ever present mantra in nuclear arms control policy and rightly so. If we are to achieve the long-standing goal of nuclear disarmament, then we must create and implement the most robust and effective verification regimes in history. Ambassador Paul Nitze defined effective verification as follows: "if the other side moves beyond the limits of the treaty in any militarily significant way, we would be able to detect such violations in time to respond effectively and thereby deny the other side the benefit of the violation." That concept is the foundation of verifying compliance with modern arms control and nonproliferation agreements.

Over the decades, verification efforts were focused on specific treaty requirements and not necessarily designed with future agreements in mind. Faced with the continuing threats posed by nuclear weapons—and all weapons of mass destruction—it is important to take stock, evaluate how existing technologies and methodologies can work together, as well as identifying new opportunities for advancements in the field.

This book brings together the views of a diverse group of international experts from national laboratories, universities, and non-governmental organizations, and strives to link the nonproliferation and arms control communities with the broader international security world in a way that will foster cross-fertilization of ideas.

Even with a clear mandate for what verification regimes should do, we have yet to really harness the tools and technologies of information revolution, which can help us to make those regimes stronger.

The seed of an idea was planted in my mind in Geneva during the New START negotiations. As we considered verification mechanisms for the Treaty, it occurred to me that, by and large, we were still thinking about verification through the lens of the 1970s. The advancements in technology since then have been nothing short of revolutionary, but it wasn't quite clear how to incorporate these advancements into an effective verification regime.

What is clear is that innovation in this area cannot come fast enough. Having new tools at the ready for any future negotiations will allow us to further implement our arms control and nonproliferation obligations. Our new reality is a smaller, increasingly networked world where the average citizen connects to other citizens in cyberspace hundreds of times each day. These people exchange and share ideas on a wide variety of topics: why not put this vast problem-solving entity to good use? Societal verification can take place on a scale that moves from active participation, such as public reporting and crowd-sourced mapping and analysis, to passive participation, such as ubiquitous sensing or data mining and analytics. There is already some very interesting work going on in social media aggregation and open-source map and satellite imagery analysis, as some of the book's chapters point out.

The number of useful tools for a WMD inspector grows by the minute. Smartphone and tablet apps could be created for the express purpose of aiding in the verification and monitoring process. For example, by having all safeguards and verification sensors in an inspected facility wirelessly connected through the cloud to the inspector's tablet, he or she could note anomalies and flag specific items for closer inspections, as well as compare readings in real time and interpret them in context. In the 1990s, U.S. weapons inspectors in Russia had to be able to cross country ski to do their jobs. They had to ski around the perimeters of facilities searching for anomalies. With the kinds of tablet apps I mentioned, perhaps we could make the cardio optional.

As we think through new ways to advance verification and monitoring, we should be aware that there may be difficult issues to confront. There are technical, legal, political, military, and diplomatic barriers ahead that would need to be overcome—never easy. The issue of privacy and how personal data are used have rightfully come front and center. We cannot assume that information will always be so readily available. There are also concerns related to reliability, so looking at how to make data immutable will be another important challenge.

In the pages ahead, in what is perhaps the most dynamic collection of verification thinking I have ever seen, you will read about how some of the best and brightest minds (from both nuclear and nonnuclear weapons states) view both our achievements to date and the challenges ahead.

By taking stock of lessons learned, investigating ways to harness the power of big data, and exploring the technical, legal, political, and other barriers to innovation, the authors outline exciting new approaches to how we can verify and monitor matters of arms control and nonproliferation. I hope that they continue to press the boundaries in this field and I hope you, the reader, will do so as well. We possessed the skill and intellect to invent weapons of mass destruction, so I have no doubt we possess the skill and intellect to eliminate them.

Former Under Secretary for Arms Control and International Security
United States Department of State
December 2018

Rose Gottemoeller

Acknowledgements

The editors wish to express their appreciation to the authors and the organizations that supported their contributions to this book, as well as the German government for providing partial support for the project.

The European Safeguards Research and Development Association (ESARDA) and the Institute of Nuclear Materials Management (INMM) (particularly the ESARDA Working Group on Verification Technologies and Methodologies (VTM) and the INMM Nonproliferation and Arms Control Technical Division) for hosting the presentations, workshops, and panel discussions that contributed to the development of the concepts and information presented here.

The project on which this book is based was funded by the German Federal Ministry for Economic Affairs and Energy under the funding numbers 02W6263 and 02W6279 via the Project Management Agency Karlsruhe (PTKA), Karlsruhe Institute of Technology (KIT). However, the responsibility for the contents of this book lies solely with the authors. The views and opinions of authors expressed herein do not necessarily state or reflect those of the United States government or Lawrence Livermore National Security, LLC. Part of this work was performed under the auspices of the US Department of Energy by Lawrence Livermore National Laboratory under Contract DE-AC52-07NA27344. LLNL-BOOK-698308.

Contents

Contributors

Keir Allen King's College London, London, UK

Jürgen Altmann Technische Universität Dortmund, Dortmund, Germany

Yetunde Aregbe EC Directorate General Joint Research Centre, Directorate G - Nuclear Safety and Security, Geel, Belgium

Rudolf Avenhaus Universität der Bundeswehr München, Neubiberg, Germany

James M. Brase Lawrence Livermore National Laboratory, Livermore, CA, USA

Morton Canty Consultant, Jülich, Germany

John Carlson Vienna Center for Disarmament and Non-Proliferation, Vienna, Austria

Cliff Chen Lawrence Livermore National Laboratory, Livermore, CA, USA

Giacomo G. M. Cojazzi EC Directorate General Joint Research Centre, Directorate G - Nuclear Safety and Security, Ispra, VA, Italy

Jill N. Cooley Consolidated Nuclear Security, Y-12 National Security Complex, Oak Ridge, TN, USA

Sharon DeLand Sandia National Laboratories, Albuquerque, NM, USA

Mona Dreicer Lawrence Livermore National Laboratory, Livermore, CA, USA

Thomas Edmunds Lawrence Livermore National Laboratory, Livermore, CA, USA

Zoe N. Gastelum Sandia National Laboratories, Albuquerque, NM, USA

Alexander Glaser Princeton University, Princeton, NJ, USA

Malte Göttsche RWTH Aachen University, Aachen, Germany

Matthias Grüne Fraunhofer INT, Euskirchen, Germany

Erwin Häckel University of Konstanz, Konstanz, Germany

Magnus Hedberg EC Directorate General Joint Research Centre, Directorate G - Nuclear Safety and Security, Karlsruhe, Germany

IAEA, Vienna, Austria

Willem A. Janssens EC Directorate General Joint Research Centre, Directorate G - Nuclear Safety and Security, Ispra, Italy

Marcus John Fraunhofer INT, Euskirchen, Germany

Yan Jie China Academy of Engineering Physics, Mianyang, China

Rainer Jungwirth EC Directorate General Joint Research Centre, Directorate G - Nuclear Safety and Security, Ispra, VA, Italy

David Keir Cloverdale Scientific Limited, Newbury, UK

Lance K. Kim EC Directorate General Joint Research Centre, Directorate G - Nuclear Safety and Security, Ispra, VA, Italy

Gerald Kirchner Universität Hamburg, Hamburg, Germany

Thomas Krieger Forschungszentrum Jülich GmbH, Jülich, Germany

Nicholas Kyriakopoulos The George Washington University, Washington, DC, USA

Clemens Listner Consultant, Berlin, Germany

Klaus Mayer EC Directorate General Joint Research Centre, Directorate G - Nuclear Safety and Security, Karlsruhe, Germany

Eric G. McKinzie Lawrence Livermore National Laboratory, Livermore, CA, USA

Michelle Nalabandian Nuclear Threat Initiative, Washington, DC, USA

Samantha Neakrase Nuclear Threat Initiative, Washington, DC, USA

Ulrik Neupert Fraunhofer INT, Euskirchen, Germany

Irmgard Niemeyer Forschungszentrum Jülich GmbH, Jülich, Germany

Frank V. Pabian EC Directorate General Joint Research Centre, Directorate G - Nuclear Safety and Security, Ispra, VA, Italy

Andreas Persbo VERTIC, London, UK

Jeffrey Preston Consolidated Nuclear Security, Y-12 National Security Complex, Oak Ridge, TN, USA

Guido Renda EC Directorate General Joint Research Centre, Directorate G - Nuclear Safety and Security, Ispra, VA, Italy

Michel Richard Division Global Security and International Regulations, Fondation pour la Recherche Stratégique (FRS), Paris, France

Joshua Rutkowski Forschungszentrum Jülich GmbH, Jülich, Germany

Joachim Schulze Consultant, Bad Neuenahr, Germany

Filippo Sevini EC Directorate General Joint Research Centre, Directorate G - Nuclear Safety and Security, Ispra, VA, Italy

Gotthard Stein Consultant, Bonn, Germany

Dirk Thorleuchter Fraunhofer INT, Euskirchen, Germany

Cristina Versino EC Directorate General Joint Research Centre, Directorate G - Nuclear Safety and Security, Ispra, VA, Italy

Maria Wallenius EC Directorate General Joint Research Centre, Directorate G - Nuclear Safety and Security, Karlsruhe, Germany

Martin Williamson Consolidated Nuclear Security, Y-12 National Security Complex, Oak Ridge, TN, USA

Erik Wolfart EC Directorate General Joint Research Centre, Directorate G - Nuclear Safety and Security, Ispra, VA, Italy

Jie Yan China Academy of Engineering Physics, Mianyang, China

John J. Zucca Lawrence Livermore National Laboratory, Livermore, CA, USA

Systems Concepts: Structuring a New Verification Approach

Mona Dreicer, Irmgard Niemeyer, and Gotthard Stein

Abstract Further reductions in nuclear stockpiles, or ultimately, the elimination of nuclear weapons is not likely to occur absent the confidence that they are not necessary for national security. After 50 years of verifying the peaceful uses of the nuclear fuel cycle, the International Atomic Energy Agency (IAEA) has shown that confidence requires a coherent and comprehensive picture of a state's nuclear-related activities in addition to the evaluation of treaty compliance. Confidence in a world with low numbers (or zero) nuclear weapons will be difficult to achieve with only an incremental treaty-by-treaty approach. To do this one must consider the broad security context including complex domestic infrastructure, a range of existing international commitments (political and legal), and technical monitoring and verification capabilities. State-level transparency or verification may be achieved by piecing together, in a well-structured way, a very broad range of information related to nuclear materials and capabilities from: declared, undeclared, and international technical monitoring data; National Technical Means; open sources; state and international trade data; and diplomatic communications. Development of a systems concept that can drive the understanding of the interactions will be crucial to achieving confidence. It would make use of decades of experience verifying bilateral and multilateral arms control agreements and takes into account varying levels of information and risk. Such a systems concept could the drive the development and implementation of new verification mechanisms for future agreements.

M. Dreicer (✉)
Lawrence Livermore National Laboratory, Livermore, CA, USA
e-mail: dreicer1@llnl.gov

I. Niemeyer
Forschungszentrum Jülich GmbH, Jülich, Germany
e-mail: i.niemeyer@fz-juelich.de

G. Stein
Consultant, Bonn, Germany

I. Niemeyer et al. (eds.), *Nuclear Non-proliferation and Arms Control Verification*,
https://doi.org/10.1007/978-3-030-29537-0_1

1 Introduction

States seek security in two ways—cooperation and competition. To deal with the threat of the use of nuclear weapons, the vast majority of states, whether or not they possess nuclear weapons, cooperate to try to minimize the threat of nuclear war. Cooperation has resulted in a collection of treaties and agreements that together establish a regime that supports international security. Over the last five decades, these agreements managed to address most of the urgent security needs regarding the spread of nuclear weapons, and have been the foundation of nuclear peace. States have been sufficiently open and transparent, to enable treaty partners to verify compliance with agreements thereby achieving the level of confidence and trust needed to meet national security requirements.

Although this security regime has resulted in nuclear peace for some time, it now seems hard-pressed to maintain a lasting or universally satisfactory settlement. States with nuclear weapons hedge their bets against the breakdown of cooperation (and peace) by maintaining arsenals of nuclear weapons and extending security assurances to non-nuclear allies. The significant support for the Treaty for the Prohibition of Nuclear Weapons (TPNW) highlights the continuing disparity between nuclear weapon 'haves' and 'have nots'. This state of affairs places ever more strain on the nonproliferation regime and existing arms control agreements. Should those foundations falter, the resulting competition—a potential nuclear arms race—could cause a dramatic reversal in the level of nuclear security across the globe.

One particular challenge to state-level confidence is that many aspects of the nuclear weapons complex have no transparency or verification regimes in place due to the need to protect classified national security information. These gaps increase uncertainty and mistrust, leading to competition rather than cooperation. Competition can undermine the security gains that have already been made. Until recently, competition between states with nuclear weapons was not as overt or as urgent as it was during the Cold War, but the tides are turning. States are reconsidering their national security policy as they seek to ensure that they can deter use of nuclear weapons. Existing competition is further complicated by new evolving technologies (e.g. additive manufacturing, cyber) that can provide broader strategic offensive and defensive capabilities. A hallmark of this competition is secrecy, which is used to maintain a competitive edge. Secrecy, in turn, exacerbates uncertainty and drives further competition.

If a lasting nuclear peace is to be achieved, verification mechanisms are needed to promote confidence in treaty compliance and transparency between competitors. Such trust does not come about naturally. The TPNW, currently open for signature, does not include mechanisms to achieve such confidence. Suitable verification mechanisms will be needed to provide the transparency and/or verification at a level of secrecy needed to protect national security. A major step towards achieving more universal confidence lies in identifying exactly where verification gaps exist and clearly defining the most useful approaches to filling them.

2 A Systematic Approach

A system can be defined as "a set of interrelated components working together toward some common objective." When approaching a complex system, involving many diverse elements and comprised of intricate interconnections and influences, engineers looks towards devising a systematic approach to understand it in totality—not piece by piece. A systems approach is used to design a methodology that can take into account internal and external factors, clearly identify stakeholder needs and help set priorities for a particular system. The field of systems engineering has been devised to focus on the system as a whole and its primary purpose is to guide system design by emphasizing definition of needs and functionality early in the process. It uses traceability, verification, and validation to ensure that the system meets desired objectives by providing a common, well-understood system analysis and design process to organize decisions. This helps drive the development of the concept that ultimately functions as desired [1].

If an overarching security goal is to reduce reliance on large nuclear weapons stockpiles, then an security framework based on cooperation and further transparency must be devised to build and maintain confidence at the state-level as well as on a global scale. A systems concept is needed to support integrated analysis of varied information, encompassing state-declared information, international technical monitoring data, information obtained by national technical means, open source and trade information, and diplomatic consultations. With the ever-greater capabilities to record and share vast amounts of data, mechanisms to systematize and integrate a wide range of information is increasingly being explored both by governments, the private sector and citizens. These capabilities can be applied to verification of existing treaty-specific commitments, but might also help to highlight places in the system where transparency would yield the greatest benefit.

Systematic approaches are not new to verification but have generally been used to develop treaty-specific regimes and have not considered the state or region as a whole. The relatively recent study and implementation of a state-level approach by the International Atomic Energy Agency (IAEA), inspired the nonproliferation and arms control technical community to explore ways to advance this concept for effective nuclear arms control verification. Experts have begun to apply a systems approach to formulate an objective, standardized, transparent, and reproducible framework that can be well-documented, so that stakeholders can confidently use it to identify and address gaps in arms control verification capabilities and approaches. Such a framework could also drive the development of new research & development directions.

3 Nuclear Weapons Complex

Each state with nuclear weapons maintains an institutional infrastructure to produce and maintain its stockpile. However the details of the complicated industrial processes (and associated safety and security) of materials production, component manufacture, assembly of weapons, and types of delivery systems are different for each state's enterprise. The details are not openly available, so it can be assumed that existing asymmetries will be a complicating factor in negotiating mutually-acceptable verification measures.

Figure 1 is a schematic that illustrates a generalized nuclear weapons complex, highlighting the differences between uranium and plutonium lifecycles. This diverse interconnected system encompasses both civilian and military material production facilities and stockpiles. The degree of separation between the civilian and military sectors is highly variable in different states. The greater the overlap between sectors, the more complicated it will be to achieve transparency. IAEA safeguards is focused on the civilian sector and has established processes to protect propriety information, but these methods are not designed to stop the spread of classified weapons information from military facilities.

The scale of the facilities in a state's nuclear enterprise also varies. Some facilities are large, complex and energy intensive, while other activities (e.g. weaponization research and development) can be accomplished with fewer resources and a smaller footprint. With the rapid advances in additive manufacturing technology, the size and types of facilities that are needed continue to change. Today, the nonproliferation regime (IAEA safeguards) monitors materials in the front end of the fuel cycle (illustrated at the top of Fig. 1), and arms control treaties and agreements, mainly between US-Russia, have focused on the back end (illustrated at the bottom of the figure), with transparent destruction of delivery systems (e.g. START) and disposition of excess military materials (e.g. HEU transparency, Plutonium Management and Disposition Agreement). However, verification measures to link refurbishment, dismantlement, and disposition of weapons components and materials have not been practically implemented to date.

Effective systems-wide confidence must address the gaps between the front end (e.g. material production, weapons production stages) and the back end (e.g. weapons dismantlement and disposition of material from the weapons) accommodating the requirements for the protection of proprietary information in civilian facilities and the secrecy in military facilities.

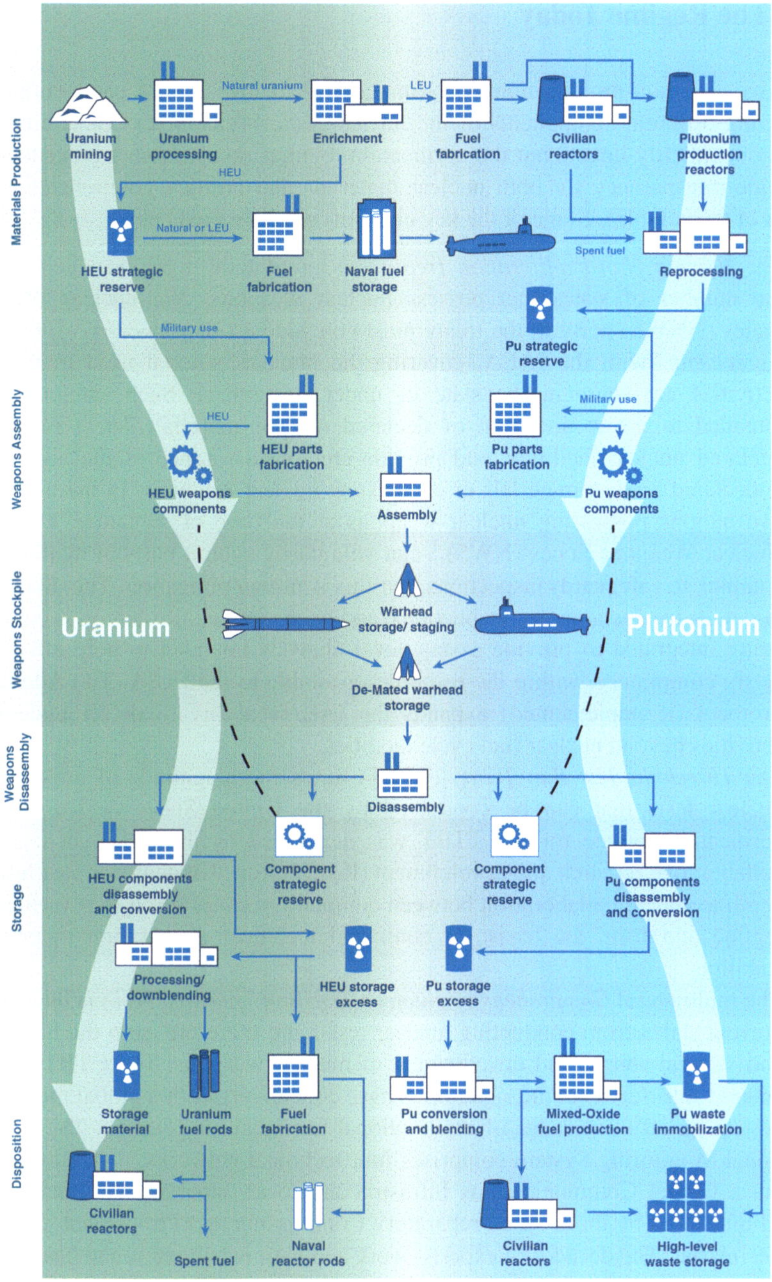

Fig. 1 Nuclear weapons complex and lifecycle

4 The Regime Today

The existing non-proliferation and arms control treaties form a patchwork, each covering different components of the nuclear weapons complex. Generally, they are not explicitly linked, but the verification regimes do establish a precedent and provide transparency for both nuclear materials and military components across parts of the complex. Some of the key elements are presented below.

- The *Nuclear Non-Proliferation Treaty* is a multilateral treaty designed to limit the number of states that possess nuclear weapons. Non-nuclear Weapons States (NNWS) party to the treaty must conclude a *Comprehensive Safeguards Agreement* with the IAEA, covering the nuclear material used in peaceful activities anywhere in the state or under its control. Such safeguards are intended to detect diversion of declared nuclear material, detect misuse of declared nuclear facilities, and provide credible assurance of the absence of undeclared nuclear materials or clandestine nuclear activities. In the treaty, the five nations possessing nuclear weapons at the time of signature, known as Nuclear Weapons States (NWS),[1] can voluntarily submit parts of their civilian complex to safeguards inspections, but this is not an obligation. The diverse set of data from declarations, inspections, and open source information is currently being integrated to provide a state-level analysis intended to help effectively verify compliance, within the resources available to the IAEA. The Additional Protocol (if implemented) expands the IAEA's ability to detect undeclared activities beyond nuclear fuel cycle facilities.
- The *Threshold Test Ban Treaty* (TTBT) limits the magnitude of underground nuclear tests that can be conducted by the United States or Russia. The verification regime for the TTBT was developed by a US–Russia technical collaborative research and development (R&D) effort that demonstrated how in-depth technical collaboration between competitors could result in a verification regime and form the basis for continued technical cooperation to promote stability.
- The multilateral *Comprehensive Nuclear-Test-Ban Treaty* (CTBT) is intended to prevent states from conducting nuclear tests, and therefore, stop the horizontal and vertical spread and development of nuclear weapons. The CTBT has not entered-into-force but the signatories have collectively built a global monitoring system based on decades of multinational technical cooperation. The International Monitoring System comprises four technical networks, transmitting data via a Global Communications Infrastructure to an International Data Center, administered by the CTBT Preparatory Commission and Provisional Technical Secretariat. The decades of experts' work and national commitment has resulted in an effective monitoring system that provides all Member States with access to monitoring and verification information.

[1]Nuclear Weapons States, or the Permanent Five, comprise US, UK, France, China, and Russia.

- Negotiations for a *Fissile Material Cut-Off Treaty* (FMCT) began in 1995 and were intended to limit the size of fissile material stockpiles—to ultimately limit global nuclear weapons stockpiles. Consensus on a technical verification regime was never achieved and negotiations are stalled. However, an FMCT-like treaty would likely employ many of the same technical verification techniques used by IAEA safeguards in NNWS and apply them in NWS and states not party to the NPT. Participation of NWS and NNWS in a common verification regime would necessitate protection against the transfer of nuclear weapons-relevant information and employment of strict mechanisms to deny any exposure of classified and national security information.
- *Strategic trade control regimes* work to control the spread of nuclear dual-use technology and associated knowledge ("deemed exports"). The evolving set of international agreements, treaties, UN Security Council Resolutions, national laws, and sanctions provide valuable information can be integrated to aid the understanding of a state's nuclear weapons intentions. Some examples are the Nuclear Suppliers Group, Missile Technology Control Regime, Wassenaar Arrangement, and UN Security Council Resolution 1540.
- The 1991 US–Russia *Strategic Arms Reduction Treaty* (START) resulted in the decrease of existing nuclear weapons arsenals. The follow-on, New START, continues to reduce and limit the number of strategic nuclear weapons and delivery systems deployed however, it will expire in February 2021. START and New START address defined treaty-accountable items in the United States and Russia, defined by mutually-agreed counting conventions. As a result an accurate count of warheads or types of warheads in each country is not known. The verification mechanisms are routine declarations and on-site inspections, largely leveraging low-tech capabilities. START implementation informed the development of the New START verification regime and the changes are the result of the state of the security environment at the time of negotiation. Even though treaty verification does not provide the specific number of warheads or types of warheads in each county, if this treaty is not expended or replaced, important transparency and consultative mechanisms will be lost. At the time of publication, it is not clear whether New START will be extended or new treaty negotiations initiated.
- Russia and the US have made some limited inroads towards reducing military fissile material stocks. The verification provisions under the 1993 US–Russia *Highly Enriched Uranium (HEU) Purchase Agreement* allowed for monitoring the down-blending of excess Russian weapons-origin HEU to low-enriched uranium. This agreement was successfully completed in 2013, but no follow-on activity was agreed upon. The 1997 *Plutonium Production Reactor Agreement* (PPRA) allowed for reciprocal monitoring and inspections of five US and five Russian shut-down plutonium production reactors, to verify the cessation of plutonium production and secure storage of material until transfer to a disposition regime. Implementation of the 2000 *Plutonium Management and Disposition Agreement* (PMDA), intended to dispose of a total of 68 metric tons of excess weapons-grade plutonium in the US and Russia, however, it has been fraught

with complications, and in October 2016 Russia withdrew, leaving the future of this effort unclear. The methods of disposition to be carried out in each country and the verification regime were never mutually agreed upon.

The changing security environment has increased the complexity of identifying and closing gaps and added stress to existing agreements. With the deterioration of US-Russia relations, the looming end date of the New START treaty and the negotiation of the TPNW, the existing degree of transparency and verification surrounding the nuclear security regime is uncertain. It is even more urgent that systematic transparent mechanisms to expand transparency and verification dialogues be devised and promoted.

5 More Pieces of the Puzzle

The variety of existing initiatives and research, previously discussed, will be the foundation in the development of a nonproliferation and arms control systems approach. The state-level framework would link the front and back end of the complex. If we assume that reductions of nuclear stockpiles could be accomplished by a network of different initiatives, a compatible set of continuity-of-knowledge regimes could begin by verifying baseline declarations and continue by monitoring movements of treaty-accountable items, any new material or weapons production or dismantlement, and transportation and storage for strategic, nonstrategic, deployed, and stored warheads [2]. This is a grand challenge because past practices have left a legacy of technical uncertainty and the level of transparency that will be needed continues to pose serious national security concerns.

The techniques used by IAEA safeguards could be applied to verifying the declared uses of nuclear materials in weapons complex facilities, even though this would not be simple. Experience gained through establishing material accountancy systems in the states of the former Soviet Union could help establish a basis for transparency regarding nuclear materials in the military sector of a state with nuclear weapons.

Over the past decades, efforts have been made to study ways to monitor the irreversible dismantlement of warheads and safeguard the resulting nuclear material. Secrecy issues are the most acute in this stage of the complex. From 1996 to 2000, Russia, the US, and the IAEA worked to develop a system for verifying nuclear weapons disarmament (the so-called *Tri-Lateral Initiative*), and there are some who believe this could be used as a basis for a future initiative [3]. The US and United Kingdom (UK) have an R&D program, conducted over the past 15 years, that was established to find technical solutions that would enable monitoring and verification of potential future arms control initiatives [4]. The unique relationship between these countries provides the security environment in which protection of classified information can be explored, tested, and evaluated.

The current trend is to expand engagement to move beyond NWS and include states that do not have nuclear weapons. The governments of the UK and Norway took the first formal step in this direction when they began an initiative in 2007 to explore how a NWS and NNWS could partner to better understand the related challenges [5]. Other ideas for addressing these issues were proposed by a diverse group of international experts and documented in a series of reports on innovating verification [6]. The recent International Partnership on Nuclear Disarmament Verification (IPNDV) [7], a public–private partnership between the US Department of State and the Nuclear Threat initiative that engages over 25 countries, is working to further understand the complexities and challenges involved in nuclear verification [8]. As it continued into its second phase, the IPNDV's three Working Groups were considering the issues of (1) verification of nuclear weapon declarations, (2) verification of reductions and (3) technologies for verification. IPNDV is now preparing for its next phase.

A systematic framework can help facilitate mutual understanding on key security issues, technical challenges, and possible solutions when nations with different levels of expertise and experience in dealing with these complex issues are part of the dialogue.

6 Innovative Systems Concepts for Nuclear Nonproliferation and Arms Control Verification

Since the publication of "Verifying Treaty Compliance" in 2006 [9], an international group of technical experts at the European Safeguards Research and Development Association (ESARDA) and the Institute of Nuclear Materials Management (INMM) organized a series of meetings and workshops to consider the following questions:

- What would an overarching framework for a network of verification regimes look like?
- Would verification standards change as stockpile reductions are implemented?
- Who are the stakeholders: multilateral or bilateral, Nuclear Weapons States or Non-Nuclear Weapons States, or open society stakeholders such as Non-Governmental Organizations, industry, and the general public?
- Can state-level confidence be achieved by linking transparency measures for nuclear materials in the civilian power and military weapons complex (past production, existing stocks, current production, dismantlement, and disposition) with transparency for the military stockpiles and delivery systems?
- Can effective arms control verification regimes still protect weapons-sensitive, proprietary, and classified information?
- How do a state's national security objectives drive the level of confidence required for transparency and verification regimes? How can these be taken into account?
- Can trade-offs or asymmetric verification regimes be designed to overcome the challenges that impede progress in certain parts of the weapons complex?

The results of the broad range of R&D projects and technical meetings and workshops organized to address these questions are presented in this book. It begins with an overview of the history and challenges of nonproliferation and arms control verification inspired by the evolution of IAEA safeguards; followed by reflections on the implications of legal, political, military and strategic export control, and issues that will influence the development of a systematic methodology. A credible framework, the models, technologies, and information analytics techniques can be down-selected and applied to monitoring or verification regimes. Due to the evolutionary nature of this technical area, only a snapshot of existing technical verification capabilities can be presented here.

The final section of this book presents the efforts of various researchers to apply a systems approach to develop a systematic, state-level concept that can take into account a broad range of existing and potential initiatives, and provide an organizing construct to consolidate the mosaic of mechanisms that will comprise a global security regime. Efforts were made to share ideas between those with diverse expertise in materials, weapons, and delivery systems verification enriched the outcomes. It is clear that in addition to analytical advantages, a systematic approach can facilitate communication among a wider variety of stakeholders and could aid in designing cost-effective research and development to advance technical monitoring and verification capabilities.

The views and opinions of authors expressed herein do not necessarily state or reflect those of the United States government or Lawrence Livermore National Security, LLC. Part of this work was performed under the auspices of the US Department of Energy by Lawrence Livermore National Laboratory under Contract DE-AC52-07NA27344. LLNL-BOOK-698308

References

1. Sage, Andrew P (ed) (2011) Systems Engineering and Practice. John Wily and Sons, page 3. https://onlinelibrary.wiley.com/doi/pdf/10.1002/9781118001028. Accessed 28 Feb 2019
2. Hinderstein C (ed) (2010) Cultivating confidence, verification, monitoring, and enforcement for a world free of nuclear weapons. Nuclear Threat Initiative, Washington DC
3. Shea T, Rockwood R (2015) IAEA verification of fissile material in support of nuclear disarmament. Harvard Kennedy School Belfer Center for Science and International Affairs, Cambridge
4. U.S. National Nuclear Security Administration Office of Nonproliferation and Arms Control and U.K. Ministry of Defence/Atomic Weapons Establishment (2015) Joint U.S.– U.K. report on technical cooperation for arms control. Washington, DC
5. United Kingdom – Norway Initiative. www.UKNI.info. Accessed 28 Feb 2019
6. Nuclear Threat Initiative (NTI) (2014) Innovating verification: New tools & new actors to reduce nuclear risks—verifying baseline declarations of nuclear warheads and materials. Four reports. http://nti.org/167R. Accessed 28 Feb 2019
7. International Partnership for Nuclear Disarmament Verification. www.ipndv.org. Accessed 28 Feb 2019

8. Hinderstein C (2018) International Partnership for Nuclear Disarmament Verification: Laying a foundation for future arms reductions. Bulletin of the Atomic Scientists 74(5):305–311
9. Avenhaus R, Kyriakopoulos N, Richard M, Stein G (eds) (2006) Verifying Treaty Compliance. Limiting Weapons of Mass Destruction and Monitoring Kyoto Protocol Provisions. Springer, Berlin-Heidelberg

Part I
Status of Verification Strategies

Verification, in the strict legal sense, is the mechanisms by which treaty partners determine if formal treaty commitments have been fulfilled. This general process is relevant for nuclear nonproliferation and arms control, as well as for other areas such as the abolition of chemical and biological weapons, and for effective global climate change measures. Existing technologies can be employed or new ones developed to verify compliance, but in each case the process must guarantee the protection of sensitive information. Strategies codified by treaties and protocols must also adapt to changing realities. This has been demonstrated for the application of international safeguards in the NPT regime. A move from mainly materials- and facility-based safeguards to a more comprehensive look at the state as a whole has occurred. A new systems approach might be the basis for other verification regimes, especially in the field of nuclear disarmament.

History, Status and Challenges for Non-proliferation and Arms Control Verification

David Keir and Andreas Persbo

Abstract This opening chapter considers the basic concepts of arms control verification, and how it works with international agreements. It shows how verification differs from, but can contribute to, general confidence-building. Later, the text elaborates on the role of technology and technological expertise in a verification enterprise, touching on the current status of technologies employed in arms control and exploring the principles on which they have been designed. The chapter highlights, in particular, the need to protect sensitive data, the need for equipment to be jointly trusted, and the requirement that equipment is designed, specified and built taking host facility considerations into mind. The text exemplifies equipment development by looking at so-called information barriers—gear intended to detect a nuclear warhead's attributes or to match a warhead against a template while protecting classified and proliferative information. Finally, the chapter points out the continuing importance of progress in arms control verification research and development, as well as in basic verification methodology.

1 On Verification

It has been said recently that we are living these days in a "post-truth" era, largely thanks to the large amount of dubious information circulating on social media and old-fashioned mass-media. But so-called "facts" have always been questionable throughout history. We all know the saying that truth is the first casualty in war, but beyond this, a healthy skepticism is more or less essential for citizens and decision-makers when interpreting statements, claims and opinions, which can range from political spin to downright lies.

D. Keir (✉)
Cloverdale Scientific Limited, Newbury, UK

A. Persbo
Verification Research, Training and Information Centre (VERTIC), London, UK
e-mail: andreas.persbo@vertic.org

I. Niemeyer et al. (eds.), *Nuclear Non-proliferation and Arms Control Verification*,
https://doi.org/10.1007/978-3-030-29537-0_2

In fact we must be aware that if we choose to believe information without checking its origin, or somehow verifying it, then we run a risk. At the very least, the risk is that we may be surprised when later on the facts are revealed to be false. At the worst, if we choose to act upon un-verified information, we may be aiding and abetting a group, an individual or an organization with the ability to cause real harm.

In the case of international agreements and treaties, this is so well recognized that there is a consensus that verification needs to be built into an agreement. This is the modern norm; but there are important exceptions where verification has been set aside as too difficult, too intrusive or actually too much trouble to be worth the benefit.

Often, in public life, there are very respectable data that can be used to cross-check "facts", opinions and claims, but they are not used, even by journalists presenting articles on the very subjects that the data relate to. The work of the late Hans Rosling has shown this, in a most accessible and entertaining form, in his openly-published presentations and video pieces. Here "big data" are animated to graphically illustrate changes and trends in worldwide populations, longevity, wealth and many other characteristics (see for example [1]).

However, and turning now to the main subject of this book, in the case of international arms control agreements and initiatives, the data required for checking statements, declarations and promises are not generally available in the public domain.

This opening chapter of the book reviews the basics of how these types of agreements might be verified—how each party could assure itself that the others are keeping their promises, across the weapons landscape of nuclear, chemical and biological weapons of mass destruction. We look at what has been achieved so far, particularly in the sphere of inspections and monitoring and what the main sticking points and challenges are.

Arms control verification has, throughout the history of arms control, been one of the most important, yet one of the most controversial, issues in negotiation. Any agreement is by nature entered into voluntarily. Verification provisions are thus a form of consensual surveillance that agreed measures are being implemented as envisioned. The word verification itself draws its meaning from a Latin root, meaning truth. Verification, from a dictionary perspective, is no more complicated than being the process of establishing the truth, accuracy, or validity of something [2]. It follows that verification, when used as a verb, must as a minimum have an object associated with it.[1] In this chapter, we are interested in the verification of an agreement, or of parts of an agreement.

[1]We tend to undergo several acts of verification on a day-to-day basis, mostly proving our identity to someone. This happens when you enter your password into your computer, onto websites, or your smartphone, or when you enter your Personal Identification Number into an Automated Teller Machine.

Functionally, arms control verification can be seen as "the process of gathering and analysing information to make a judgement about parties' compliance or noncompliance with an arms control agreement" [3]. In most cases this will be done by a team of trained inspectors. This definition is commonly used and emphasises verification as a process. In principle, verification consists of three steps:

1. Gathering of information relevant to the fulfillment of the obligations under the arms control treaty (*information gathering stage*);
2. Review of the information collected against the norms against which compliance is judged (*analytical stage*); and
3. As a result of the review, a determination of whether the rules of the treaty have been met (*compliance determination stage*).[2]

These steps may be cycled through daily, or even more frequently, in a real-life verification application—especially one which follows a process (such as receipt of a nuclear warhead for decommissioning, staging, disassembly and components consignment). In a future nuclear warhead dismantlement treaty verification, the data would need to be analysed and conclusions drawn as part of the inspectors' need for continuity-of-knowledge.

Much literature on the topic of verification focuses on the first of these stages—the gathering of information. Key questions are: What technologies are used to do this? What are the conditions for inspector access to sites relevant to the agreement? Under what circumstances can parties to the agreement restrict, limit or prevent certain types of information being transmitted to the verifiers? How much information does an inspector need to collect on certain aspects of state behaviour?

A relatively under-studied topic is how the information gathered fits with the norms. A compliance determination will be constrained by the norms, no matter what the evidential strength of information gathered. Whether a rule is applicable can be in question despite partial or full agreement *as to the facts*. For instance, a swipe sample taken a particular location during an inspection, once analysed by a specialist laboratory could point to the fact that highly enriched uranium exists in that area. However, even if this is the case, the existence of HEU may not be in contravention of, say, a nuclear safeguards agreement. The enriched material could, as has happened in real life, have been present on contaminated inspection equipment. It could also have been part of a previous enrichment campaign at a time when such production was not strictly prohibited. Finally, and quite obviously, the production could have been lawfully declared.

In other words, the state where the sample was taken might answer that while they do not deny the presence of highly enriched uranium at the site, the particles were transferred there on contaminated equipment, and therefore were not indicative of undeclared production (this would constitute a *partial agreement* with the facts). Alternatively, the state could say that the swipe analysis is correct, but that they are under no obligation to declare the activity that produced the uranium (this would be *full agreement* with the facts).

[2]For a particularly cogent explanation of verification theory, see [4], especially on pp. 104.

Treaty text can sometimes be vague, and this in itself gives rise to difficult questions. The parties to the agreement can be agreed that data collected means the same thing, but differ on whether the data points to non-compliance. One party may think that a particular behaviour is non-compliant, while the other believes that it is compliant. Often, compliance determinations are handled by the state parties themselves—and disagreements *as to the law* are supposed to be handled by international arbitration. While divorced from the technical process of gathering and analysing information, the *compliance determination stage* is perhaps the most significant stage of a verification process.

Most existing verification regimes are iterative, meaning that the process of gathering data, analysing it, and making a compliance determination is continuous and in many cases overlapping in time. New data is being collected while old data is being analysed. The most pertinent example of this is the International Monitoring System run by the Comprehensive Nuclear Test Ban Treaty Organization, where a web of monitoring stations collect near real-time data on state compliance for subsequent computerised and human analysis.

2 On Verification and Confidence-Building

A common view is that a "verification system cannot ... legally verify compliance by States that have chosen not to become a party to a treaty" [3]. The statement emphasises that verification arrangements are for the most part consensual. However, there are instances of non-consensual verification. The United Nations (UN) Security Council is free to impose legally binding rules on UN member states.[3] This has been done in the past, through for instance Security Council resolution 1284 of 17 December 1999, which created the United Nations Monitoring, Verification and Inspection Commission (UNMOVIC). Non-consensual verification is, however, a rare exception in arms control. The fact that arms control verification in the vast majority of instances is consensual also means that it is mostly a cooperative endeavour. For the most part, the inspector and the inspected have a joint interest in concluding the inspection successfully. Some, therefore, highlight the confidence-building aspect of the verification enterprise. In 1978, for instance, the UN General Assembly resolved that "disarmament and arms limitation agreements should provide for adequate measures of verification satisfactory to all parties concerned in order to create the necessary confidence and ensure that they are being observed by all parties" [5].

[3]The powers of the Council derives from articles 24 and 25 of the UN Charter. Security Council decisions will prevail over member states obligations under international agreements, and this follows from article 103 of the Charter. If this provision had not been there, UN member states could technically opt out of Council jurisdiction by means of international treaty.

If the treaty or agreement concerns nuclear weapons, warheads, components or fissile material—or the facilities and equipment used to produce these—a great deal is at stake.[4] So much so that it would be extraordinary to contemplate merely trusting signatories to carry out their obligations without some formal system for checking that those obligations are being carried out exactly as agreed. There are three main reasons for this:

1. To detect slow, inadequate or inappropriate implementation of declared obligations—allowing states party to the agreement to become aware of that type of non-compliance and therefore be able to take action on that information;
2. To detect deliberate cheating as soon as possible—allowing states party to the agreement to become aware of that type of non-compliance and therefore be able to take action on that information;
3. Because of the detection capabilities described above; to deter non-compliance with the actions agreed in the treaty or agreement.

Part of what this book tries to do is to examine the ways in which that could, or should, be accomplished, in a variety of future situations.

3 On Verification Provisions

Verification provisions are either negotiated into the main text of the treaty or separately concluded in a protocol.[5] This is usually conducted in the same negotiation so that the treaty and the protocol can be signed and ratified as one package. It is rare to negotiate and conclude verification provisions after a treaty has entered into force. A promising attempt to separately agree on verification arrangements for the 1972 Biological Weapons Convention met with failure in the early 2000s.

If relations between contracting parties are healthy, or if the subject matter is of little military importance to them, verification provisions are sometimes omitted. In seeking the US Senate's advise and consent for the 2002 Treaty of Moscow, the US administration argued that verification was unnecessary. The government highlighted that "neither side has an interest in evading the terms of the treaty". Why did the US reach that conclusion? Namely because the agreement "simply codifies unilaterally announced reductions—and gives both sides broad flexibility in implementing them [7]." As noted above, verification arrangements have an assumed objective of building confidence between parties—it follows that they will

[4]Much has been written about how verification requirements change over time, and how verification requirements are likely to be stricter in a regime focusing on nuclear explosive devices. For a reasonably recent overview of the debate, see [6].

[5]The multilateral practice in the 1990s was to give effect to verification procedures in specially designated Verification Protocols. This practice was kept in the negotiation of the New START arms reduction agreement between the Russian Federation and the United States.

be viewed as unnecessary if the contracting parties believe that they are in a trustful relationship—however misguided that belief may be.

Verification provisions attached to a treaty or agreement are married to the language of the main commitment or prohibition. If, for instance, an agreement stipulates that a state is to reduce its weapons holdings by a defined number of explosive devices, the verification system would logically examine whether each device has been decommissioned, as promised. The verification task can focus exclusively on verifying the *correctness* of the promised dismantlement action. If—on the other hand—an agreement stipulates a ceiling for holdings, the task becomes more complicated. Verifying that a country does not have more than, say, 200 specified types of explosive device, will require a well-defined baseline declaration. It would also need to assess the *completeness* of the promised dismantlement action.[6]

The verifiable aspects depend, in each case, on the aims and the specific language of the treaty or agreement. Often it will be that parties agree that specific actions will be undertaken under that agreement, in specific locations, by specified dates.[7] Often this will come down to the *treaty-accountable items*.[8]

In the case of fissile materials—and the associated components which are required to make nuclear weapons—the primary aims of verification to date have ranged across the following:

- For non-nuclear weapon states, the "timely detection of diversion of significant quantities of nuclear material from peaceful nuclear activities to the manufacture of nuclear weapons or of other nuclear explosive devices or for purposes unknown, and deterrence of such diversion by the risk of early detection" [8].
- In the case of future nuclear disarmament activities "... the confirmation, by a monitoring party, that declared treaty-limited items (TLIs) are consistent with the declaration made by the host country" [9].

The verification process is limited to monitoring compliance with the agreement itself. It is therefore considered a principle that requests for inspections or information should be used **only** for the determination of compliance, with care being taken to avoid abuses [10]. Thus, if we consider the point of view of any party to such an agreement who will be subject to verification, in the form of monitoring, inspections, inventory-taking, document checking, etc., they will want to limit and define the rights of other (monitoring or inspecting) parties to specific verification activities. Otherwise, items belonging to the monitored state, unrelated to the

[6]If, for instance, a state has 260 explosive devices and promises to have no more than 200, it only need to declare that it in fact has 230 devices and submit 30 to verified dismantlement. After dismantlement, the country has 200 devices 'on the books' but 230 in reality. The country would be non-compliant, but this cannot be verified by simply counting the number of dismantled devices.

[7]Or, of course, sometimes parties will have agreed to refrain from specified actions—as in the case of the Comprehensive nuclear Test Ban Treaty.

[8]In the safeguards context, for instance, the treaty accountable item is often a specified mass of a metal (specifically plutonium or uranium) and various source, or precursor, materials (such as uranium yellowcake).

treaty, might come under scrutiny and investigation—this could be an unnecessary intrusion into the affairs of the inspected state or, worse, an abuse of inspection rights.

One way of limiting the rights, activities and access of the monitoring parties is to define the scope of verification very rigorously. So, for instance, in the case of a disarmament action, the host state could take the position that the only thing that will be verified will be a **declaration** they have made under the treaty or agreement. Everything else would then be agreed as outside the scope of the verification regime.

4 Verification in the Nuclear Field: The Role of Technology and Technological Expertise

The job of carrying out effective verification monitoring and measurements starts well before any inspector sets foot in the monitored state. It is vital, at the very beginning of the technical part of the treaty or agreement, to have all the negotiators well briefed and well-supported on what is possible technically—and what the implications are for each and every technical detail that is proposed and negotiated over.

Technical input will be required throughout negotiations to increase the prospects of reaching an agreed verification protocol able to provide desired levels of detection and deterrence. Technical expertise will also be important to keep negotiators up to breast with language covering access, allowed equipment, operating procedures, data gathering protocol, as well as data transmission and storage.

The verification of nuclear materials (in particular military metals and components) is likely to be subject to constraints. To neglect careful negotiation of any one aspect may result in an impotent verification system, with inspectors unable to provide assurance of compliance however hard they try. By the time inspectors or monitors arrive at a host facility, it will be too late to realise, for example, that the equipment specified and measurement procedures agreed are inadequate, inappropriate, or downright misleading.

5 Current Status of Verification Technologies in Arms Control Verification

Since the 1980s, much work has been devoted to the development of equipment designed for nuclear materials and nuclear weapon and component verification measurements (plus the associated chain-of-custody requirements). This activity has mostly been centred in the United States, largely at national laboratories funded by the Department of Energy. Most of the measurement technologies which have shown promise are based on the fact that fissile materials (and therefore nuclear weapons and their fissile components) emit gamma and neutron radiation of very specific types and energies. This provides the possibility of identification 'fingerprints' and in some approaches, 'templates' which could allow unambiguous identification of treaty-accountable items.

Except in the case of safeguards measurements on civil nuclear materials, there is no agreed set of equipment for the various tasks and no agreed best practices for the operation of equipment that has shown promise. There is also no model protocol for important activities such as verifying baseline declarations of warheads, components and military fissile materials, let alone dismantlement and decommissioning. The last issue was one of the main things addressed in a study led by the Nuclear Threat Initiative (NTI), which gathered expert opinion on verification during a series of workshop meetings from 2012 to 2014 [11].

There are three important and well-known requirements for technologies applied to verification of military fissile materials, warheads and components:

1. The absolute need to protect sensitive data while obtaining good enough information to contribute to decision-making about compliance
2. The condition that the equipment is jointly-trusted (authenticable and authenticated);
3. The requirement that the equipment is designed, specified and built in such a way that the host facilities will allow it access and deployment in nuclear facilities (i.e. it must be safe and certifiable).

The various technical reports and reviews of weapons verification technologies, dating from the 1980s up to the present day, sometimes lack the sort of detailed error and uncertainty analysis that would usually form part of a scientific report involving measured data. Particularly in the case of technologies being applied experimentally to nuclear arms control-related measurements, the trials and demonstration of the equipment are often reported at a summary level, leaving the reader uncertain as to whether the measurements are accurate to ±1% or ±50%—and how that changes as measurement conditions change.

To be useful for prospective inspector use in future verification schemes, each type of verification technology needs to be subjected to a systematic study of the circumstances in which measurement uncertainties vary. This mean that as factors such as shielding of the radioactive emissions, geometry, distance and other features of the measurement environment change, the way in which the results and their uncertainties change, are essential information for planning and writing operational procedures. Of course, some of the experimental and development work in this particular field has been done in the classified realm; and this may be the reason for the lack of detailed information on capabilities and limitations. It should be pointed out that, by contrast, in the world of civil fissile material safeguards measurement, the opposite is true and a comprehensive guide to accuracy and standard uncertainty for a range of measurement types is in routine use [12].

There is another significant problem in nuclear arms control verification. While there is general agreement that no one technology or measurement type is going to provide the verification confidence required to make the decision of compliance with declarations, or to draw any broader conclusion; there is no consensus about how to combine or 'fuse' different kinds of data. Nor is there any alternative best practice on how to assemble and present data from many different measurement types to reach any such conclusion.

Almost as important is how uncertainties in such an array of measurements should be propagated through information-combination, to give an overall uncertainty in the final result—for decision-makers.

6 Information Barriers: Attributes and Templates

Returning to the issue of multiple requirements, listed in Sect. 5; one way in which scientists in this field have attempted to satisfy the requirement for high-fidelity data for decision making while not accessing sensitive data (about nuclear weapons design for instance) has been to invent the 'information barrier'.

The barrier is a piece of equipment which acts as the data handling and analysis recipient for various detector types. The classified measured data is used by the information barrier to internally compare attributes (such as isotopic ratios) with set values or *thresholds* for those attributes, and then to output the result of that comparison a green (for agreement) or red (for disagreement) light. Thus no inspector or monitoring personnel have sight of the classified data.

In another approach, the information barrier collects a detailed set of classified data—amounting to a fingerprint, or image of the first target object it is presented with. That image or fingerprint is classified and is never seen but is stored securely (possibly encrypted) under joint custody. Later, other target objects can be exposed to the same detector/information barrier and internally compared to the first template.

If the first template was assuredly an image or fingerprint of a genuine treaty-accountable item, then conformity of succeeding target objects with it implies that they too are the same type of treaty-accountable item.[9] This information can be used to verify continuity of knowledge of a single item through a process flow or, additionally can be used to build confidence in the identity of a collection of separate items.[10]

[9]Establishing, beyond all reasonable doubt that the first item used to set up the template really is the treaty accountable item—and that this can be relied on as the 'gold standard' measurement is not easy. this has become known as the 'initiation problem'.

[10]There are subtler arguments to be made about the value of taking template measurements when large numbers of declared identical warheads or components are being dismantled and yielding significant amounts of product (fissile material leaving the military inventory). These are based on the philosophical discussion of confidence and likelihood of falsification of many target objects at enormous cost. One thing is clear however: If measurements (template or attribute-type) are not taken of declared warheads prior to the dismantlement process, there is no going back to take them later when the value of doing so becomes clearer.

Information barriers, of course, are yet more items of equipment which, if they are to be used, must also satisfy items 2 and 3 of the constraints list above, namely, be certifiable and authenticable and also have their inbuilt data uncertainties characterised for all relevant deployment scenarios.

7 Fixed and Potentially Movable Constraints

In truth, there are some real physical boundary conditions which current technologies can not overcome. Highly-enriched uranium, for instance, is a key fissile material to be detected, identified and quantified in nuclear arms control verification. However, it barely emits radioactivity. The emission rates for both gamma and neutron radiation are extremely low, and so collecting statistically significant data using so-called passive NDA (non-destructive assay) techniques is very challenging. When shielding of the fissile material is present, when distances from the detector to target item are large and when allowed counting times are short, it may become impossible to gather data with any significance for decision-making. These challenges may mean that the information barriers could cease to function meaningfully.[11]

On the other hand, there are other constraints which are there not because of physical laws but because of security and other political concerns. In these cases, were it possible to relax the level of information protection required, then more verification confidence might be achievable, through access to better raw data.

It is worth noting that the direction of progress in the last couple of decades, as far as US-Russian bilateral disarmament agreements is concerned, has been from no verification technology for SALT, to delivery system verification in START, to limited warhead verification for New-START, utilising radiation detectors to verify absence of fissile materials in declared non-warhead objects. The next step expected would be for an increasing role for technology. However, it is difficult to predict when and what the next step will be.

8 The Continuing Importance of Progress

The most influential efforts in this field have come from the era of START 3 preparative research at the US national laboratories, from supporting research to the US-Russian Trilateral Initiative. That was the time when significant resources were

[11] We note that there are a number of non-destructive assay technologies which can be referred to collectively as ‘active interrogation’, which could potentially be the answer to the poor inherent radiation signatures emanating from low-radiation level objects. Lack of space prevents us from exploring these in the current chapter.

devoted to technology development, prototyping and demonstration of capabilities. These efforts have been supplemented in the 2000s the UK-Norway Initiative and the (mostly classified) UK-US initiative, which has been in operation for more than a decade.

A fairly recent pooling of expert opinion—particularly focused on the requirements for verifying baseline amounts of fissile material and numbers of warheads, for hypothetical future disarmament activities—was the NTI Pilot Project on Verification [11]. The authors were both involved in this exercise over a 2 year period up to 2014. The document set produced by NTI Pilot Project on Verification will likely continue to influence policy in the field of arms control verification research for the next few years, particularly in US public sector funded programs. More importantly, the foundation of the International Partnership for Nuclear Disarmament Verification (IPNDV) has resulted in a strong on-going initiative that features non-nuclear weapon states as full and influential members.

More than 20 countries are part of the IPNDV and its work is done partly by Working Groups, each with an agreed Terms of Reference. Each is tasked with looking in detail at such issues as technologies for verification and on-site inspections. The IPNDV has entered Phase Two of its activities in 2018 and proceeds now with some restructuring of the Working Groups. There is an easily-accessible website for IPNDV for readers who wish to follow its progress.

Less policy-focussed, smaller and more 'agile' is the QUAD group. This has grown out of the UK-Norway Initiative and is now an active group consisting of laboratories in UK, Norway, Sweden and the US, with backing from the governments of each. The most striking achievement so far has been the "LETTERPRESS" exercise, an experiment in multilateral nuclear arms control verification. Held in late 2017 at a former UK weapons facility, the initial stages of nuclear weapons being taken out of active service, but before dismantlement, were studied. Training versions of the UK's 1970s nuclear warhead (minus any proliferative parts or information) were used as trial objects. A selection of monitoring and inspection approaches and technologies were deployed and evaluated. A presentation of this was given at the 2018 NPT Review Conference, and a pamphlet for public release, with colour photographs, has also been distributed.

In introducing this book, we have focused on technologies, along with the constraints operating in these treaty-type environments. We have set a scene that could be viewed as somewhat negative in outlook. However, the challenges provided by this indivisible mélange of scientific, policy and political conditions are both fascinating and technically rich. Furthermore—even in the current era of little arms control activity—progress in this area is urgent and important, if the world is to move towards the situation envisaged in the Nuclear Non-Proliferation Treaty—of a future universally disarmed world.

References

1. https://www.ted.com/speakers/hans_rosling. Accessed 28 Feb 2019
2. Oxford Dictionaries, British & World English (2016) Definition of Verification in English. https://en.oxforddictionaries.com/definition/verification. Accessed 28 Feb 2019
3. United Nations Institute for Disarmament Research (UNIDIR), The Verification Research, Training and Information Centre (VERTIC) (2003) Coming to Terms with Security: A Handbook on Verification and Compliance. UNIDIR/2003/10:1 Geneva
4. Dekker GD (2001) The Law of Arms Control International Supervision and Enforcement, Martinus Nijhoff Publishers, The Hague, p 104
5. United Nations General Assembly (1978) Resolutions and Decisions Adopted by the General Assembly during Its Tenth Special Session, New York, paragraph 31
6. Perkovich G, Acton JM (eds) (2014) Abolishing Nuclear Weapons: A debate, Carnegie Endowment for International Peace. https://carnegieendowment.org/files/abolishing_nuclear_weapons_debate.pdf. Accessed 28 Feb 2019
7. Rumsfeld, D (2002) Prepared Testimony for the Senate Foreign Relations Committee regarding the Moscow Treaty, US Department of Defense, 17 July 2002. https://www.gpo.gov/fdsys/pkg/CHRG-107shrg81339/pdf/CHRG-107shrg81339.pdf. Accessed 28 Feb 2019
8. International Atomic Energy Agency (IAEA) (1972) The Structure and Content of Agreements Between the Agency and States in Connection with the Treaty on the Non-Proliferation of Nuclear Weapons, INFCIRC/153 (Corrected), para 28
9. McArthur D, Hauck D, Smith M (2013) Confirmation of Nuclear Treaty Limited Items Pre-dismantlement vs. Post-dismantlement. ESARDA Bulletin 50:116–123
10. United Nations (2008) Verification in All Its Aspects, including the Role of the United Nations in the Field of Verification. United Nations, New York, p 25
11. Nuclear Threat Initiative (NTI) (2014) Innovating Verification: New Tools & New Actors to Reduce Nuclear Risks. Four reports. NTI, Washington DC. http://nti.org/167R. Accessed 28 Feb 2019
12. International Atomic Energy Agency (IAEA) (2010) International Target Values 2010 for Measurement Uncertainties in Safeguarding Nuclear Materials, STR-368

The Evolution of Safeguards

Jill N. Cooley

Abstract Implementation of IAEA safeguards and the drawing of safeguards conclusions has changed dramatically from an ad hoc arrangement in 1962, to an approach focused on verifying nuclear material at declared facilities and drawing safeguards conclusions at the level of individual facilities, to one that assesses the consistency of all safeguards relevant information regarding a State's nuclear programme and draws a safeguards conclusion for the State as a whole. The notion of implementing safeguards that considers a State's nuclear and nuclear-related activities and capabilities, as a whole, is referred to as the State-level concept. The State-level concept is applicable to all States with a safeguards agreement in force. Through implementing safeguards in the context of the State-level concept, a number of benefits in terms of effectiveness and efficiency have been realized, including taking better account of State-specific factors which allow for the development and implementation of customized State-level approaches (SLAs). SLAs provide options for safeguards measures to be implemented in the field and at Headquarters, allowing the IAEA to compare their cost-effectiveness and providing greater flexibility in safeguards implementation. Instead of mechanistically applying the activities listed in the safeguards criteria, the implementation of SLAs is focused on attainment of technical objectives. By doing so, safeguards implementation is more performance-oriented and is helping the IAEA to avoid spending resources on doing more than is needed for effective safeguards. The objectives are established by the Secretariat using structured and technically based analytical methods conducted according to uniform processes and defined procedures. Utilizing the same technically-based processes and procedures for developing SLAs for all States helps to ensure consistency and non-discrimination in safeguards implementation, efficient performance of the work, and more soundly based safeguards conclusions.

J. N. Cooley (✉)
Consolidated Nuclear Security, Y-12 National Security Complex, Oak Ridge, TN, USA
e-mail: jill.cooley@ib3global.com

I. Niemeyer et al. (eds.), *Nuclear Non-proliferation and Arms Control Verification*,
https://doi.org/10.1007/978-3-030-29537-0_3

1 Background

In 1962, the International Atomic Energy Agency (IAEA) conducted its first safeguards inspection at a research reactor in Norway. From that beginning, the IAEA's safeguards legal authority and implementation practices have continued to evolve to address technological advances in the nuclear fuel cycle as well as proliferation challenges. This evolution has also been driven by technological advances in the measures and tools used for safeguards. In 1970, when the Treaty on the Non-Proliferation of Nuclear Weapons entered into force, independent State regulatory authorities, statistical techniques for assessing material balances and nondestructive assay as a measurement tool barely existed.[1] The foundations of modern safeguards were only then being developed. Comprehensive safeguards agreements (CSAs) based on INFCIRC/153 [1] established requirements for systems of accounting for and control of nuclear material (para 32), safeguards measures (e.g. statistical techniques and random sampling) and the use of State factors (para 81), all of which were in their infancy. Under the initial concept for implementation of CSAs, IAEA safeguards activities were focused primarily on verifying nuclear material at declared locations in States with 'significant' nuclear activities.[2]

By 1990, the number of States with safeguards agreements in force (104), the number of nuclear installations subject to inspection (912), and the quantities of nuclear material under safeguards (60,000 significant quantities[3]) had grown considerably. With this growth, came an increase in the number of inspectors in the IAEA Department of Safeguards (188 full-time inspectors compared to 44 in 1977). To ensure adequate effectiveness and consistency in the implementation of safeguards, safeguards criteria were developed. The criteria specified the verification activities to be conducted for each type of nuclear facility and material to detect the diversion of a significant quantity of nuclear material and misuse of the declared facility to produce undeclared nuclear material. Safeguards conclusions were drawn and reported regarding the non-diversion of nuclear material placed under safeguards at declared nuclear facilities. The shortcomings of this approach to safeguards implementation became evident in the early 1990s with the IAEA's inability to detect Iraq's clandestine nuclear weapons program.

---

[1] In 1970, safeguards were being applied to 82 nuclear facilities in 32 States.

[2] States with significant nuclear activities were defined to be those with a facility or a quantity of nuclear material exceeding the INFCIRC/153, paragraph 37 limits for exempting nuclear material from safeguards and thus were subject to routine in-field inspection activities.

[3] A significant quantity is the approximate amount of nuclear material for which the possibility of manufacturing a nuclear explosive device cannot be excluded.

2 Early Strengthening Measures

Efforts to strengthen the safeguards system, in particular the IAEA's ability to detect undeclared nuclear material and activities in States with CSAs, began almost immediately. The need to give greater consideration to a State as a whole, rather than focusing primarily on nuclear material and facilities declared by the State, was clearly recognized. In 1992, the IAEA Board of Governors affirmed that the scope of a CSA was not limited to nuclear material actually declared by a State, but included any material that is required to be declared. Thus, the Board confirmed that the IAEA has the right and obligation, under such agreements, to verify not only that State declarations of nuclear material subject to safeguards are 'correct' (i.e. they accurately describe the types and quantities of the State's declared nuclear material holdings) but that they are also 'complete' (i.e. that they include all nuclear material and activities that should have been declared). In this same time period, the Board also approved the early provision of design information and the voluntary provision of other information relevant to a State's nuclear activities.

In order to address the detection of undeclared nuclear material and activities in a State, different tools and techniques were required from those needed for the timely detection of the diversion of declared nuclear material. Strengthening measures, which were progressively adopted, especially those of the Model Additional Protocol [2], involved:

- acquisition of a broader range of safeguards relevant information[4];
- more emphasis on information analysis;
- broader IAEA inspector access to locations in States beyond declared facilities;
- use of advanced technical verification measures; and
- a more investigative approach in implementing safeguards.

Measures that can be implemented under the IAEA's existing legal authority provided for in CSAs include environmental sampling, remote data transmission, the use of satellite imagery, and collection and analysis of safeguards relevant information from all sources. Additional protocols (APs) concluded with CSA States provide the IAEA with complementary legal authority and equip the IAEA with important additional measures that provide for broader access to information and locations. The implementation of an AP significantly increases the IAEA's ability to verify the peaceful use of all nuclear material in a State with a CSA (i.e. the correctness and completeness of the State's declarations).

Strengthened safeguards also requires emphasis being placed on considering the entire nuclear fuel cycle of a State (i.e. the State 'as a whole') rather than on individual facilities. This shift in focus is important, marking a 'State level'

[4]Safeguards relevant information is information relevant for the implementation of IAEA safeguards and which contributes to the drawing of soundly based safeguards conclusions. It is collected, evaluated and used by the IAEA in exercising its rights and fulfilling its obligations under safeguards agreements.

approach towards safeguards implementation rather than, as previously, one with a nuclear facility focus. The purpose of all of these measures has been to increase knowledge and understanding about a State's nuclear material, activities and plans.

3 State Evaluation

The framework for this move from safeguards implementation and conclusions drawn at the facility level to implementation and conclusions drawn for a State as a whole is the safeguards State evaluation process. The State evaluation conducted for an individual State seeks to answer the interrelated questions of whether all safeguards relevant information about a State's nuclear program is consistent, whether the 'picture' of the State's present and planned nuclear program is complete, and whether sufficient information is available about a State's nuclear activities and plans to enable the IAEA to provide credible assurance, through its safeguards conclusions, that the State is complying with its safeguards obligations.

For all States with safeguards agreements in force, the IAEA collects and processes three types of safeguards relevant information:

1. information provided by the State itself (e.g. declarations and reports, including voluntarily-provided information);
2. information from safeguards activities conducted by the IAEA in the field and at Headquarters (e.g. inspections, design information verification, complementary access, material balance evaluations); and
3. other relevant information (e.g. from open sources, including commercial satellite imagery, or provided by third parties).

The State evaluation process involves the on-going evaluation of all safeguards relevant information available to the IAEA about a State aimed at assessing the consistency of that information in the context of a State's safeguards obligations. Information provided by the State is reviewed for internal consistency, for coherency with results of safeguards verification activities and for compatibility with all other available information. Critical to the State evaluation process is the identification of anomalies or inconsistencies requiring follow-up through, for example, the acquisition of further information or the conduct of additional in-field verification activities. Defining and conducting appropriate follow-up activities is essential to ascertain whether the identified anomalies or inconsistencies indicate the possible presence of undeclared nuclear material or activities or the diversion of nuclear material from peaceful activities. The comprehensive State evaluations conducted for individual States provide the basis for a State-level approach to safeguards implementation where verification activities can be planned and conducted, results evaluated, and follow-up actions identified for each State individually.

State evaluations also provide the basis on which the IAEA draws it safeguards conclusions reported annually to the IAEA Board of Governors and the international community. For States with CSAs and APs in force, a safeguards conclusion that all nuclear material has remained in peaceful activities in a State (i.e. the 'broader conclusion') is based on the Secretariat's findings that there is no indication of the diversion of declared nuclear material from peaceful nuclear activities and no indication of undeclared nuclear material or activities in the State. Because the information and access provided under an AP are essential for the IAEA's ability to provide credible assurance of the absence of undeclared nuclear material and activities in the State, the safeguards conclusion drawn for a State with a CSA alone relates only to the non-diversion of declared nuclear material from peaceful activities.[5]

Work began within the IAEA in the mid-1990s to establish the internal infrastructure to support the State evaluation process, including the collection, processing, analysis and storage of safeguards relevant information from a broad range of sources. Subscriptions to news agencies and information collection services were set up to obtain, for example, scientific and technical literature, government reports, and trade publications; contracts with commercial satellite imagery providers were put in place to acquire imagery; and capabilities were established to analyze procured imagery and collect data transmitted from unattended safeguards instrumentation in the field. Software tools were implemented to facilitate the management and analysis of the information, guidance was developed for conducting and documenting State evaluations, and training in all of the new tools and techniques was implemented. Small teams of safeguards inspectors were formed to evaluate all the information about individual States and an interdepartmental review committee was established to provide management oversight.

4 Integrated Safeguards

The implementation of the strengthened safeguards measures provided the IAEA with increased information about a State and added to the IAEA's ability to consider the State as a whole. This was particularly true for those States with an AP in force. The IAEA began to consider how this information could be used in the determination of safeguards activities both in the field and at Headquarters. The best opportunity for the IAEA to maximize the effectiveness and efficiency of safeguards implementation was for those States for which the broader conclusion had been drawn. Through an optimized combination of safeguards measures provided for under a State's CSA with those provided for under its AP, by taking into account State-specific factors,[6] and through the evaluation of all safeguards relevant information, the IAEA sought to enhance the efficiency of safeguards implementation for

[5]The safeguards conclusion for States with voluntary offer agreements or item-specific safeguards agreements relates only to the non-diversion of nuclear material or non-misuse of facilities or other items to which safeguards had been applied.

[6]At that time, State-specific factors were referred to as 'State-specific features and characteristics'.

such States, without compromising effectiveness. The implementation of safeguards in this manner was called 'integrated safeguards'. The framework for this effort was described in reports by the Director General to the Board of Governors in the early 2000s.

In 2001, in the context of integrated safeguards, the IAEA began implementing individual safeguards approaches for States ('State-level safeguards approaches' or 'SLAs') for which the broader conclusion had been drawn. Although some considerations relating to the State as a whole were reflected in these approaches, the primary basis for determining safeguards activities at declared facilities in these States remained the safeguards criteria, albeit their application adjusted to take into account the broader conclusion for such States.

In 2004, in an effort to distinguish the general State-level approach to safeguards implementation from the customized SLAs being developed for individual States, the Secretariat introduced the term 'State-level concept' to refer to the State as a whole perspective in implementing safeguards. During the 2000s, IAEA efforts were concentrated on implementing APs being brought into force, conducting verification and evaluation activities necessary to draw broader conclusions, and progressively developing and implementing SLAs for States under integrated safeguards. The Secretariat, however, was clear in reports to its policy making organs that SLAs were to be developed and implemented for all States.

5 Furthering the Development of the State-Level Concept

By 2010, the State evaluation process was well established within the Department of Safeguards. State evaluations were being conducted for all States with safeguards agreements in force and providing the basis for safeguards conclusions. Additionally, 47 SLAs were being implemented for States under integrated safeguards. In-field verification activities, however, were still largely based on implementation of the safeguards criteria or the criteria-based SLAs. Driven by the desire to take full advantage of the increasing amount of safeguards relevant information available and the growing verification workload resulting from the increasing number of nuclear facilities and quantities of nuclear material under safeguards,[7] combined with budgetary constraints, internal efforts were refocused to integrate the State evaluation process with safeguards verification activities and to move from criteria-based safeguards to an objectives-based approach. The goal was to implement SLAs more focused on the attainment of technical objectives with the appropriate level of safeguards activities instead of mechanistically carrying out the activities listed in the safeguards criteria. A fundamental principle in this further development was

[7] In 2010, there were 175 States with safeguards agreements in force, and 1175 nuclear installations and 172,180 significant quantities of nuclear material under safeguards.

maintaining safeguards effectiveness while using resources more effectively and efficiently.

Efforts to further develop and implement elements of the State-level concept included preparing guidance for establishing State-specific technical objectives and developing SLAs, further consideration and use of State-specific factors and strengthening the process for conducting State evaluations. In 2012, driven by Member States' requests for more information on the development and implementation of the State-level concept, the Secretariat embarked on what would be a 2-year communication effort to describe and document in detail the elements of the State-level concept. The following provides a description of those elements.

6 The State-Level Concept

The State-level concept refers to the general notion of implementing safeguards that considers a State's nuclear and nuclear-related activities and capabilities, as a whole. The State-level concept is applicable to all States with a safeguards agreement in force, taking into account the objectives deriving from their respective safeguards agreement and the rights and obligations of the parties to that safeguards agreement.

In applying the State-level concept, the IAEA implements safeguards in a holistic manner, using integrated processes to support safeguards implementation (Fig. 1). Associated with these processes are the elements of the State-level concept, namely:

- establishment and pursuit of safeguards objectives for a State (i.e. generic objectives that are established on the basis of the scope of a State's safeguards agreement and related technical objectives that are developed on the basis of either acquisition or diversion path analysis);
- development of a customized safeguards approach for a State and its execution through an annual implementation plan;
- consideration and use of State-specific factors in the implementation of safeguards;
- evaluation of all safeguards relevant information available to the IAEA about a State; and
- drawing and reporting of a safeguards conclusion for a State each year.

6.1 Establishment of Safeguards Objectives

The purpose of IAEA safeguards is to verify a State's compliance with its safeguards obligations. Safeguards implementation is governed by the safeguards agreement and, where applicable, the AP, between the IAEA and the State or States concerned. To implement effective safeguards, the Secretariat pursues State-level objectives on

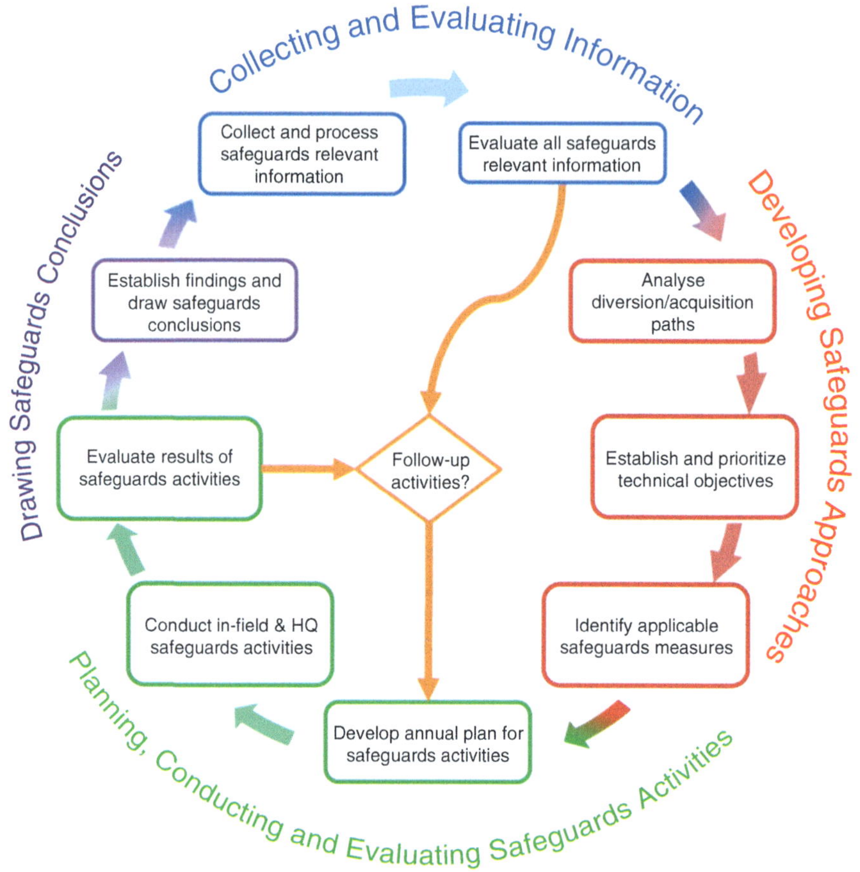

Fig. 1 Processes supporting safeguards implementation (Source: IAEA)

the basis of the State's safeguards agreement. These are generic objectives that are common to all States with the same type of safeguards agreement.

Under a CSA, the IAEA seeks to verify that all nuclear material required to be safeguarded is declared by the State (i.e. the correctness and completeness of the State's declarations). To do so, safeguards activities are conducted to address the following generic objectives:

- to detect any diversion of declared nuclear material at declared facilities or locations outside facilities where nuclear material is customarily used (LOFs);
- to detect any undeclared production or processing of nuclear material at declared facilities or LOFs; and
- to detect any undeclared nuclear material or activities in the State as a whole.

Under an item-specific safeguards agreement,[8] the IAEA seeks to verify that no items subject to safeguards are used for the manufacture of any nuclear weapon or to further any other military purpose and that such items are used exclusively for peaceful purposes and not for the manufacture of any nuclear explosive device. To do so, safeguards activities are conducted to address the following generic objectives:

- to detect any diversion of nuclear material subject to safeguards under the safeguards agreement; and
- to detect any misuse of facilities and other items subject to safeguards under the safeguards agreement.

Under a voluntary offer agreement (VOA), the IAEA seeks to verify that nuclear material in selected facilities or parts thereof is not withdrawn from safeguards, except as provided for in the agreement. To do so, safeguards activities are conducted to address the following generic objective:

- to detect any withdrawal of nuclear material from safeguards in selected facilities or parts thereof, except as provided for in the agreement.

Acquisition/Diversion Path Analysis to Establish Technical Objectives

To address the generic objectives for a State, the Secretariat establishes technical safeguards objectives ('technical objectives') to guide the planning, conduct and evaluation of safeguards activities for that State. Technical objectives are established through the conduct of either an acquisition path analysis (for States with CSAs) or a diversion path analysis (for States with item-specific safeguards agreements or VOAs). The technical objectives are focused on detecting and deterring any proscribed activity (e.g. removal of nuclear material or items placed under safeguards, undeclared production or processing of nuclear material at declared facilities) along a technically plausible path. Unlike the generic objectives, which are common to all States with the same type of safeguards agreement, technical objectives may differ from State to State, depending in particular on the nuclear fuel cycle and related technical capabilities of a State. The technical objectives form the basis for identifying safeguards measures and conducting safeguards activities for a State.

Acquisition path analysis is a structured method used to analyse the plausible paths by which a State, from a technical point of view, could acquire nuclear material suitable for use in a nuclear weapon or other nuclear explosive device. Each plausible acquisition path has multiple steps at which potential indicators[9] or

[8]Item-specific safeguards agreements are based on INFCIRC/66/Rev.2 [3].

[9]Indicators are inputs to a process that suggest that a State may have a capability in that process. Examples of indicators are attempts by the State to purchase equipment contained in Annex II of the AP and other equipment from the export control 'trigger list', and certain dual use items.

signatures[10] could be detected. Technical objectives are established for each of these steps (e.g. processes or activities) to identify what must be detected with respect to diversion of declared nuclear material, misuse of facilities or other undeclared activities. For example, for a State with a CSA, a technical objective under the generic objective of detecting any undeclared nuclear material and activities in the State as a whole might be to detect any undeclared research and development associated with reprocessing. The analysis takes into account the scope of a CSA (i.e. all nuclear material in all peaceful nuclear activities in the State) in order to address the three generic objectives for the State.[11]

The nuclear fuel cycle and related technical capabilities of the State are a key factor considered in the acquisition path analysis. Assessment of a State's nuclear fuel cycle capabilities include consideration of the State's existing nuclear fuel cycle, knowledge of and expertise in nuclear fuel cycle technologies (including through past and current research and development activities), experience with operating related processes and facilities (e.g. enrichment of non-nuclear isotopes), capacity to develop or import the technology and/or expertise needed, ability to produce or import the feed material for the processes, and time and resources needed to develop the necessary capabilities.

The technical objectives are prioritized based on the technical plausibility of the acquisition path for the State and the relative importance of the objective with respect to covering a plausible acquisition path. Factors taken into consideration in prioritizing the technical objectives include (1) the stage of the nuclear fuel cycle the technical objective addresses with priority focused on nuclear material that could be directly used for a nuclear explosive device; (2) the IAEA's ability to detect the associated indicators and signatures; and (3) the number of paths covered by the technical objective. For example, for a State with nuclear power reactors but with no declared spent fuel reprocessing capabilities, a high priority technical objective might be to detect research and development activities related to spent fuel reprocessing.

Diversion path analysis is used to establish technical objectives for States with an item-specific safeguards agreement or a VOA. Diversion path analysis is a structured method to analyse the paths by which, from a technical point of view, nuclear material subject to safeguards could be diverted from a facility, or by which facilities or other items subject to safeguards could be misused.[12] Hence, the technical objectives are limited to detecting steps on a path to divert safeguarded material or to detecting the misuse of facilities or items under safeguards. The objectives

[10]Signatures are emanations from a process that provide evidence that an undeclared activity may be occurring. An example of a signature of undeclared production of high enriched uranium would be high enriched uranium particles in a low enriched uranium enrichment plant.

[11]Acquisition path analysis does not consider weaponization nor does it involve judgements about a State's intention to pursue any such path.

[12]Diversion path analysis does not involve judgements about a State's intention to pursue any such path.

are prioritized based on an assessment of the State's technical capabilities to divert nuclear material or misuse a facility.

6.2 Development of a State-Level Safeguards Approach

The basis for optimized safeguards implementation for a State is a customized SLA, focused on the attainment of safeguards objectives. An SLA, detailed in an internal safeguards document, identifies the generic and technical objectives for the State and the applicable safeguards measures to address them. The SLA is executed through an annual implementation plan.

Applicable Safeguards Measures

Safeguards measures comprise both in-field and Headquarters measures, including inspections, design information verification, complementary access, review of reports and declarations, open source information collection and analysis, satellite imagery analysis, sample taking and analysis, and the use of unattended monitoring systems and remote data transmission. The identification of applicable measures takes into consideration the scope of the IAEA's legal authority (e.g. if the State has an AP in force) and other State-specific factors (e.g. the possibility for the IAEA to carry out unannounced inspections effectively, the technical capabilities of the State or regional safeguards authority). Where possible, an SLA identifies more than one measure that could be used to address each technical objective to provide for flexibility in implementation as well as the comparison of the cost-effectiveness of the different measures. The selection of measures should ensure that an appropriate balance is maintained between activities to verify declarations, to gain a better understanding of a State's nuclear programme, and to detect undeclared activities, as well as between activities conducted in the field and at Headquarters.

Annual Implementation Plan

To execute the SLA, the IAEA develops an annual implementation plan. The plan specifies the safeguards activities to be conducted in the field and at Headquarters in a given calendar year for a State in order to meet the technical objectives. The selection of the activities, and their focus and intensity, take into consideration State-specific factors and results of the State evaluation. An annual implementation plan for a State is revised as necessary during the course of the year in order to respond to new information received and findings from the on-going State evaluation. When a planned safeguards activity cannot be performed in accordance with the annual implementation plan, alternative means for meeting the relevant technical objective are to be considered. The effectiveness of safeguards implementation is evaluated to ensure that the activities conducted achieved the technical objectives specified in the SLA. The evaluation includes assessments to ensure that all activities were conducted in a manner consistent with the IAEA's established procedures and guidelines and that the rationale for any changes to the plan was appropriate and properly documented.

6.3 State-Specific Factors

In developing and implementing an SLA for a State, the IAEA takes into account State-specific factors. This practice is long-standing; indeed the use of State factors in safeguards implementation is called for in all types of safeguards agreements. More systematic consideration and better use of State-specific factors facilitates the further optimization of safeguards implementation.

The term 'State-specific factors' refers to safeguards-relevant features and characteristics that are particular to a State which are used in the development of an SLA for a State, and in the planning, conduct and evaluation of safeguards activities for that State. State-specific factors are based on factual information about a State, are objective and are objectively assessed by the IAEA in the implementation of safeguards. The State-specific factors used by the IAEA are: (1) the type of safeguards agreement in force for the State and the nature of the safeguards conclusion drawn by the IAEA; (2) the nuclear fuel cycle and related technical capabilities of the State; (3) the technical capabilities of the State or regional system of accounting for and control of nuclear material (e.g. does the State authority conduct national inspections or audits or does it possess and use its own verification equipment); (4) the ability of the IAEA to implement certain safeguards measures in the State (e.g. remote data transmission or unannounced/short notice inspection schemes); (5) the nature and scope of cooperation between the State and the IAEA in the implementation of safeguards (e.g. the timeliness and completeness of State reports; facilitation of inspector access; responsiveness to addressing anomalies, questions, or inconsistencies); and (6) the IAEA's experience in implementing safeguards in the State (e.g. the number and type of unresolved anomalies or local security conditions impeding IAEA access).

State-specific factors are considered at different stages of developing an SLA, and in the planning conduct and evaluation of safeguards activities for a State. Certain State-specific factors, such as the State's nuclear fuel cycle and related technical capabilities, are taken into account in acquisition or diversion path analysis and in the determination of technical objectives—key elements in developing an SLA. Other factors, such as the nature and scope of cooperation between the State and the IAEA in the implementation of safeguards, are taken into account in the planning, conduct and evaluation of safeguards activities for the State.

6.4 *Strengthening the State Evaluation Process*

As described above, the State evaluation process is fundamental to safeguards implementation and the application of the State-level concept. The State evaluation process has been strengthened in the recent development efforts. State evaluation groups (SEGs) have been established within the Department of Safeguards for every State with a safeguards agreement in force to conduct the evaluations. SEGs consist of staff members with the appropriate expertise (e.g. on the specific nuclear fuel cycle in the State and associated verification activities) to evaluate all safeguards relevant information for the State. The SEG conducts on-going evaluation of all the safeguards relevant information available to the Agency using structured processes and Departmental methodologies to minimize biases and ensure objectivity. It evaluates the results of the IAEA's safeguards activities to determine whether follow-up activities are needed or whether the information changes the IAEA's knowledge about a State or the assessment of the State's capabilities—which could affect the acquisition path analysis, prioritization of technical objectives or the selection of safeguards measures—thereby ensuring that safeguards implementation is responsive to any changes. The SEG is responsible for annually documenting the results of its evaluation in periodic State evaluation reports and for developing the SLA and annual implementation plan for the State. The proper assessment of the consistency of all safeguards relevant information about a State requires a team of staff members with the requisite knowledge and expertise to conduct collaborative analysis. Training in collaborative analysis for SEGs has been instituted and oversite mechanisms have been put in place to ensure that State evaluation is a well-defined, thorough and robust process.

6.5 *Drawing Safeguards Conclusions*

The final 'product' of safeguards implementation is the IAEA's annual safeguards conclusion on a State's compliance with its safeguards obligations. In order to draw an independent and soundly based safeguards conclusion, the IAEA needs to have conducted a sufficient level of safeguards activities and a comprehensive evaluation of all safeguards relevant information available to it about a State, including the results of its verification activities. It also needs to have addressed all anomalies, questions and inconsistencies identified in the course of its safeguards activities, and assessed whether there are any indications that constitute a safeguards concern. A safeguards conclusion is drawn when all the necessary safeguards activities have been completed and no indication has been found (i.e. there are no 'findings') by the Secretariat that, in its judgement, could constitute a proliferation concern.

The internal process for drawing safeguards conclusions is designed to ensure that the conclusions are valid and based on an adequate level of safeguards activities and follow up activities, and supported by thorough evaluation. The annual State evaluation reports prepared by the SEGs document the results of the consistency analyses and safeguards findings and propose safeguards conclusions and follow-up activities. Approved by the relevant Operations Director, the State evaluation report is reviewed by an internal committee that makes a recommendation on the safeguards conclusion to the Director General. The Director General in turn reports the Secretariat's findings and conclusions for all States to the Board of Governors through the annual Safeguards Implementation Report.

7 Conclusions

Implementation of IAEA safeguards and the drawing of safeguards conclusions has changed dramatically from an ad hoc arrangement in 1962, to an approach focused on verifying nuclear material at declared facilities and drawing safeguards conclusions at the level of individual facilities, to one that assesses the consistency of all safeguards relevant information regarding a State's nuclear programme and draws a safeguards conclusion for the State as a whole. Through implementing safeguards in the context of the State-level concept, a number of benefits in terms of effectiveness and efficiency have been realized, including taking better account of State-specific factors which allow for the development and implementation of customized SLAs. SLAs provide options for safeguards measures to be implemented in the field and at Headquarters, allowing the IAEA to compare their cost-effectiveness and providing greater flexibility in safeguards implementation. Instead of mechanistically applying the activities listed in the safeguards criteria, the implementation of SLAs is focused on attainment of technical objectives. By doing so, safeguards implementation is more performance-oriented and is helping the IAEA to avoid spending resources on doing more than is needed for effective safeguards. The objectives are established by the Secretariat using structured and technically based analytical methods (i.e. acquisition or diversion path analyses) conducted according to uniform processes and defined procedures. Utilizing the same technically-based processes and procedures for developing SLAs for all States helps to ensure consistency and non-discrimination in safeguards implementation, efficient performance of the work, and more soundly based safeguards conclusions.

This work of authorship and those incorporated herein were prepared by Consolidated Nuclear Security, LLC (CNS) as accounts of work sponsored by an agency of the United States Government under Contract DE NA0001942. Neither the United States Government nor any agency thereof, nor CNS, nor any of their employees, makes any warranty, express or implied, or assumes any legal liability or responsibility to any non governmental recipient hereof for the accuracy, completeness, use made, or usefulness of any information, apparatus, product, or process disclosed, or represents that its use would not infringe privately owned rights. Reference herein to any specific commercial product, process, or service by trade name, trademark, manufacturer, or otherwise, does not necessarily constitute or imply its endorsement, recommendation, or favoring by the United States Government or any agency or contractor thereof. The views and opinions of authors expressed herein do not necessarily state or reflect those of the United States Government or any agency or contractor (other than the authors) thereof.

References

1. International Atomic Energy Agency (IAEA) (1972) The Structure and Content of Agreements Between the Agency and States in Connection with the Treaty on the Non-Proliferation of Nuclear Weapons, INFCIRC/153 (Corrected)
2. International Atomic Energy Agency (IAEA) (1997) Model Protocol Additional to the Agreement(s) between State(s) and the International Atomic Energy Agency for the Application of Safeguards, INFCIRC/540 (Corrected)
3. International Atomic Energy Agency (IAEA) (1968) The Agency's Safeguards System (1965, as provisionally extended in 1966 and 1968), INFCIRC/66/Rev.2

Part II
Factors Influencing the Development of a Systems Concept for Verification

Verification of compliance is a legal determination made by governments that is essential to building confidence in agreements made between nations. The verification of declarations is important to the detection of treaty violations but of equal importance is the ability to confirm the absence of non-complaint activities. This is carried out in the context of legal, political and military considerations—in an ever-changing environment. Risk and systems-driven verification makes use of more quantitative methods and judgments but uncertainty will always exist. A review of the complex factors that influence verification and some lessons-learned from existing nonproliferation, arms control and export control regimes, help to identify problems and sensitivities and can offer approaches for new initiatives.

Legal Issues Relating to Risk and Systems Driven Verification

John Carlson

Abstract Broadly speaking, current verification approaches—exemplified by *traditional* IAEA safeguards—seek to show whether a declaration made by a state under a particular treaty is true or false. The IAEA safeguards experience shows the limits to this approach—not all situations are so neatly bivalent. A major focus for safeguards development over the last 20 years or so is the problem of detecting, and drawing conclusions about, *undeclared* nuclear material and activities. This involves the difficulty of proving a negative—can *absence of evidence* be taken to be *evidence of absence*? As the scope of arms control is extended and nuclear disarmament progresses, future verification missions will be dealing with increasing degrees of *uncertainty*, having to make judgments about unknowns. Risk and systems driven verification, involving greater use of qualitative methods and judgments, will be an essential aspect of this. With greater uncertainty, the *confidence-building* objective of verification will be more difficult, and at the same time more important. A less definitive verification environment has major implications for the application of legal precepts such as the standard of proof, the duty to cooperate, and non-discrimination, and related policy issues such as the use of information. Compared with the traditional emphasis on the right of the state, more focus will be required on the international interest—emphasising the need for cooperation and transparency. The need for an effective approach to managing uncertainty should also be reflected in the way new treaty regimes are developed, for example in decision-making processes, non-compliance determinations, adaptability of verification procedures to meet evolving circumstances, dispute resolution, enforcement, and so on. In the development of risk and systems driven verification, and the associated legal and institutional aspects, the IAEA verification system has a deep and rich experience to draw on.

J. Carlson (✉)
Vienna Center for Disarmament and Non-Proliferation, Vienna, Austria

I. Niemeyer et al. (eds.), *Nuclear Non-proliferation and Arms Control Verification*,
https://doi.org/10.1007/978-3-030-29537-0_4

1 Introduction

In thinking about a new verification philosophy, and the legal issues involved, the first question that comes to mind is, what is different about risk and systems driven verification, compared with established verification approaches?

Broadly speaking, the basis of current verification approaches could be described as the *falsification* (that is, showing to be false) of declarations made by states. These declarations relate to treaty-accountable materials or items, and include details such as quantities, locations, physical and chemical forms, and so on. Verification measures to confirm the accuracy or otherwise of a state's declarations include on-site inspections, visual observation, and measurement and analysis. Verification can be complemented by measures to maintain continuity of knowledge (such as containment and surveillance).

Applying this approach, verification seeks to show whether a state's declaration is false (that is, not true). It follows that a declaration cannot be considered *meaningfully verifiable* unless in principle it could be shown to be false (if in fact this *is* the case). Verification measures and strategies are required that can provide a definitive outcome. With this approach verification findings are *bivalent*, that is, the declaration is either false or it is true.

A familiar example of this approach is *traditional* IAEA safeguards. The IAEA safeguards experience shows the limits to this approach—not all situations are neatly bivalent. The obvious example is the problem of detecting, and drawing conclusions about, *undeclared* nuclear material and activities (that is, nuclear material and activities which a state has a legal obligation to declare but has not done so). This has been a major focus for safeguards development over the last 20 years or so. This problem involves the difficulty of proving a negative—can *absence of evidence* be taken to be *evidence of absence*?

If we thought too rigorously about verification, we might conclude that the absence of something is *inherently unverifiable*—it is not something that can be shown to be false, therefore it is not meaningfully verifiable. Yet, there is international consensus that the safeguards system *must* endeavour to provide assurance about the absence of undeclared material and/or activities.

Furthermore, it is obvious that in the real world *at least some* undeclared materials and activities *can* be detected. The key factors in this regard include:

(a) authority—the access to locations, information and people available to the verifier;
(b) capability—verification technologies and resources; and
(c) methodology—how to design and implement effective detection strategies, and how to draw valid conclusions.

Conclusions about *absence* will be *qualitative*, rather than bivalent—and this brings us to an answer to the opening question: risk and systems driven verification is distinguished by the greater use of qualitative methods and judgments.

This is not to say that qualitative judgments were absent from traditional safeguards—for instance, decisions on safeguards intensity have always taken account of risk analysis—but qualitative methods and judgments will become more prominent.

While the main focus on qualitative judgment concerns managing *uncertainty*—the challenge of undeclared material and activities—the IAEA safeguards experience shows that qualitative judgment is also important in meeting another contemporary challenge: how to manage verification workloads that are growing faster than available budgets and resources.

2 Future Verification Missions

It is likely that extending the scope of arms control, and progress towards the eventual elimination of nuclear weapons, will proceed in a stepwise manner, involving a range of new verification missions. At this stage the exact steps and sequence are speculative, but the new missions are expected to be along the following lines:

- fissile material cut-off—proscribing production of fissile material for nuclear weapons, with verification to confirm that all production after entry-into-force is for peaceful or non-proscribed purposes;
- fissile material disposition—transparency measures for fissile material inventories, and verification of progressive, irreversible transfer of fissile material to peaceful or non-proscribed purposes;
- nuclear weapon limitations—verification that deployed nuclear weapon numbers are within agreed limits;
- nuclear weapon dismantlement—verification that nuclear weapons are dismantled as agreed, and the resulting fissile material is irreversibly transferred to peaceful or non-proscribed purposes;
- nuclear inventory baselines (involving *nuclear archaeology*)—to help assure there is no nuclear material outside the purview of verification arrangements.

To varying degrees these new verification missions can be expected to involve features similar to those of current IAEA safeguards, such as:

- legally binding (that is, treaty-based) commitments to undertake, or refrain from, specified actions and to accept verification measures;
- a multilateral inspectorate;
- declarations—requirement for parties to declare relevant materials, items and activities, and to maintain records and reporting;
- ongoing inspections and other verification measures to confirm parties' declarations;
- verification activities aimed at detecting possible *undeclared* materials, items and/or activities—including information collection and analysis, monitoring and

other detection techniques, and necessary additional access rights for inspectors; and
- procedures for determining compliance and taking enforcement action.

Entrusting these new missions to the IAEA would take advantage not only of the IAEA's unparalleled verification experience, but also its institutional and legal arrangements in decision-making, establishing technical criteria and methods, determination of treaty compliance and so on. If states decide that new verification agencies are more appropriate, they will need to ensure the new agencies are given similar authorities and capabilities.

3 Verification Objectives

In developing and implementing any verification system it is essential to have a clear idea of what the objectives are. The objective of verification may seem self-evident: to detect violations (material breaches) of, or non-compliance with, the provisions of the relevant treaty. The statement of objectives for IAEA safeguards [1, para 28] elaborates on this, referring to:

(a) timely detection of diversion of significant quantities of nuclear material from peaceful nuclear activities to the manufacture of nuclear weapons or of other nuclear explosive devices or for purposes unknown; and
(b) deterrence of such diversion by the risk of early detection.

There are some important qualifications built into this statement—for example, *timely*, *significant* quantities, *other nuclear explosive devices*, *purposes unknown*—that help interpret the objective and apply it to actual situations. In particular the statement avoids making the objective *too rigorous*, and therefore too difficult to achieve in practice. Thus, the IAEA does not have to prove a device is a nuclear *weapon*, only that it is a nuclear explosive device; and inspectors don't have to discover the actual device (in practice the state is likely to obstruct inspectors from doing this), it is sufficient to show that *diversion* has occurred (nuclear material is unaccounted for or undeclared nuclear material has been found) and that the *purpose* for the diversion is unknown.

The IAEA safeguards statement also goes beyond the basic objective of detection, and points to the important *deterrent* value of detection risk (or more accurately, *detection probability*).

There is a third objective of verification—providing confidence. If a verification system is perceived as being *effective*—having a sufficiently high detection probability—it will reinforce the confidence of each party to the treaty that the other parties are meeting their treaty commitments, and will thereby reinforce commitment to the treaty. Where detection probability is easy to calculate (for example, verification of a declaration of a specified number of treaty-accountable items), confidence is readily established. In situations of greater uncertainty (more unknowns)—which is the likely operating environment for risk and systems driven verification—confidence-building will be more difficult, and at the same time more important.

These considerations underlie a number of legal issues relating to verification, such as standard of proof, duty to cooperate, non-discrimination, and use of information.

4 Some Key Legal Considerations

4.1 Non-compliance: Standard of Proof

It is unrealistic to require the IAEA to *prove* the existence of a nuclear weapon program. It is most unlikely inspectors will find a nuclear weapon or nuclear weapon components. A state facing exposure in such an obvious way is likely to deny inspectors access and try to maintain ambiguity. The drafters of the IAEA's standard comprehensive safeguards agreement (IAEA document INFCIRC/153) [1] understood that the standard of proof should not be set impracticably high, hence the agreement refers to diversion for *purposes unknown* (para 28). Further, the agreement provides for the IAEA Board of Governors to report non-compliance to the Security Council if the IAEA is *not able to verify* that there has been no diversion to nuclear weapons (para 19). The effect of these provisions is that the standard of proof for safeguards is the *balance of probabilities*. This is consistent with general international practice, which favours the balance of probabilities where this is practical or realistic.[1]

Just as the standard of proof should not be too high, it is also essential not to set it too low, or this will impact on the credibility of the verification conclusions. There needs to be sufficient evidence to show that a weapon purpose is plausible in the circumstances. The IAEA has not attempted to articulate rules or guidelines for this, but relevant factors would include: the materials and quantities involved; whether there is a systematic pattern of safeguards breaches; whether the breach or breaches could be part of an overall program aimed at acquiring weapons; whether inspectors are being obstructed; and so on.

If new verification agencies are established, it will be important to drawn on IAEA experience to ensure that the process for determining non-compliance is not overly rigorous and allows for qualitative judgments.

[1]As a general rule international tribunals are not bound by strict rules of evidence or standards of proof. International practice is to take into account the circumstances of the case, particularly what is practical or realistic. This has led to the International Court of Justice and other tribunals adopting a *preponderance of evidence* test, drawn from the Civil Law system, in appropriate circumstances.

4.2 Obligation to Cooperate

A state cannot be allowed to frustrate the verification agency in the performance of its responsibilities. Cooperation with the verification process is an essential aspect of compliance. This is reflected in the standard IAEA safeguards agreement, which provides as one ground for a non-compliance determination that the Agency is *not able to verify* that there has been no diversion. The phrase *not able to verify* can cover a number of circumstances, but inability to conduct verification because of obstruction or lack of cooperation by the state is an obvious example. It is important that new verification treaties include non-cooperation as a case of non-compliance.

Looking at cooperation more broadly, states should view verification activities positively, as presenting an opportunity for them to demonstrate that they are meeting their treaty obligations. It is important not to view the relationship with inspectors as an adversarial one. Cooperation with inspectors is not just an obligation, but an essential part of a collaborative activity towards a common goal, namely, the successful operation of the particular treaty.

Cooperation and confidence are closely related. If a state cooperates fully with inspectors this adds to confidence about the state's commitment to treaty objectives. An adversarial approach to verification has the opposite effect. If a state asserts a legalistic interpretation, attempting to narrow down the inspectors' remit, this will raise suspicions that the state is trying to hide something. In the context of IAEA safeguards, it has been noted that lack of full cooperation with inspectors will, at the least, result in a *confidence deficit*, which will be counter-productive to the inspected state's own interests. Safeguards provide an objective mechanism for resolving suspicions—cooperation with the IAEA is clearly in the interest of a state that has nothing to hide.

As verification moves into areas of greater uncertainty, ambiguity and lack of conclusive evidence are likely to become more prevalent. In these circumstances the full cooperation of the state assumes even greater importance. This applies not only to issues of inspector access, but also to provision of information, including establishment of transparency arrangements.

4.3 Non-discrimination

Non-discrimination is essential not only in meeting states' concerns about equality and fairness, but especially in avoiding biases that could result in unsound verification conclusions (in particular, that a state has complied with a treaty when in fact it has not).

An essential feature of risk and systems driven verification is that a verification approach will be designed to meet state-specific factors, that is, the specific circumstances of each state. This is exemplified by the State Level Concept (SLC) now used in IAEA safeguards. In the development of individual State Level

Approaches (SLAs) under the SLC, the IAEA has emphasised that *differentiation* amongst states, taking account of their differing factual circumstances, does not equate to discrimination provided an objective process, applicable to all states, is used.

Several years after the introduction of the State Level Concept, some states have questioned whether bias may have crept into the SLC process. As a consequence, over the past 4 years or so the Board of Governors has reviewed the SLC process. The Board's close interest indicates, not that there is a problem inherent in the SLC as a concept, but rather the importance of the IAEA secretariat clearly articulating how the objective process is applied in practice. In particular, it is essential to show there is a quality assurance process that counters potential bias—accordingly the secretariat has given giving priority to establishing a robust quality assurance process [2]. Demonstrating non-discrimination is an essential aspect of building confidence in new verification approaches.

4.4 Use of and Access to Information

Information collection and analysis are central to risk and systems driven verification. Major issues involved here include: whether there are limitations on the kind of information that can be used by a verification agency; protection against false information; whether information available to a verification agency can be shared with states; and information-sharing between verification regimes.

Type of Information IAEA practice has established that there are no *a priori* limitations on the kind of information that can be collected and analysed for verification purposes. For example, the IAEA often receives information from states about nuclear activities in other states, including information collected through national technical means (*intelligence*). This is consistent with the IAEA Statute [3], which provides that each member should make available such information as would, in the judgment of the member, be helpful to the Agency (Article VIII.A).

Some states have expressed concerns about unreliability or bias in information provided by one state about another, but the IAEA's response is that every state with information on matters under investigation should contribute such information to the Agency[2]—bias can be countered through comparing information from diverse sources, and unreliability can be resolved through investigation, including inspections. In practice, the main problem in this area is not that inaccurate or biased information is being supplied to the IAEA, but that states remain reluctant

[2]The final documents of the 1995, 2000 and 2010 NPT Review Conferences affirm that states should direct concerns regarding non-compliance by others, along with supporting evidence, to the IAEA to investigate and decide on necessary actions (there were no final documents agreed at the 2005 and 2015 Review Conferences).

to share intelligence information, notwithstanding that the Agency has rigorous confidentiality procedures to protect sensitive information.

The IAEA is well aware of the risks in being dependent on information provided by states, and has established a program to collect and analyse open source information. This helps the Agency to independently assess information provided by states, and also identifies matters warranting further investigation, including through commissioning commercial satellite imagery. Information collection and analysis has become an essential aspect of the design of a State Level Approach and verification plan for each state.

An issue receiving increasing attention is the possible use of *societal verification* for treaty verification purposes. Some states are hostile to this on grounds of unreliability. On the other hand, studies have shown this is a potentially valuable source of information as part of a holistic approach to information analysis. As thinking about the use of such information develops, it should be possible to build in quality checks against the problem of unreliable information.

Protection Against False Information Some concerns have been expressed that the IAEA could be given inaccurate or even deliberately false information which could result in adverse safeguards conclusions. However, the IAEA would not base its conclusions solely on such information but would seek to test it through consistency checks against information from other sources and, where warranted, through inspections.

IAEA safeguards agreements give the state the opportunity to refute information used by the Agency as the basis for an inspection or complementary access. In the case of a request for a special inspection, the Agency is required to consult the state [1, para 77]. If there is any disagreement the state has the right to refer the issue to the Board of Governors [1, para 22]. In the case of complementary access under the additional protocol to resolve a question or inconsistency, the IAEA must first provide the state with an opportunity to clarify and facilitate the resolution of such question or inconsistency [4, Article 4.d]. Other verification treaties have similar provisions. For example, both the Comprehensive Nuclear-Test-Ban Treaty (CTBT)[3] and the Chemical Weapons Convention (CWC) provide for consultation and clarification prior to any on-site or challenge inspection [5, Article 29 and following]; [6, Article IX.2 and following].

If a state maintains that information provided to the IAEA or other verification agency is inaccurate, it is in the interest of that state to cooperate and help the agency establish the true facts. Assuming the information is location-specific, the verification agency can if necessary undertake an inspection. The state should have no objection to an inspection: all verification treaties have managed access provisions to protect national-security or other sensitive information. If the information

[3]The CTBT has not yet gained the necessary ratifications for entry into force. Pending entry into force, the CTBT's International Monitoring System is operating on a provisional basis, but the treaty's on-site inspection provisions are not in operation.

is not location-specific, therefore not readily resolvable by inspection, the state can cooperate with the verification agency in finding a way of resolving outstanding concerns.

Confidentiality of Information The standard IAEA safeguards agreement requires the IAEA to maintain confidentiality of information obtained in connection with the implementation of the agreement [1, para 5]; see also the Statute [3, Article VII.F]. This blanket proscription on information sharing contrasts with more recent treaty-based regimes. Under the CWC's provisions, much of the information in parties' declarations and summary data from inspections is made available to the parties at large. The CTBT takes information sharing further: parties are given equal access to all data collected by the CTBT's International Monitoring System. Any party is able to analyse the data for itself and to seek clarification of a suspect event, including through calling for an on-site inspection.

Thus under both the CWC and the CTBT parties are in a position to cross-check the information available to the verification agency—to identify gaps in that information where they may be able to assist, and to reach an informed assessment on how well the verification system is operating. For future verification missions consideration should be given to the confidence-building benefits of sharing information among treaty parties.

Information-Sharing Across Verification Regimes Currently there are political and legal barriers to information-sharing: for example, the IAEA considers that the requirement for confidentiality of safeguards information precludes it from sharing such information with the CTBT's Provisional Technical Secretariat even where there would be a clear mutual benefit, in terms of verification effectiveness, in doing so. IAEA member states should consider modifying the proscription on sharing safeguards information, to enable information sharing with other treaty verification agencies on a confidential basis where this would contribute to better verification outcomes.

In systems driven verification, where effective use of a broad range of information is essential, better arrangements are required to ensure that information from the different verification agencies is available where it is needed.

4.5 *Verification Tools: Inspections and Monitoring*

Risk and systems driven verification will make less use of regular inspections. As is the case today with strengthened safeguards and the additional protocol, there will be more emphasis on *unpredictability* in the timing of inspections and on access by inspectors to a broader range of locations. The IAEA's additional protocol provides for access to a range of nuclear-related locations, as well as a right of access anywhere in a state if required to resolve a *question or inconsistency* concerning information provided by the state.

Challenge Inspections An important aspect of inspector access is the availability of challenge inspections, that is, non-routine inspections aimed at investigating suspected treaty violations.

IAEA safeguards agreements allow for the Agency to request access of this kind under the *special inspection* provisions. Unfortunately member states have regarded special inspections as being too accusatory, with the consequence that the IAEA has invoked its special inspection authority only twice, in Romania (1992) by invitation, and in North Korea (1993).[4] It seems there is too high a political bar to use of special inspections, the cost of "failure" being considered too high. This attitude is mistaken: if a special inspection finds nothing, and in so doing shows that suspicions were unfounded, that should be considered a *success*—the inspection has served its purpose (assuming of course that the inspection was competently conducted at the correct location!).

Contention over special inspections was one of the influences in the development of the *complementary access* concept contained in the IAEA's additional protocol. As noted above, complementary access may be available anywhere in a state. Complementary access, however, is no substitute for a special inspection: only specified verification activities (primarily environmental sampling) may be conducted; and for certain types of location, if the state claims it is unable to provide access, verification activities may be limited to *adjacent locations*. So a special inspection may well be required to conclusively resolve the matter.

Other treaty regimes employ a different approach to challenge inspections. In the CWC and the CTBT, such inspections may be initiated by treaty parties, rather than the inspectorate. The CWC uses the specific term *challenge inspection*. The inspectorate (the OPCW—Organisation for the Prohibition of Chemical Weapons) conducts routine inspections at declared sites, but does not attempt to detect undeclared activities away from declared sites. A treaty party may request a challenge inspection, if it has reason to believe there may be a prohibited activity. To guard against frivolous or vexatious inspection requests, a request for a challenge inspection can be disallowed by a three quarter majority vote of the OPCW Executive Council (a *red light* filter).

The CTBT has no routine inspections, and no standing inspectorate. Instead, an International Monitoring System is operated under the treaty, looking to indicators of a prohibited event, namely, a nuclear explosion. After the CTBT enters into force, a party believing that a nuclear explosion has taken place may call for an *on-site inspection* of the location concerned. The request must be approved by at least 30 of the 51 members of the CTBTO Executive Council (roughly a 60% majority—a *green light* filter).

In the case of new verification treaties, unless they adopt the IAEA safeguards system, it will be necessary to provide for inspections similar to challenge inspections. The CWC approach with its *red light* filter would seem to be an appropriate

[4]North Korea refused to allow the special inspection, leading to a non-compliance finding by the Board of Governors.

model. The approach of the CTBT *green light* filter may be too difficult to achieve in practice—as inspections provide the means of resolving suspicions, it would be counterproductive to set the bar too high.

In the case of the IAEA safeguards system—whether for its current remit or for future verification responsibilities that might be entrusted to the Agency—there is a case for having, in addition to the special inspection mechanism, a challenge inspection mechanism that can be initiated by a state if the secretariat does not initiate a special inspection.

Timeliness of Access All of the challenge access mechanisms outlined above have some procedural delays. In the case of a special inspection, the IAEA must first consult the state and seek agreement. If the state does not agree, the state can invoke arbitration [1, para 77]. However, if the secretariat considers that access is essential and urgent, it can request the Board to call upon the state to take requested action without delay [1, para 18]. It is hard to say how long this process could take, but even without delay for arbitration it could be several days before an inspection could proceed.

In the case of complementary access under the additional protocol, the state is to be given an opportunity to clarify the matter before access is requested. Advance notice of at least 24 h is required for access—so here too the overall process could take several days. Under the CTBT, a state that is subject to an on-site inspection request must be given 72 h to clarify the matter. The Executive Council must take a decision on the inspection request within 96 h of receipt. The state is to receive at least 24 h notice of the arrival of the inspection team. Thus at least 5 days could elapse, probably rather longer, before the inspection commences.

These time frames operate in the context that evidence of nuclear activities will be difficult to remove and traces could be detected for many months or longer even if a deliberate effort was made to sanitise the location. Obviously this depends on the subject of the inspection—if it is a relatively small discrete item such as a nuclear weapon, successful removal and concealment may well be possible within a very short time. Future verification missions are likely to require development of new procedures to ensure that rapid access is possible.

Monitoring Depending on the development of suitable technologies, in the future some form of wide area monitoring may well play a key role in detecting prohibited activities away from known (that is, *declared*) locations. The CTBT's International Monitoring System (IMS) is in provisional operation—this deploys a number of verification technologies looking for indicators that a prohibited activity (a nuclear explosion) has occurred. Data from the IMS will show if there has been a prohibited activity and should enable its location to be identified, enabling investigation through an on-site inspection.

Extensive experience of wide area environmental monitoring was gained in Iraq after the first Gulf War, when air and water sampling were deployed to detect any indicators of undeclared chemical or nuclear activities. Subsequent studies into possible use of wide area environmental monitoring by the IAEA have concluded that the technologies are not yet sufficiently developed to be cost-effective for safeguards, but this is expected to be just a matter of time. The IAEA's additional protocol anticipates the use of *wide area environmental sampling*, but requires the Board of Governors' approval of the technique before the IAEA can request access to a location in a state to carry out such sampling [4, para 9]. This procedural requirement applies only to the particular circumstances set out in the additional protocol—that is, that the IAEA seeks access to locations through the protocol. If the Agency does not require access, or does not use the additional protocol as a basis for seeking access, Board approval to deploy wide area environmental monitoring is not a legal requirement.

4.6 Dispute Resolution

Under the IAEA's standard safeguards agreements, including the additional protocol, a state with any question as to the interpretation or application of its safeguards agreement may take the issue to the Board of Governors. If not satisfied with the Board's response, the state can request arbitration, except in the case of a finding of non-compliance. The Board may call upon the state to take action that is essential and urgent to ensure verification that diversion has not occurred, irrespective of whether the dispute settlement procedures have been invoked—that is, the state cannot refuse to take action (such as facilitating an inspection) on the basis that there is a dispute.

The CTBT and CWC have similar, albeit simplified, provisions allowing for urgent action to be taken to uphold the objects of the treaty notwithstanding that the state concerned may have invoked the dispute settlement procedures. In new verification regimes it will be essential to include provisions to deal with this issue.

4.7 Treaty Structure

For substantially new verification missions—as distinct from further development and strengthening of the current IAEA safeguards system—it is assumed that the parties will require these missions to have legal force, and therefore will require new treaties. This does not preclude the IAEA from the responsibility for implementing new missions. On the contrary, the IAEA Statute authorises the Agency to apply safeguards pursuant to unilateral, bilateral or multilateral arrangements where the parties so request [3, Article III.A.5].

The Statute gives the term *safeguards* a broad meaning, basically to ensure that *materials, services, equipment, facilities, and information are not used in such a way as to further any military purpose* [3, Article XII], with the details to be specified in an agreement between the Agency and the state(s) concerned [3, Article XII.A.6]. In the case of NPT safeguards, the details are set out in a bilateral agreement between each NPT party and the IAEA.

Executive Authority Adding new verification missions to the IAEA's responsibilities would enable the parties to take full advantage of the Agency's technical capabilities. Depending on what the parties want, they could also take advantage of the Agency's institutional arrangements and working procedures, including those for dealing with the legal issues discussed above. In this respect, the main issue for the parties to decide is where they want decision-making authority to reside. This is of critical importance for decisions on compliance, remedial action, disputes, and so on. Are the parties prepared to have the Board of Governors take these decisions or do they want a separate executive body, specific to the particular treaty?

The NPT is an example of a major treaty that does not have an executive body: in effect the IAEA Board of Governors serves this purpose, at least for decisions on safeguards compliance. In the case of new treaties, for example the proposed Fissile Material Cut-off Treaty (FMCT), it is possible that the parties may wish to entrust the verification function, but not the executive function, to the IAEA. This is because not all the nuclear-armed states are permanent members of the Board of Governors: unless this is changed, those that are not permanent Board members may insist on a separate FMCT Executive Council which guarantees them a seat. While it would be complicated to have two separate executive bodies involved, this is workable: it would require detailed provisions on the relationship between the Board and the new treaty body.

Verification Adaptability Risk and systems driven verification approaches can be expected to develop in an iterative manner as they have to adapt to new and unexpected situations. Ready adaptability will be an important attribute. The IAEA safeguards system has shown itself to be capable of evolutionary development. If new verification missions are built on the IAEA system, it should not be difficult to ensure adaptability. However, if new treaties establish new verification bodies, it will be important to establish procedures that work efficiently for approving and updating the operational details of the verification system.

The IAEA safeguards system has a complex multi-layer framework. The first level is the NPT, a multilateral treaty which sets out the basic political commitments under the non-proliferation regime. The second level comprises safeguards agreements, which are bilateral treaties between each state and the IAEA. The safeguards agreements and additional protocols set out the legal rights and duties of the IAEA and the state with respect to implementation of safeguards, including: the national accounting system; information to be provided to the IAEA; rights of access by inspectors; types of inspections; use of safeguards equipment; determination of compliance; and so on.

The third level comprises documents of less-than-treaty status, setting out working procedures: *subsidiary arrangements* for the state, and *facility attachments* for individual facilities. While these documents are not legally binding, they are given legal effect through their direct relationship to the implementation of particular provisions of the safeguards agreement. Finally, there are the IAEA's internal working documents: technical criteria and the like.

This multi-level approach, with a separation of mainly political from mainly technical subject matters, has proven highly effective, ensuring a verification system that is readily adaptable to changing circumstances and new challenges.

The negotiation of the model additional protocol (INFCIRC/540) [4], agreed in 1997, illustrates how this basic approach allows flexibility for major updates of the verification system. Another advantage of having separate negotiations on the technical details is that such negotiations can proceed quite expeditiously. Despite their complexity, negotiation of the standard safeguards agreement (INFCIRC/153) and the model additional protocol (INFCIRC/540) each took only about 18 months to complete.

Other treaties have taken a different approach. For example, the CWC comprises a single treaty, which sets out the basic political commitments of the parties, establishes the treaty organization, including executive and verification bodies, and details the verification system in an annex, the Verification Annex.

The Verification Annex is very detailed, totalling 94 pages. A potential difficulty in having a substantial level of detail on the verification system as part of the principal treaty is that any change to these details will involve a formal treaty amendment. The CWC contains a (relatively) simplified approval process for *technical* amendments (if an amendment is recommended by the Executive Council and no party objects within 90 days), but the process takes some 320 days, and presumably will involve formal treaty amendment procedures in member states. If an amendment does not pass the simplified process (for example, if one party objects), it must be treated as a substantive amendment, which requires an Amendment Conference with a majority of all parties voting in support and no party casting a negative vote. Updating the verification system could be a major political exercise, and could prove very difficult in practice.

The CTBT has a two-level framework: the Treaty contains the political commitment and sets out the principal elements of verification. Verification modalities are contained in a Protocol to the Treaty. These modalities are to be supplemented by detailed working-level operational manuals.

The Protocol has a simplified process for *changes*, as distinct from the amendment procedure for the main Treaty provisions. Changes may be proposed by any party: if the change is recommended by the Executive Council it shall be adopted unless a party objects within 90 days. If there is any objection, however, the change becomes a matter of substance for consideration at the next meeting of the Conference of Parties. Matters of substance require a two-thirds majority of parties present and voting. As the Protocol is part of the Treaty, any change might involve formal treaty processes within the national laws of the parties. As the operational manuals are of less-than-treaty status, amendments to the manuals should be easier to deal with.

5 Conclusions

No major legal obstacles are foreseen in the development of risk and systems driven verification. However, risk and systems driven verification is likely to be addressing increasing degrees of uncertainty, having to draw conclusions about unknowns. With greater uncertainty, the *confidence-building* objective of verification will be more difficult, and at the same time more important. A less definitive verification environment has major implications for precepts such as the standard of proof, the duty to cooperate, non-discrimination, and the use of information. Compared with traditional verification practice, there will be greater emphasis on the need for inspected states to demonstrate cooperation and transparency.

New verification missions will require appropriate legal frameworks for determining treaty compliance, and so on. Verification objectives will need to be carefully defined, and it will be important to ensure that verification modalities and arrangements can be readily updated as verification approaches undergo continuous improvement.

The IAEA already has a substantial level of experience with new verification approaches. The current IAEA verification system is the product of decades of evolution. It will be difficult for entirely new verification systems to replicate the way the IAEA system works. This is a good reason to build on the IAEA system, or at least to learn from it, rather than trying to “re-invent the wheel”.

References

1. International Atomic Energy Agency (IAEA) (1972) The Structure and Content of Agreements Between the Agency and States in Connection with the Treaty on the Non-Proliferation of Nuclear Weapons, INFCIRC/153 (Corrected)
2. International Atomic Energy Agency (IAEA) (2014) Supplementary Document to the Report on The Conceptualization and Development of Safeguards Implementation at the State Level (GOV/2013/38), GOV/2014/41, pp 46–48
3. International Atomic Energy Agency (IAEA) (1989) Statute
4. International Atomic Energy Agency (IAEA) (1997) Model Protocol Additional to the Agreement(s) between State(s) and the International Atomic Energy Agency for the Application of Safeguards, INFCIRC/540 (Corrected)
5. Comprehensive Nuclear-Test-Ban Treaty (CTBT)
6. Chemical Weapons Convention (CWC)

Verification: The Politics of the Unpolitical

Erwin Häckel

Abstract The nuclear safeguards system of the International Atomic Energy Agency—the world's most widely accepted scheme of multilateral verification—entails an inescapable paradox. While it is based on strictly scientific terms and methods of data collection and evaluation, equality of treatment and meticulous observance of non-discrimination among members of the Agency, its frame of reference is avowedly political, shaped by historical contingencies, steeped in the gradients of power and influence, prejudiced on mutual distrust and jealousy as is typical of international politics. The evolution of the safeguards system over time has brought this paradox more sharply to the fore. As demands on the technical accuracy and reliability of IAEA verification were being continuously increased, so was the ambition of Member States to bring their influence to bear on the definition and operation of safeguards. Efforts to resolve the paradox have resulted in an increasingly intimate meshing of politics and verification. By way of shifting the focus of safeguards from nuclear material accountancy and facility monitoring to "the State as a whole", and more recently the adoption of the "State-level concept", it is now the correlation of international power itself which has become dominant in the direction of the Agency's safeguards effort. Balancing these competing interests will remain a big challenge for the IAEA. 2018 afterthoughts on current developments at the end of the chapter reconfirm the inherent dilemma of international verification.

1 Introduction

Verification is defined as a procedure to authenticate conformity with established standards; international verification is understood to monitor and evaluate compliance with international treaty obligations. The concept and practice of

E. Häckel (✉)
Department of Political Science and Public Administration, University of Konstanz, Konstanz, Germany
e-mail: erwin.haeckel@t-online.de

I. Niemeyer et al. (eds.), *Nuclear Non-proliferation and Arms Control Verification*,
https://doi.org/10.1007/978-3-030-29537-0_5

international verification has been developed and refined over several decades since the Second World War, mainly in the context of arms control and disarmament efforts, in particular in the nuclear field. Controls over nuclear technology have become the mainstay and driving force in the evolution of international verification. Nuclear nonproliferation was and continues to be the first and foremost concern of international verification; from this origin the verification approach has spread to other fields of transnational regulation such as chemical and biological weapons, environmental change, or drug control. In this chapter, the focus is set primarily in the framework of the nonproliferation role of the International Atomic Energy Agency (IAEA), with occasional glances at other areas.

Verification is a political project, but there is an inherent contradiction between politics and verification. Politics is concerned with essentially unregulated matters such as power, conflict, interests and beliefs. Verification is concerned with regulated matters such as standards, measurements, proofs and consistencies. Politics tends to be subjective, voluntaristic and contentious; verification strives to be objective, truthful and consensual. Politics is about shaping the future, verification about evaluating the past. Where politics claims to possess normative autonomy and legitimate authority, verification invokes scientific analysis and incontrovertible evidence. Politics and verification appear like fire and ice, working in principle to the exclusion of each other.

So it seems. But, of course, things are not as clear and simple as that, most obviously in the context of nuclear nonproliferation where the stakes for politics are extremely high and the requirements for verification extremely strict. International verification and nuclear policies are dependent upon each other and inseparably intertwined. If nonproliferation is taken as an important political objective, binding international agreements are required to that effect, the faithful observance of which needs to be assured by verification. If assurance through verification is assumed to be indispensable for nonproliferation, unequivocal political commitments are necessary in support of its implementation. What is more, the two seemingly contrary principles tend in practice to reinforce each other. The higher the aims and ambitions of nonproliferation policy are raised, the stricter and more demanding the standards of verification need to become. It is this spiralling connection of politics and verification which has inspired the nonproliferation conundrum from its earliest beginning—and is continuing presently.

2 Lessons of History Forlorn

A brief historical overview illustrates this pattern of interaction. In the first half of the twentieth century various schemes of arms control and disarmament were introduced such as the Hague Convention of 1907, the Versailles Treaty of 1919, the Washington Naval Agreement of 1922 and the Geneva Protocol of 1925. While these multilateral agreements had different objectives (like humanitarian rules of armed conflict, containment of a defeated enemy country, prevention of a global arms race, prohibition of chemical and biological agents in warfare), they had two common characteristics: first, they were initiated and framed by the most powerful nations of their time, and second, they were concluded without any serious verification mechanism. The latter shortcoming, which turned out in the end to be an essential weakness, was partly due to the belief that international verification was unnecessary or impracticable or unsustainable, but more fundamentally to the prevailing opinion that it would be incompatible with the basic tenets of national sovereignty and international order. Only at the Versailles Conference and, later, the 1932 Disarmament Conference under the League of Nations, was there some extended discussion of verification. But it proved to be highly divisive between the Great Powers and ended inconclusive [1]. Consequently, many of the agreed treaty provisions were willfully ignored, secretly evaded or blatantly violated. If there was a lesson of the inter-war period, it was a dual one: no dependable compliance without verification; no international verification without a broad and solid consensus at the highest level of politics.

By the end of the Second World War, the tables had been turned. Compliance with the rules of the game was henceforth considered of utmost importance. The United States, now beyond question the foremost power of all and the sole nation in possession of the atomic bomb, set about to construct a new world order on the basis of this unique status. In the proposal presented by the U.S. chief representative Bernard Baruch (hence thereafter called the "Baruch Plan") in mid-1946 to the newly created United Nations, there was envisaged the establishment of a specialized agency under the U.N. Security Council, called the International Atomic Development Authority (IADA), with a monopoly to hold and control all nuclear material and technology around the world [2]. With apparently unparalleled generosity, the U.S. was prepared to relinquish its atomic weapons and submit its secret know-how to IADA as soon as the prospective agency was fully set up. IADA would allow and encourage the use of nuclear technology exclusively for peaceful purposes in all member states and would closely monitor their compliance with treaty obligations. At the same time, the Security Council would act as the supreme guardian and stand by—with veto powers suspended—to chastise any perpetrator with "condign punishment" [3]. It was expected that, under such intimidating conditions, no member state would dare to cheat and rebel against IADA, the atomic juggernaut.

The Baruch Plan was initially welcomed by a vast majority of United Nations member states, traumatized as they were by the devastation of the World War. But then it foundered on the obstinate resistance of the Soviet Union and a few others. Although the plan is now only of historical interest, it remains highly relevant as a conceptual yardstick for any global verification regime. It combined the strictest type of control over a dangerous technology with severe disciplinary powers vested in an international authority. At a closer look, it was a scheme invalidating all claims of national sovereignty under the umbrella of a peaceful world order. Seemingly driven by an irresistible logic, the Baruch Plan was bound to fail because it was based on the assumption of an overriding political consensus that was simply not there. Even just a few months after the destruction of Hiroshima, not all survivors of the World War were ready to sacrifice great power ambitions of their own for the promise of a world without nuclear weapons.

Whatever came about in the following decades in terms of an international verification project stood in the shadow of this seminal failure. Never again has something as high-flown as the Baruch Plan been attempted to achieve. Never again was one nation, the United States (or, for that matter, anyone else), in such an extraordinary position to lead the world community towards an aspiration of its own definition. Never again was the fear of nuclear weapons so deeply felt and the concomitant eagerness to share the benefits of nuclear energy so enthusiastically embraced as in the immediate post-war period. Most importantly, in the second half of the twentieth century the number of states anxious to retain their sovereignty grew threefold while the number of states having nuclear weapons in their possession increased to nine. The historic moment, if there ever was one, to shape a consensual world order around the atom had passed. All subsequent pursuits of a more modestly designed nuclear order had to steer a precarious course between those stumbling blocks.

3 Lessons of History Regained

The demise of the Baruch Plan's grand design has spawned, over the course of the last half-century, an array of more or less isolated, more or less unsystematic attempts to combine international security arrangements with verification provisions. Leaving a large number of dead-born arms control and disarmament projects aside, it is useful to take a closer look at those undertakings that were actually transformed into treaty-based instruments during this period. Only a few among them, prohibiting the introduction of nuclear explosives into uninhabited locations like Antarctica, outer space and the sea-bed, entrusted observance of treaty provisions to the collective watchfulness of member states. More frequently, though, arms control treaties were based on carefully crafted terms of verification.

Two categories can be distinguished. There was, on the one hand, the exclusive series of Soviet/Russian-American agreements to regulate their nuclear arms race on a strictly bilateral basis, ranging from the Hot Line Agreement of 1963 through the ABM, PNE, INF, various SALT and START Treaties up to 2010. Most of these contained highly sophisticated verification mechanisms which were, however, designed and administered on their strictly bilateral terms alone and therefore required only technical elaboration within a pre-ordained framework of mutual political understanding of common interests. Although these arrangements commanded the bulk of public attention during the Cold War (and rightly so, as the balance of nuclear terror rested upon them in the bipolar era), they were of transient temporal significance and contributed to the evolution of verification concepts only in a limited way. Questions of compliance could be easily settled through bilateral exchanges and negotiations or, alternatively, unilateral dismissal.

On the other hand, there was brought into existence a series of multilateral agreements where verification assumed an essential and increasingly prominent role. They were based on the participation and adherence of as many member states as possible, designed for (or, in a few cases, later adapted to) unlimited duration, and in most cases equipped with or closely coupled to an administrative body, usually a permanent international organization with extraterritorial status and specialized expert staff, plus a provision for review conferences of member states at regular intervals. The International Atomic Energy Agency (IAEA), founded in 1956, served as a model and grew up to be the central reference point of nuclear policies worldwide. Its "safeguards" system of standardized inspections and data analysis evolved into the most highly developed and most widely accepted instrument of transnational verification, was adopted and adapted for monitoring compliance with treaty obligations under the Nonproliferation Treaty (NPT) of 1968 and various regional nuclear-weapon-free zones, and provided benchmark precedents for further verification agencies like the Organization for the Prohibition of Chemical Weapons (OPCW) of 1993 and the Comprehensive Test Ban Treaty Organization (CTBTO) of 1996.

The distinction between the former (bilateral) type of an international arrangement and the latter (multilateral) type is more than an arithmetical difference. It goes to the bottom of the verification syndrome. In a bilateral constellation, each of the two parties is in full control of the relationship. Each is equally free and simultaneously responsible for defining and authorizing, upholding or ignoring, breaking or sanctioning the rules of the game. Verification of compliance between two parties is a self-reliant activity, and cooperation a self-determined allowance. In a multilateral constellation, however, each of the numerous parties is bound into a collective discipline and holds not more than a share (possibly a small, although not necessarily an equal one) in those common responsibilities. Rules apply to all parties but are not necessarily shaped and approved by each of them. Verification of compliance is removed from members' individual authority, and cooperation can strain the limits of one's tolerance. Enforcement of sanctions against rule-breakers can expose them to severe deprivation. In such a setting, verification becomes an all-important instrument of distributive justice and the definition of compliance a matter of supreme interest for everyone.

4 The Evolution of Safeguards

The elevated importance of verification in a collective security arrangement touches upon three fundamental questions. First, what is the appropriate methodology for certifying treaty compliance? Second, who shall determine whether a treaty obligation has been violated? Third, how shall a certified treaty violation be dealt with? While the first question is mostly a technical one, the other two are obviously political. What is more, improvement of verification mechanisms tends to increase also their political content and sensitivity.

The historical development of IAEA safeguards illustrates this point. The Agency's verification system has undergone significant changes and adjustments over the last half-century.[1] Building on President Eisenhower's "Atoms for Peace" Plan of 1953, it was originally conceived as an international watchdog over the peaceful use of nuclear facilities supplied to foreign countries by the United States, then the one and only nuclear exporter of any significance. When the safeguards system INFCIRC/66 became fully operational in the mid-1960s, still on a modest scale, it continued to concentrate on inspections of U.S.- delivered facilities abroad. The NPT, which came into force in 1970, required a new system of safeguards for an era that was expected to see a rapidly expanding use of nuclear energy worldwide at the same time as the U.S. dominance of the market was anticipated to be increasingly challenged by foreign competitors. The new system INFCIRC/153, agreed in 1971 after lengthy international discussions, sought to strike an equitable balance between a precise and objective method of verification on a non-discriminatory basis and due respect for commercial interests and technological competition. The method chosen was the application by the IAEA of nuclear material accountancy in NPT member states together with inspections of nuclear installations at selected "strategic points".

This system continues to be basically in operation up to this day. However, within a short time external factors such as the increasing globalization of nuclear trade and technological innovation called for further adjustments to make NPT safeguards more ambitious and demanding, more stringent and fool-proof. In pursuit of this purpose, what was previously a carefully restricted, lucid and unobtrusive concept of verification gradually changed its character: it became, step by step, more expansive, more comprehensive, more intrusive, more selective and more prohibitive. In other words, more controversial and more political.

1. The *expansion* of IAEA verification responsibilities was driven by the inherent dynamics of nuclear technology. Although the attractiveness of nuclear electricity generation was sometimes vastly overrated, it resulted in a growing number of countries exhibiting keen interest in this source of energy. In order to gain access to nuclear technology, most of them found it necessary, more or less willingly, to join the nonproliferation regime and accept IAEA safeguards. With the steady increase of NPT membership to near-universality, and with the inexorable spread of nuclear commerce, technology and know-how beyond national borders, the nuclear world, which used to comprise merely a handful of industrial countries in the northern hemisphere, became a truly global one. More nuclear activities than ever before came into the purview of safeguards. A net effect of these trends has been the growing salience, disparity and diversity of vested interests in nuclear affairs in more and more countries and, consequently, the steadily increasing number of countries joining the IAEA (now standing at 170), where they strive to be seated in the Agency's central policy-making body, the 35-member-states Board of Governors. The Board is responsible for defining and

[1] The best and most authoritative accounts of the early years and subsequent evolution of IAEA safeguards continue to be [4–6].

approving the concepts, methods and instruments of IAEA safeguards. Disputes about nuclear verification are a direct reflection of the expansive engagement of political interests in the Agency.

2. The *comprehensiveness* of IAEA safeguards increased as it became evident that monitoring the peaceful use of nuclear facilities and fissile material in member states alone was insufficient for nonproliferation purposes. Nuclear trade with other countries and sensitive activities within and beyond treaty-bound states had to be included and extended into the domain of "comprehensive safeguards" and "full-scope safeguards". Dual-use items and immaterial assets were to be covered by "strengthened safeguards". Undeclared facilities had to be taken into account; the intake of unofficial information into safeguards assessments came into practice. Safeguarding the atom is now an infinitely more complex and demanding task than in the early years. The increasingly ambitious definition and progressive broadening of the legitimate concerns of international verification has turned IAEA safeguards into a pervasive presence in many fields of nuclear affairs that were previously regarded as an undisputed preserve of national sovereignty.

   As the Agency's safeguards methodology came to embrace since the early 1990s the concept of "the State as a whole", it has become standard practice for the IAEA Secretariat to look beyond the narrowly defined field of nuclear affairs in order to establish the trustworthiness of individual countries' commitment to their nonproliferation obligations. Each of these steps towards a more comprehensive safeguards system has, however, been bitterly contested among member states, mostly upon allegations that the IAEA tends to overstretch the limits of its statutory authority [7, 8].

3. The *intrusiveness* of verification was increased as a preventive countermeasure against illicit nuclear activities by member states and criminal networks on the sub-national level. Following the discovery of various instances of disloyal behavior, clandestine weapons development and conspiratorial collusion among governments and profiteers, the Additional Protocol of 1997 granted the IAEA the authority to investigate inside member states to seek assurance of the absence of illegal nuclear activities. The earlier function of safeguards as a retrospective instrument for certifying past treaty compliance was thereby complemented and to a certain extent substituted by a prospective capacity to detect future possibilities of non-compliance. Innocuous nuclear activities and even states without any nuclear programme can now, under conditions set up by the Additional Protocol, be required to submit to inquisitive IAEA inspections and suffer a significant degree of enforced transparency. While this may be justified as a logical consequence of the Agency's verification mission, it reaches far beyond what was originally assumed to be standard safeguards practice under the NPT. Not surprisingly, a considerable number of states (mostly in the Middle East, Latin America and Sub-Saharan Africa) have so far refused to accept the Additional Protocol for themselves.

4. The *selectiveness* of safeguards has been an extant feature all along, but more recently has assumed a new dimension. While unequal treatment of nuclear-

weapon states and non-nuclear-weapon states was accepted (grudgingly, to be sure) as a fundamental distinction of the international nonproliferation regime from its beginnings, strict observance of "the equality of misery" among non-nuclear-weapon states used to be regarded as a cornerstone of fairness in the application of NPT safeguards. An unintended consequence of this rule was that for several decades the bulk of the IAEA's safeguards effort was spent (or, as critics argued, wasted) on a handful of industrial countries with large nuclear energy programmes such as Japan, Germany and Canada. Resources in finance and personnel being scarce in the IAEA, and the Agency's budget being held at near-zero real growth by member states in spite of a growing volume of responsibilities, safeguards policy has lately shifted towards a more economical pattern of allocation. Starting with EURATOM and then extending to other countries with impeccable nonproliferation records, notably in connection with the Additional Protocol, the IAEA has come to streamline routine inspections in some places while at the same time seeking to concentrate its limited resources in other places for more thorough scrutiny of suspicious targets and activities of immediate concern.

The recently proclaimed "State-level approach " (and, more lately, the "State-level concept") goes one step further in this direction. Spurred on by the detection in 2003 of a clandestine nuclear programme (presumably for weapons purposes) in Iran, the IAEA has sought to refine the methodology of its safeguards system in such away that henceforth it should be able to identify at a very early stage any country which might set out on the path towards illicit nuclear activities.[2] Based on a comprehensive assessment of a state's nuclear profile, the Agency now seeks to focus its investigative spotlight upon those nations (rather than specific nuclear facilities or activities) where, in its judgment, special vigilance is warranted.

While the IAEA insists that this approach is entirely objective, unprejudiced and non-discriminatory, it walks a precarious tightrope of equity. The "State-level concept" entails, in effect, an incentive for national governments to earn and maintain a reputation of faithful treaty observance through loyal cooperation with the IAEA. Uncooperative behaviour is requited with frequent visits and insistent inquiries from safeguards inspectors. Whatever the merits of this approach, it has the net effect of treating some non-nuclear-weapon states differently from others in such a way that the Agency discriminates between states that are considered trustworthy and those that are not. Not all member states consider this to be an appropriate procedure for an international organization. Administration of the "State-level concept" is likely to remain contentious within the IAEA.

5. The *prohibitiveness* of safeguards is to a certain extent inherent in their ascribed function of "deterrence" (of which, more later). Nuclear weapons are, by definition, clearly and categorically prohibited for non-nuclear-weapon states under the NPT, and the avowed purpose of verification is to "to deter the proliferation

[2]For an analysis of the connection between the Iran crisis and the IAEA's safeguards reform effort see [9].

of nuclear weapons by the early detection of the misuse of nuclear material or technology" [10]. While the principle is beyond dispute, its practical application is not. Key definitions and conceptual determinants that are fundamental for verification purposes such as "detection time", "significant quantities" of fissile material, legitimate uses of such material, "sensitive" capabilities and the like are neither self-evident nor immutable, they need to be specified and from time to time refined and readjusted. There has been a tendency in the IAEA to redefine the technical requirements for effective safeguards in such a demanding manner as to make them (in practical if not in legal terms) next to prohibitive for certain nuclear activities. For instance, if large reprocessing or enrichment plants in national possession, laboratory experiments with metallic HEU or plutonium, tritium production, high-speed implosion technology and similar cutting-edge research projects are denounced by renowned experts as being "unsafeguardable",[3] it becomes well-nigh impossible to uphold them in the civilian environment of a non-nuclear-weapon state.

At the same time, the Agency is confronted with limitations of its own. The safeguards inspectorate does not (and should not) normally possess genuine knowledge and practical experience of nuclear weapons or their manufacture itself. When it comes to the task of verifying ostensibly peaceful nuclear activities in a suspicious military compound (e.g., probing into possible military dimensions of Iran's or Syria's earlier nuclear programmes), IAEA inspectors may reach the limit of their professional competence. The Agency may in certain instances hire such expertise from nuclear-weapon states—as was actually done, e.g., for recruitment of qualified personnel to the Iraq Action Team probing into Saddam Hussein's clandestine nuclear programme in the early 1990s.[4] But this is a delicate point as it may compromise the reputation and self-esteem of the IAEA inspectorate, which rests on its independent professionalism. Over-extension of the Agency's safeguards ambition can in fact damage its own standing and integrity.

Altogether, the historical evolution of IAEA safeguards has brought politics and verification more closely together and their interaction into sharper relief. If verification has become more politicized, how has this affected the veracity of politics?

[3]IAEA officials would, of course, never speak of things nuclear being "unsafeguardable". Others have used the term more freely in public; see [11–16].

[4]Significantly, though, the U.N. Security Council mandated the IAEA Director General personally with this task, not the IAEA itself as an institution.

5 Authority and Authenticity in International Verification

Verification is often discussed in terms of its technical details and scientific concepts. But it is no less an institutional and organizational problem. Who decides? How are decisions made?

In the IAEA, the Department of Safeguards with its corps of inspectors and technical expert staff under the overall responsibility of the Director General enjoys a high degree of autonomy and professional self-regulation. Its findings and conclusions in verification matters command such an amount of confidence and authority that they are usually taken as the Agency's decisive judgment. However, this is not really correct. It is the Board of Governors where the statutory authority for certifying the evidence of verification properly lies. While the Director General presents to the Board of Governors the findings of the Secretariat in his annual Safeguards Implementation Report (SIR), the Board is the place where, when the chips are down, the validity of verification is ultimately determined. The Board can accept and endorse the judgment of the inspectorate, as it ordinarily does, or reject and override it. Although such a conflict of judgments will rarely come into the open, the possibility looms in the Agency's institutional makeup. As Thomas Hobbes observed in his classic treatise on sovereignty: *auctoritas, not veritas, facit legem*. Authority, rather than authenticity, decides what is right or wrong.

In regard of verification the Board has a dual function: a rule-making and an adjudicating one. Both are thoroughly political, although the former is couched in technical and the latter in quasi-judicial terms. In the first function, the Board sets the rules and regulations for member states to act in accordance with their treaty obligations and for the inspectorate to verify compliance. In the second function, the Board deliberates for itself whether or not to support the inspectorate's assessment and what kind of consequences to draw from it. Most important, of course, is this function when it comes to decide a question of non-compliance.

Significantly, the Board has never set up rules to define exactly what non-compliance means, thus leaving for itself a large measure of flexibility and discretion [17–19]. Only on five occasions—Iraq (1991), Romania (1992), North Korea (1993), Libya (2004) and Iran (2006)—has it ever determined that a state was in non-compliance with its NPT safeguards agreement. Much more frequent are those cases where the Agency's inspectors detect irregularities in a member state's records which need to be clarified and set straight through further investigation and thorough discussion with national authorities. Such follow-up inquiries are sometimes quite lengthy affairs and can extend over several years. The Board has wisely refrained from declaring any irregularity a breach of treaty obligations when it may have resulted simply from negligence of local nuclear operators or incompetence of national administrators, and brought to public attention only those few cases which it deemed to be a serious challenge to the Agency's authority and integrity.

The inspectorate's expertise and the Board's authority rarely clash because they serve different functions. The Department of Safeguards collects and interprets evidence, the Board considers and authorizes conclusions. Both bodies stand firmly on their own ground, but in the end the Board prevails. Safeguards inspectors are technical experts and scientific professionals in the service of an international agency; Board members are diplomats in the service of national governments. Inspectors owe their allegiance to a global mission, governors theirs to the political interests of individual member states. Under normal circumstances they tend to cooperate with each other to the collective advantage of all. When the IAEA as a whole speaks out with one voice (usually through the person of the Director General), its official opinion reflects the interaction of all these factors.

However, international organizations being what they are, harmony and unanimity are not normally the order of the day. The 35-member Board of Governors is legally a society of equals where formal votes are rarely taken and decisions are mostly made by consensus, and its composition is newly apportioned among the Agency's member states every year. Nonetheless, power and influence are distributed quite unevenly among Board members. Fully a dozen of them (including the five recognized nuclear-weapon states plus all the major industrial countries and nuclear technology holders) are de facto permanent members, having held a seat on the panel for more than a half-century. This makes for a high degree of professionalism and continuity in the IAEA policy-making process. But it leaves all other members at a permanent disadvantage in terms of personal experience, organizational insight, information access, networking skills and administrative leverage within the Agency. Informal bureaucratic precedence of this kind can yield tangible dividends when, for instance, the Director General consults with the Board, as he regularly does, before filling senior positions in the IAEA Secretariat, including the safeguards department.

Weak and strong member states have different means to bring their unequal weight to bear—in the IAEA as in any other international organization. There is on the Board of Governors always a certain number of delegates who represent desperately poor and wretchedly dysfunctional states, mostly in the developing world, who have no genuine interest in nuclear matters, let alone the intricacies of safeguards, but are eager to trade their country's vote as a diplomatic bargaining chip in exchange for financial aid, technical assistance or other compensation from the richer colleagues. Deals of this kind tend to strengthen even further the position of the mightier states. Strong states, in particular the quasi-permanent Board members, can exert their influence in a more varied and subtle fashion. Possibly the most effective way is to offer voluntary contributions in support of the IAEA's underfunded programmes, be it through direct financial transfers, in the form of advanced technical equipment, secondment of specialized personnel or cost-free services in national laboratories and nuclear facilities. The Agency's safeguards budget is, in fact, significantly relieved by such voluntary contributions from a number of member states, notably the United States [20]. But it would be naive to assume that such generosity comes for utterly unselfish reasons.

More recently, the intake of informal information (published and unpublished sources, national intelligence reports, commercial area surveillance, etc.) by the safeguards department and the adoption of a broader "State-level approach" have nudged the IAEA closer and closer to a degree of dependence on externally generated data in its task of gathering and evaluating evidence, the verification of which may be beyond the Agency's genuine capacity. There have been warnings from insiders that safeguards inspectors might make themselves vulnerable to "fabricated evidence".[5] In a similar vein, misgivings have been expressed from the middle of non-aligned member states (the majority of whom has never had a chance to sit on the Board of Governors) to the effect that the IAEA must not, in its quest for nuclear "transparency", neglect "confidentiality", "objectivity" and "respect for Member States' sovereignty" as the basis of legitimate verification [21].

The IAEA could indeed find itself on the horns of a dilemma. Demands come from among its member states for more ambitious safeguards, but at the same time for more safeguards restraint. Both demands cannot be reconciled easily. On the one hand a powerful group, led by the nuclear-weapon states and dominant in the Board of Governors, insists on having a much stricter safeguards regime installed before nuclear disarmament can be seriously considered. On the other hand a large number of developing countries, orchestrated by the Non-Aligned Movement (NAM) and enjoying a comfortable majority in the Agency's General Conference, suspects the IAEA safeguards system to be all too easily misused for the purpose of protecting and perpetuating the Western powers' global hegemony.

What these contrary positions reveal (regardless of any intrinsic merits) is fundamentally a mood of distrust between major groupings of states. More specifically, though, it is a display of no-confidence in the IAEA and its safeguards system. Many developing countries share an experience of inferiority and marginalization in the IAEA where they feel the goals of nonproliferation to be systematically overvalued at the expense of nuclear disarmament and technological co-operation. Safeguards appear to them as an expensive nuisance rather than an effective measure to fulfill the IAEA's statutory promise "to accelerate and enlarge the contribution of atomic energy to peace, health and prosperity throughout the world".[6] The nuclear-weapon states and their allies in the Board of Governors, on the other hand, tend to believe that the Agency's safeguards routine may be insufficient to prevent the spread of nuclear weapons in the international system. Uncompromising insistence on the latter view, if carried to the extreme, can actually discredit the validity of international safeguards and undermine the institutional authority of multilateral systems of verification altogether.[7]

[5]An explicit warning against "fabricated evidence" (in scarcely veiled allusion to the Iraq and Iran cases) was voiced by Peter Jenkins, former British ambassador to the IAEA, in a public panel discussion during the CENESS conference "Nuclear Energy, Disarmament, Nonproliferation", Moscow, 21 November 2014. (Author's personal notes).

[6]The formula, taken from Article II of the IAEA Statute of 1957, is traditionally invoked to emphasize the Agency's obligation to work to the practical benefit of all its members.

[7]A gross example of such intransigence can be seen in the decision of the United States under George W. Bush to go to war against Saddam Hussein's Iraq in 2003 in open defiance of the findings of the IAEA and the United Nations Monitoring, Verification and Inspection Commission (UNMOVIC) that there were at that time no weapons of mass destruction left in Iraq; see [22]. In

Nowhere is this more apparent than in the controversy over Iran's nuclear programme, which has been agitating international nuclear diplomacy for more than a decade ever since an Iranian opposition group uncovered its secret components in 2002. While the IAEA safeguards inspectors tried over this period—more or less successfully—to verify its clandestine details, the P5+1, a group of six states comprising the recognized nuclear-weapon states plus Germany, took it upon themselves to bring Iran back into the fold of rule-abiding non-nuclear-weapon states. A key element in their negotiations was the demand that Iran should forego an indigenous fissile material production capability and be subject to a special verification regime for an extended period of time. There is a significant message in such peremptory requirements: IAEA safeguards alone, even when they go beyond those of the Additional Protocol, are not accepted by the leading members of the Board of Governors as being adequate enough to restore confidence in the loyalty of a non-compliant NPT member state. In blunt terms, this raises the question whether in the absence of trust anything short of an effective denial of access to sensitive technologies for non-nuclear-weapon states will pass for a really reliable safeguard.

The Joint Comprehensive Plan of Action (JPCOA), which was at long last concluded by the negotiating parties in June, 2015, leaves this question open.[8] It is an uneasy compromise, combining face-saving allowances and time-begging restrictions with precarious assurances of good will for a period of 15 years, during which a more permanent solution needs to be found. It puts an extraordinary burden of verification upon the IAEA for the coming years and, as such, is neither fit nor intended to serve as the model for a general balance of interests among the world's nuclear community.

6 Verification and Deterrence

Verification is not an end in itself. The purpose of IAEA safeguards, according to the Agency's model NPT safeguards agreement of 1971, is "deterrence" of prohibited nuclear activities "by the risk of early detection" [25]. This carefully phrased formula combines two important notions. "Deterrence" means that the risk of detection alone is not enough, it must be really fearful for the perpetrator. "Early" means that the fact of detection alone is not enough, it must happen at a point of time long before corrective action against the perpetrator might become self-defeating. To illustrate, if an disloyal member of the NPT could acquire secretly an operational nuclear arsenal before being detected, any punitive action of adequate

a characteristic fashion of self-assurance, the Bush Administration accepted the essence of these findings as soon as they were in effect confirmed after the war by the report ("Duelfer Report") of the Iraq Survey Group (ISG), an investigative body unilaterally installed by US Government agencies; see [23].

[8]A thorough exposition and analysis of the JCPOA can be found in [24].

severity ("condign punishment", in the famous words of Bernard Baruch) would come only at a prohibitive cost: possibly nothing less than a nuclear war.

Before taking a closer look at the interplay of those two concepts of "deterrence" and "early detection", it should be noted that they ring familiar to historians of the nuclear age. Both have played a central role in the evolution of nuclear strategic thinking during the Cold War, culminating in the doctrine of "mutual assured destruction" (MAD) between deadly enemies. It is not without irony to point out that the same conceptual toolbox is used simultaneously to make credible the threat of nuclear war, the prevention of nuclear war and the prevention of nuclear proliferation. There may be more common ground in this frame of reasoning than a mere coincidence. Nuclear strategists have long been pointing out the frightening ambiguities and fragility of the theory of deterrence. Suffice it here to say that deterrence (and to a lesser extent, early detection) is neither a proven fact nor a scientific concept. In an interpersonal perspective it is a psychological construct. In an international perspective it is a political scheme, if not an act of belief. In practical terms, cynics would argue, it works as long as it works.

Earliness (or "timeliness") is considered important to forestall a sudden "break-out" of dishonest treaty members and to allow as much space as possible for clarification, negotiation and reconciliation before harsh compulsory measures are taken against an uncooperative state. Since the IAEA lacks any serious compulsory instruments of its own, it needs to rely on the leverage of more powerful external actors to bring pressure to bear on an unrelenting offender. Consequently, the IAEA Statute provides for recourse to the U.N. Security Council in case of an irreconcilable conflict.

The definition of timeliness has been a perennial point of contention inside and outside the IAEA. It does make a lot of difference whether a suspected would-be nuclear power is days or weeks or years away from "the bomb". An accurate estimation of break-out time is therefore a highly desirable element of verification. But it would be a futile attempt (although it has been made again and again) to define it in terms of objective, quasi-scientific conditions. What matters instead is the willingness and ability of the international community to deal with cases of non-compliance in an adequate manner within a time frame that may or may not be foreseeable. It is, once again, a political question, not a technical one.

Dealing with non-compliance is further complicated by the fact that a sudden break-out (such as North Korea's abrupt renunciation of the NPT in 2003) is not what the international community is prepared for. More common is the problem of countries with an advanced nuclear infrastructure (presently more than a handful come to one's mind) that are, technically speaking, very close to a nuclear weapons capability. Again, there is no preparation in sight for this problem of latent (or "creeping") proliferation. The "risk of early detection" is obviously not a solution if, for example, a country like Japan began taking nuclear arms seriously into consideration.

For the IAEA to report a case of non-compliance to the Security Council is certainly a decision of momentous proportion as it signals the transition of responsibility from a specialized technical watchdog agency to the supreme guardian of world peace. The Board of Governors has always been rather hesitant to take this final step of disciplinary escalation. In reality, though, the step is not such a big one. In the IAEA, as has been emphasized above, no important decision is made without the leadership (or at least the consent) of the quasi-permanent group of states on the Board. In the Security Council, the same applies to the Permanent Five (P5) with their individual veto power. In both bodies, those groups are in effect identical. Therefore, if the Agency reports a case to the Council, it can be sure that its findings will be respected in New York. And if the Council makes a decision on the case, it can be sure that it will be heeded and faithfully executed in Vienna. Although the two institutions are only loosely connected in legal terms, this means for all practical purposes that they live in a symbiotic relationship where the IAEA serves as the Security Council's verification body and the Council as the Agency's disciplinary arm.

What, then, about "deterrence"? In principle, the Security Council has the authority to make life really bad for a country that has been caught cheating on the IAEA. In practice, its members are rarely ready to impose upon such a country severe sanctions that would involve great costs for themselves. Viewpoints tend to be more diverse and national interests more complex within the Security Council than on the Board of Governors because nuclear affairs are only one item of concern beside others in the U.N. framework, and normally not the most urgent or rewarding ones for all members. In particular the great powers with veto status are prone to weigh their multiple relationships and engagements specifically against each other. Nuclear nonproliferation may rank high on their scale of national interests but not highest in all cases. Under such circumstances it is not unusual for multilateral sanctions to be imposed by the Security Council but simultaneously undercut by individual members; or even more bizarre, U.N. sanctions (as presently still operating against Iran) being unanimously supported by the Permanent Five but overlaid by crosscutting sanctions directed by some members against others of the same group.

U.N sanctions against a non-compliant NPT member state can be quite onerous for the targeted country. But they have rarely had an immediate and incisive effect of sufficient magnitude to bring a stubbornly defiant perpetrator to his knees. Once a national government has made the enormous decision to go for nuclear weapons, willfully in violation of international law, it will not be easily swayed by outside pressure even if this entails economic hardship and international isolation. Countries like South Africa, North Korea or Iran have suffered greatly from international sanctions but withstood them for many years.

One major reason is, of course, that sanctioned countries never stand entirely alone. They have always more or less numerous sympathizers or powerful supporters in the international system. Thus, in an exemplary show of self-confidence and defiance, the 120-nation Non-aligned Movement (NAM), meeting in Tehran in mid-2012, unanimously declared its support for the disputed Iranian nuclear energy programme, including the right to ownership of a full nuclear fuel cycle (meaning uranium enrichment, an activity which the Security Council had repeatedly demanded Iran to suspend), and expressly condemned the American-led effort to isolate and punish Iran with economic sanctions.[9] Election of the host country to the NAM chairmanship for a 3-year period at the same meeting may not really have helped Iran in its further dealings with the great powers. But it served a clear message to the Security Council that "condign punishment" is not on the minds of the majority of NPT member states.

The deterrent effect of IAEA safeguards, if there is any, does not work with automaticity and direct causality. Detection of disloyal behaviour in the NPT community does not trigger all by itself a forceful response on the multilateral level. Multilateralism works effectively only if sufficient numbers of stakeholders with widely divergent views and interests are brought together into a stable coalition through skillful diplomacy and dynamic leadership. Such favourable conditions are not easily available in the Security Council. More often than not they are stifled by the natural inertia of the international system.

And yet, safeguards can carry a deterrent effect on certain occasions, if only in an indirect and skewed fashion. When the IAEA has found a country to be in violation of its treaty obligations, and corrective action from the Security Council is not forthcoming or not seen to be effective, some countries (individually or together with like-minded ones) may feel challenged and entitled to take things in their own hands. These countries—more ready than others to intervene in imminent cases of nuclear proliferation—invariably come from a core group of half a dozen, plus an occasional outsider, neighbour or concerned ally, usually acting under the leadership of the United States. Not surprisingly, the core group consists of the nuclear-armed Permanent Five—the well-known dominant powers in the Security Council and on the IAEA Board of Governors. In their typical mode of intervention, demonstrated in the critical cases of North Korea, Libya and Iran, they have leaned together on the culprit governments, employing a combination of diplomatic negotiations, economic sanctions and co-operative incentives—but never excluding veiled or blunt threats of coercive military action against illicit nuclear installations. While these threats have until now not been carried out (except unilaterally by the United States and Britain against Iraq and by Israel, notably a nuclear-armed non-NPT state, against Iraq and Syria), they were meant to be taken seriously and were probably taken as such. Their deterrent value remains therefore open to dispute—as deterrence remains altogether.

Interventions of this sort were undertaken without a clear legal mandate (although some were later approved by the Security Council), and their legitimacy continues to be hotly debated in international fora. It is certainly very doubtful whether enforcement of international treaty obligations should be left to the discretion of a self-appointed posse of nuclear-weapon states. But it is clear at the same time that responsible and active engagement for the preservation of the nonproliferation order is lacking elsewhere, notably on the side of non-aligned countries where indignation about nuclear weapons is as widespread as resentment

[9]For a detailed report see [26].

against intrusive safeguards. Indolence and complacency among a majority of the world's nation-states are a worrisome sign of dysfunctional multilateralism.

7 Reconciling International Politics and Verification

It has been argued in this chapter that verification, in particular the nuclear safeguards system of the IAEA, is a political project within a political setting, shot through with political interests and interventions, framed in political terms for political purposes. It would therefore be dishonest and erroneous to conceive and describe it as unpolitical, objective and purely scientific.

But it would be equally erroneous to dismiss it as a sham, created and upheld simply to disguise the unfettered rule of hegemonial powers or the arbitrariness of an hermetic bureaucracy. Suspicions of this kind seem to ring in the background whenever advanced safeguards concepts like the Additional Protocol or the "State-level concept" are denounced for being unfair, excessively intrusive and burdensome. When such insinuations are raised against the IAEA, it appears that the Agency is held victim to its own propaganda which has insisted for many years that safeguards are entirely objective, egalitarian, unobtrusive and, well, unpolitical.

What went wrong in this line of reasoning (and was perhaps wrong from its earliest beginning) resulted probably from a confusion of the methodology and application of safeguards with their intention and utility. The former aspect is dependent on truthfulness and integrity, and the safeguards inspectorate has rightly earned for itself a reputation of impeccable professionalism. The latter aspect is thoroughly political, and it should be displayed openly for what it is.

Safeguards inspectors are neither entitled nor qualified to decide themselves what should be allowed or not in the nuclear world. Theirs is a task of faithfully attending to certain practices of good behaviour among states, the rules for which are made elsewhere. Their findings are neither designed nor weighty enough for the purpose of deterrence (the claim of the model safeguards agreement notwithstanding). Nor is the IAEA fit to define and guarantee a nuclear world order. All such expectations are overly demanding and addressed to the wrong recipient. Often they seem to serve as a convenient excuse to shunt responsibility for dealing with grave problems away from political authorities to seemingly impartial institutions. Grave problems, however, are not suitable for shunting, they just stay in place. What graver problem could there be than the role of nuclear weapons in the world?

One possible conclusion at this point might be to simply acknowledge that politics reigns supreme, even in the field of verification. But again, this would be only half the truth. If the noblest purpose of politics is to organize the survival of mankind, it needs clearly to improve upon the past. The looming presence of nuclear weapons is a permanent reminder of the imperfection and fragility of the international order, and multilateral verification helps raising awareness of this dilemma. Inasmuch as the "State-level concept" of IAEA safeguards promises to contribute to a more widely shared appreciation of the value of fairness and civility in international relations, it is good politics.

8 Afterthoughts: The Crisis of Multilateralism

It was argued in this chapter that the system of nuclear safeguards, as defined and applied in the framework of the International Atomic Energy Agency, has evolved over more than six decades in a dyadic process of international cooperation and institutionalized verification. Cooperation to establish common rules of the game among the Agency's member states and verification to establish trust in their compliance with the agreed rules resulted in a more and more refined toolbox of monitoring instruments for the legitimate uses of nuclear technology. On top of this, the Joint Comprehensive Plan of Action (JCPOA), commonly known as the "Iran nuclear deal", put into effect an exceptionally strict and ambitious scheme of international supervision to curb one nation's suspicious nuclear program at least for a limited number of years. The JCPOA, concluded on 14 July 2015 by six leading nations together with Iran, was widely hailed as a highlight of multilateral consensus-building diplomacy. It was acclaimed and endorsed by national governments around the world and by several international bodies, including the Security Council, NATO and the European Union (and, of course, the IAEA itself, which stepped in promptly to apply the agreed safeguards provisions and subsequently certify their correct implementation in Iran).

Three years later, the situation had changed dramatically. When President Donald Trump announced for the United States on 8 May 2018 to withdraw from the Iran deal ("the worst deal ever", as he declared on the Twitter network), he dealt a severe blow not only to the JCPOA but in fact to the whole edifice of multilateral agreements that make up the fragile stability of the nuclear nonproliferation regime. The regime is based on a sense and practice of shared responsibility among the world's major powers. If one of them breaks away from their common commitment, the global arrangement immediately lacks one of its essential supports. Legitimacy and sustainability of a multilateral undertaking is dependent on loyal adherence of all contracting parties.

Not all contracting parties are equally important, however, and the United States is undeniably the most important in nuclear matters. It has been in this position ever since the Second World War, initiating, shaping and upholding nuclear rules and institutions worldwide, including the IAEA and the NPT. Multilateralism in cooperation with like-minded allies was the prevailing mode and persistent vehicle of American global influence all along. To denounce multilateral obligations and instead proclaim bilateral deal-making, diplomatic bullying, economic arm-twisting or crude power play as the preferred method in pursuit of "America first", as Donald Trump has sounded out with a flourish, amounts to nothing less than a slap in the face of international institutions like the IAEA—the organization that was entrusted with safeguarding the JCPOA. More broadly, it displays a fundamental determination to turn away from more than 75 years of successful American globalism.

What does all this mean for verification? Significantly, President Trump did not mention the IAEA or its safeguards system when he dumped the Iran deal. He simply ignored the Agency, adding insult to contempt for the legacy of his predecessor Barack Obama. Whether or not the IAEA affirmed the proper execution of safeguards obligations in Iran (as it had indeed done regularly throughout the previous years) was shown in a blunt and blatant manner to be totally irrelevant for Trump's decision. This went far beyond George W. Bush's insistence in 2003 to wage war against Iraq on the assumption that there were weapons of mass destruction deployed in Saddam Hussein's army (in defiance of international inspection teams who argued that there was no evidence to such effect); whereas later, to his credit, Bush came around to accept the Duelfer Report's conclusion in 2004 that his assumption had been wrong. Trump, in contrast, has never bothered with the validity of facts other than those of his own making. By refusing to consider the weight of the IAEA's opinion, he snubbed the Agency as well as the other JCPOA signatories who expressed their confidence in its professional judgment. More profoundly, Trump laid open his ignorance of the important role and function of international verification.

Verification is not an end in itself. In the context of nonproliferation its main purpose is to create confidence and reduce the amount of mutual distrust and anxiety among an unruly crowd of sovereign states, nuclear and non-nuclear. In order to be accepted and trusted, verification needs to be carried out by a multinational institution, such as the IAEA, and in order to be effective it needs to be respected and supported by the most powerful international actors. The stability of the nonproliferation system depends on their allegiance. If one of the major actors such as the United States breaks away from the consensus (for instance, by seeking a better "deal" for its own purposes), the multilateral linkage is broken, and verification runs idle. When it comes to the creation of a stable nuclear order, unilateral disruptive actions or bilateral deals without internationally agreed verification are bound to be deficient, if not worthless. They lack the legitimacy and sustenance flowing from multilateral verification.

Promises and shortcomings of bilateralism were brought into sharp relief when Donald Trump, shortly after his repudiation of the Iran deal, met with North Korean leader Kim Jong Un on 12 June 2018 in Singapore to strike a new "deal" over Pyongyang's nuclear program. What had been an impossible task for the Six-Party Talks of regional powers in previous years suddenly appeared to be resolved during an afternoon session by two maverick heads of government. In their joint statement celebrating the Singapore summit as "an epochal event of great significance ... for the promotion of peace, prosperity and security of the Korean Peninsula and of the world", Trump and Kim announced their determination to build "a lasting and robust peace regime on the Korean Peninsula". To this end and in true fashion of cunning deal-makers, Trump pledged to Kim "to provide security guarantees to the DPRK", whereas Kim "reaffirmed his firm and unwavering commitment to complete denuclearization of the Korean Peninsula" [27]. Of course, no mention was made of verification or other implements to fulfil and assure compliance with such grand designs. Instead, the two summiteers reiterated again and again (in fact, eight times in their short one-page declaration) their "commitment" to make good on all that. "Commitment" is a customary diplomatic formula for engagement without legally binding effect. The verbiage rings familiar to anyone who has in mind article VI of the Nonproliferation Treaty where all parties agreed to move forward "in good faith" towards nuclear and general disarmament, 50 years ago.

Not much scrutiny is required to realize that the Singapore statement is, at best, an easy-going declaration of good intentions and, more likely, a frivolous media stunt for the personal grandstanding of two eccentric leaders. Both sides of their pretended bargain turn out to be a bluff when taken seriously. Trump's offer of "security guarantees" is insincere and void unless it entails a life insurance for the Kim dynasty and its totalitarian rule in North Korea, and unless it was fully underwritten by other countries in the region; besides, who should trust a warrant from someone who just recently tore apart his predecessor's with Iran? On the other side, Kim's "commitment to complete denuclearization" is an empty formula unless it entails acceptance of IAEA safeguards that would need to be far more intrusive and irreversible than anything previously imposed upon Iran or Iraq; and why should the North Korean leader earnestly consider giving up his precious nuclear doomsday machine and submit to an international (let alone a unilateral American) inspection regime that would be incompatible with his country's sovereignty? In either case, it is strange to note that the Singapore statement, if its wording were taken to all its consequences, would boost the IAEA to a level of extended responsibilities which the Agency has never experienced before, and would at the same time enmesh and entangle the United States in a network of multiple accountabilities and obligations that could surely not be handled in the Trumpian style.

The Iranian and North Korean cases, contradictory as they are, both indicate that President Trump's personality and worldview do not easily go along with the established principles and practices of the nuclear nonproliferation regime. The IAEA's verification system has as yet escaped his wrath (which up to now was directed at other areas of international collaboration, such as trade, climate policy, regional integration, cultural exchange, or migration), but this could change. Trump's deep-rooted displeasure with multilateral rules and institutions may yet lead him to realize that the United States, traditionally the biggest contributor to the Agency's activities, has still more unexplored possibilities to throw its weight around, even at the cost of its own standing.

What about the future? There are some who tend to expect that after Trump's incumbency of 4 or possibly 8 years in the White House, things will swing back to the normality of America's benevolent leadership role at the helm of a liberal world order. But what is normal? Donald Trump is not alone, neither at home nor abroad. False claims of easy solutions to complex problems have more recently come into vogue, as for instance the adoption by more than 70 states of a Nuclear Ban Treaty without proper verification. In an era of ready-made "fake news", considerable damage has already been done to the reputation and influence of professional expert communities, notably in the field of nuclear and non-nuclear arms control. Trust in the integrity of scientific inquiry and conscientious evaluation of facts and evidence is not something that can be switched on and off as it pleases. The truthfulness of politics, precarious as it may be, remains the essential premise for verification to count.

References

1. Goldblat J (1994) Arms Control: A Guide to Negotiations and Agreements, International Peace Research Institute (PRIO), Oslo, pp 11–20
2. United States Proposal for the International Control of Atomic Energy, June 14, 1946. In: US Department of State (1985) A Decade of American Foreign Policy: Basic Documents 1941–1949, rev. edition, Washington DC, pp 865–871
3. United States Proposal for the International Control of Atomic Energy, June 14, 1946. In: US Department of State (1985) A Decade of American Foreign Policy: Basic Documents 1941–1949, rev. edition, Washington DC, p 868
4. Szasz PC (1970) The Law and Practices of the International Atomic Energy Agency. Legal Series No. 7. IAEA, Vienna, pp 531–657
5. Scheinman L (1987) The International Atomic Energy Agency and World Nuclear Order. Resources of the Future, Washington DC, pp 121–172
6. Fischer D (1997) History of the International Atomic Energy Agency: The First Forty Years. IAEA, Vienna, pp 243–324
7. Rockwood L (2014) The IAEA's State-Level Concept and the Law of Unintended Consequences. Arms Control Today 44(7):1–7
8. Joyner D (2014) A Response to Laura Rockwood. In: Arms Control Law. https://armscontrollaw.com/2014/09/14/a-response-to-laura-rockwood/. Accessed 28 Feb 2019
9. Hibbs M (2015) Iran and the Evolution of Safeguards. In: The Verification Research, Training and Information Centre (VERTIC) (eds) (2015) Verification & Implementation. A biennial collection of analysis on international agreements for security and development. VERTIC, London, pp 1–25
10. International Atomic Energy Agency (IAEA) (2015) The Annual Report for 2015, GC(60)/9. IAEA, Vienna, p 95
11. Gilinsky V, Sokolski H (2014) Is the IAEA's Safeguard Strategic Plan Sufficient? Proc. IAEA Symposium on International Safeguards, Vienna, 2014. http://npolicy.org/article.php?aid=1262&rtid=6. Accessed 28 Feb 2019
12. Sokolski HD (2005) After Iran: Keeping Nuclear Energy Peaceful. NPEC Policy Paper. http://npolicy.org/article_file/After_Iran-Keeping_Nuclear_Energy_Peaceful.pdf. Accessed 28 Feb 2019
13. Carrigan AL (2015) Can the IAEA Verify the Iran Deal? In: Bulletin of the Atomic Scientists. http://thebulletin.org/can-iaea-verify-iran-deal8302. Accessed 28 Feb 2019
14. Leventhal PL (1991) The Nuclear Watchdogs Have Failed. In: The New York Times, 24 September 1991
15. Cochran TP, Paine CE, Fettus GH (2007) NRDC comments on the Energy Department's Notice of Intent to Prepare a Programmatic Environmental Impact Statement for the Global Nuclear Energy Partnership. https://www.nrdc.org/resources/nrdc-comments-energy-departments-notice-in-prepare-programmatic-environmental-impact. Accessed 29 Feb 2019
16. Allison G, Setter O (2014) Blocking All Paths to an Iranian Bomb: How the West Can Avoid a Nuclear Maginot Line. Discussion Paper, Belfer Center for Science and International Affairs, Harvard Kennedy School. https://www.belfercenter.org/sites/default/files/files/publication/BlockingAllPaths.pdf. Accessed 29 Feb 2019

17. Carlson J (2009) Defining Non-Compliance: NPT Safeguards Agreements. Arms Control Today 39(4):22–27
18. Goldschmidt P (2009) Exposing Nuclear Non-Compliance. Survival 51(1):143–164
19. Findlay T (2015) Proliferation Alert! The IAEA and Non-Compliance Reporting. Project on Managing the Atom, Report 2015–04, Belfer Center for Science and International Affairs, Harvard Kennedy School. http://belfercenter.ksg.harvard.edu/files/proliferationalert-web.pdf. Accessed 29 Feb 2019
20. Findlay T (2016) What Price Nuclear Governance? Funding the International Atomic Energy Agency. Project on Managing the Atom, Report 2016-01, Belfer Center for Science and International Affairs, Harvard Kennedy School. http://belfercenter.ksg.harvard.edu/files/WhatPriceNuclearGovernance-Web.pdf. Accessed 29 Feb 2019
21. Preparatory Committee for the 2015 Review Conference of the Parties to the Treaty on the Non-Proliferation of Nuclear Weapons (2014) Working Paper submitted by the Group of Non-Aligned States Parties to the Treaty on the Non-Proliferation of Nuclear Weapons, NPT/CONF.2015/PC.III/WP.1
22. Blix H (2004) Disarming Iraq: The Search for Weapons of Mass Destruction. Pantheon Books, New York
23. Duelfer CA (2009) Hide and Seek: The Search for Truth in Iraq. Public Affairs Books, New York
24. Samore G (ed) (2015) The Iran Nuclear Deal: A Definitive Guide. Belfer Center, Harvard Kennedy School. http://belfercenter.ksg.harvard.edu/files/IranDealDefinitiveGuide.pdf. Accessed 30 Apr 2018
25. International Atomic Energy Agency (IAEA) (1972) The Structure and Content of Agreements Between the Agency and States in Connection with the Treaty on the Non-Proliferation of Nuclear Weapons, INFCIRC/153 (Corrected), para. 28
26. Erdbrink T (2012) Nonaligned Nations Back Iran's Bid, but not Syria. In: The New York Times, 31 August 2012
27. Joint Statement of President Donald J. Trump of the United States of America and Chairman Kim Jong Un of the Democratic People's Republic of Korea at the Singapore Summit. https://www.whitehouse.gov/briefings-statements/joint-statement-president-donald-j-trump-united-states-america-chairman-kim-jong-un-democratic-peoples-republic-korea-singapore-summit/. Accessed 29 Feb 2019

Military Dimensions

Michel Richard

Abstract The military dimensions associated with nuclear arms control, disarmament and non-proliferation verification of a state's commitments are key for any significant and solid progress to be made. Verification of compliance and assurances of the absence of cheating are very sensitive issues, in particular with respect to the military capabilities that form the foundation of strategic and political posture at a national, regional and global level. This article sets the scene through a brief historical review of the influence that verification has on the credibility of arms control measures, the effectiveness of nuclear non-proliferation efforts and the importance of the nuclear disarmament process. In the framework of nuclear disarmament verification the main features of the military nuclear system are the doctrines, research & development, production, deployment and elimination of nuclear weapons, and the verification measures associated with disarmament commitments. Finally, a brief description of the complexity of regional, multinational and international instruments is provided. Using a systems concept to establish a verification mechanism, taking into account the specificity and the issues impacting individual states, could overcome the global complexity and national antagonisms and provide the confidence needed to enhance world global security.

1 Introduction

Since the United States dropped the atomic bomb on Japan to end World War II, the possibility of use and the deterrence of use of nuclear weapons have been at the heart of military strategic doctrine. This is true for countries that possess nuclear weapons or rely on an ally's nuclear umbrella, those that do not currently possess nuclear weapons but may consider doing so in the future, and those that wish for global elimination.

M. Richard (✉)
Division Global Security and International Regulations, Fondation pour la Recherche Stratégique (FRS), Paris, France
e-mail: micheljrichard@orange.fr

I. Niemeyer et al. (eds.), *Nuclear Non-proliferation and Arms Control Verification*, https://doi.org/10.1007/978-3-030-29537-0_6

In the 1950s and 1960s, the rapid growth of the United States (US) and the Soviet Union (USSR) nuclear arsenals, was followed by United Kingdom (UK), France, and China as they achieved nuclear status. This formed the club of five Nuclear Nonproliferation Treaty (NPT) nuclear weapon states. Later India (1974 and 1998) and Pakistan (1998), non-NPT parties, conducted nuclear weapons tests, more recently joined by North Korea's (2006, 2009, 2013, 2016, 2017) tests which withdrew from the NPT (2003). The threat of continued nuclear proliferation in Iraq, Libya, Iran and the increasing pressure for reduction and dismantlement of nuclear capacities, remain very sensitive issues and major factors in the political and strategic standing of states.

Nonetheless, significant progress has been made since the Cuban missile crisis (1962) and especially since the end of the Cold War in the early 1990s. More than 70 years after Hiroshima and Nagasaki and 25 after the end of the Cold War, the world's nuclear arsenals are still estimated to total around 13,900 warheads at June 2019, nearly 91% of which are in the hands of the US and Russia [2]. This still remains a major concern but a real improvement compared to the 70,300 weapons active in 1986 [3].

In the aftermath of the disintegration of the USSR, the end of the Cold War and the new strategic context, which began with the perspective of a more peaceful world, brought new hopes and gave a new momentum to the nuclear disarmament process. Alas, at the turn of the century, the resumption of antagonism between the western bloc, young Russia, and emerging China plus regional disputes (i.e. India and Pakistan) have soon slowed down the pace of nuclear disarmament process.[1]

Nuclear weapons remain a major component of the security posture and backbone of military political and alliance strategy for those who wish to possess them. This includes the countries that rely on the umbrella of US deterrence. Some states see nuclear weapons as a global strategic stability factor and others view them as a regional power vector. If strategic, regional and national conditions are met, progress on nuclear disarmament would be an important enhancement of global and regional security. This progress is not possible without the confidence that security it not undermined. The key challenge is to set the right balance between efficiency of disarmament instrument's verification systems and the legitimate rights of states to protect sensitive national security information. Transparency and effective verification will be essential to build confidence between states or group of states and allow progress towards that goal. Confidence building measures, such as:

- verifiable declarations,
- voluntary actions,
- changes in state doctrines,
- irreversible destruction of equipment,
- facilities or testing sites,

[1]For a discussion on how to advance in nuclear disarmament, see [1].

- disposition of fissile material stocks no longer needed for defense purpose, and
- removal of nuclear warheads, with ad hoc visits

could all efficiently contribute to the establishment of a climate conducive to progress, as stated by UNSCR 1887 Preamble, unanimously adopted by heads of state.[2]

Over the last 30 years, a renewed nuclear landscape has emerged in uncertain strategic environment, with new actors and new threats and a higher risk of nuclear use. Global Strategic Stability is more and more elusive and fragile.[3] Any progress expected in nuclear disarmament should proceed stepwise carefully taking into account the strategy, policy, doctrines, regional or international context, culture, military capacity, compliance with existing instruments, and nuclear capacities, both military and civilian, of the states involved. This does not mean that ultimately, everything should or could be verified but the whole system should be considered.

If these conditions are met, nuclear disarmament will proceed steadily, slowly and stepwise. Unfortunately, these conditions are not met currently and there is little hope they will be in foreseeable future, as relations with Russia are worsening. China continues to build up its military nuclear capacities and to manifest tendencies to regional hegemony. Tensions between India and Pakistan did not subside. In the Middle East, the nuclear deal with Iran (Joint Comprehensive Plan of Action) and its verification protocol bring some hope of removing the risk of a nuclear capable Iran (temporarily?). But tensions are still very high and Israel does not appear to be ready to make any move to drop its nuclear guard.

This pessimistic picture of the international context prevents progress in nuclear disarmament and related verification. Progress will only occur if policy-makers feel it's in their country's national interest. It will only happen when political and security conditions enable it to happen [6]. Transparent, verifiable and irreversible nuclear disarmament complemented with credible disarmament measures in all other fields, such as conventional forces, missile defense and space and backed up by a fair and efficient verification regime taking into account all aspects of state's security, complemented by confidence building measures, is the key to movement towards disarmament.

Today the landscape is a mix of hope and pessimism. On one hand, no new progress advancing international instruments has been seen. We are still waiting for some major steps towards a future global nuclear disarmament, such as entry-into-force of the Comprehensive Nuclear Test Ban Treaty (CTBT) and the opening of negotiations for a Fissile Material Cut-off Treaty (FMCT). The Conference on Disarmament (CD) in Geneva, the multinational fora for disarmament negotiations, has produced no major progress for almost two decades. The fact that some nuclear

[2]"Resolving to seek a safer world for all and to create the conditions for a world without nuclear weapons, in accordance with the goals of the Treaty on the Non-Proliferation of Nuclear Weapons (NPT), in a way that promotes international stability, and based on the principle of undiminished security for all." [4].

[3]For a comprehensive discussion on the future of doctrines and deterrence, see [5].

powers keep modernizing their strategic forces and some continue to increase their nuclear arsenal (India [7], Pakistan [8], China? [9] shows more is still needed to achieve the common goal of a global nuclear disarmament. CTBT cannot enter-into-force until major nuclear players as USA, China, India, Pakistan Iran, Israel and North Korea (DPRK) ratify the treaty.[4] Although important work was conducted in 2014–2015, negotiations for a treaty banning the production of fissile materials for nuclear weapon (FMCT or cut-off treaty) have not yet re-started at the CD.

On the other hand, there has been significant progress in recent years. France and UK announced reductions of their nuclear arsenals and put forward transparency measures.[5] France closed the South Pacific test site and begun dismantling its nuclear weapons fissile material production facilities. New START, the Treaty between the US and Russian on *Measures for the Further Reduction and Limitation of Strategic Offensive Arms* has entered into force. No further reduction of fissile material available for nuclear weapons has been announced, after the significant reductions undertaken by US and Russia.

2 Setting the Scene

The linkage between nuclear disarmament verification and strategic postures, nuclear deterrence doctrines and military capabilities is evolving so the strategic context becomes more elusive and complex. Since the end of World War II we have seen that military dimensions, verification processes, improvement in the security context and confidence between states are closely interlinked and can only proceed together.

2.1 Growth of Nuclear Arsenals

In the aftermath of the Second World War, the direct confrontation of both Western and Communist blocs triggered and fueled considerable growth in the US and former Soviet Union's nuclear arsenals. This was seen during Cuba missile crisis which brought the world to the edge of a nuclear war. During this period, Britain, France and China acquired nuclear status and began to develop their own nuclear military capabilities, albeit much smaller than those of the US and USSR. After several setbacks (e.g. Vietnam, Suez, etc.) these states chose not be completely dependent on the two superpowers. The possession of nuclear weapons gave them

[4]For any information on CTBT see the CTBTO Preparatory Commission website: http://ctbto.org/.

[5]Discours sur la dissuasion nucléaire—Déplacement auprès des forces aériennes stratégiques. Istres (13). https://www.defense.gouv.fr/content/download/352889/5043677/file/discours-sur-la-dissuasion-nucleaire-deplacement-aupres-desforces-aeriennes-strategiques-istres-3.pdf.

the political and strategic stature needed, in particular to become members of the UN Security Council.

At the same time, several other states, some very close to the nuclear threshold, concluded that their security would be better without nuclear weapons and renounced them.[6] Others renounced possession of nuclear weapons and campaigned for nuclear disarmament while choosing to be sheltered under the nuclear umbrella of one or other of the superpowers (e.g. NATO members, Japan, Korea under the American umbrella, members of the Warsaw Pact under the Soviet Union). This complicates the military dimensions of nuclear arms reduction talks.

2.2 *Fissile Materials and Technology Control*

Very soon after the end of World War II, the US became aware of the risk the diffusion of nuclear material and the uncontrolled spread of nuclear technology. As the USSR developed their nuclear capability, they gradually came to share this view. The fear of the huge destructive power of nuclear weapons raised the vital need to block their dissemination. This required the control of the spread of sensitive materials and technologies without hindering peaceful applications of nuclear energy. This "fearful dilemma" inspired the speech of President Eisenhower "*Atoms for Peace*" (December 1953) [11] and gave birth to the International Atomic Energy Agency (IAEA) [12]. The treaty establishing the European Atomic Energy Community (EAEC, 1957), known as "EURATOM treaty" requires, inter alia, the application of nuclear safeguards to all members of EU whatever their nuclear status [13] (contrary to IAEA safeguards which were applied only to non-nuclear weapons states, unless volunteered by a nuclear weapons state).

2.3 *De-escalation*

After the Cuban missile crisis (1962), the world, and in particular the two superpowers became concerned with the risk of a global nuclear war being triggered by a local crisis. US and USSR began talks on mutual reduction of their arsenals and limiting nuclear testing in the atmosphere. The first agreement was a treaty banning all nuclear weapons test detonations in the atmosphere, outer space, and underwater environments, allowing continued underground testing (PTBT, 1963). Abolition of underground testing was addressed by the CTBT 30 years later. By then, the establishment of a verification system of controls and mutual inspections

[6]Sweden, Switzerland Brazil, South Africa and many other countries started to develop a nuclear weapons programs in the 1960s. Most ended by the time NPT started and some others much later; see [10].

for underground explosions had been negotiated between the USSR and Western countries. The competing requirements of security and confidentiality were already clear when considering the means to detect and evaluate suspicious events.

2.4 *Proliferation Awareness*

After the accession of UK, France and China to nuclear status, it appeared that other countries were seeking to acquire nuclear weapons increasing the risk of regional nuclear conflict.

> There are indications because of new inventions, that 10, 15, or 20 nations could have a nuclear capacity, including Red China, by the end of the Presidential office in 1964. This is extremely serious. I think the fate not only of our own civilization, but I think the fate of world and the future of the human race is involved in preventing a nuclear war [14].

Discussions in Conference on Disarmament, established under UN auspices in Geneva to prevent the proliferation of nuclear weapons, resulted in the adoption of the Treaty of the Non-proliferation of Nuclear Weapons (NPT, 1970). This is the cornerstone of nuclear disarmament and nonproliferation and intended to close the door to the "nuclear club" while allowing the peaceful uses of nuclear energy. Only the five members having already detonated a nuclear weapon would have the legal right to possess them (US, USSR (later Russia), UK, France and China). Any other nation that tested a nuclear weapon after entry into force of the NPT has been accepted as a Nuclear Weapons State.

2.5 *Post-NPT Era (1970–1990)*

During the Cold War, the main concerns were that the arms race between the US and USSR increased nuclear arsenals, improved nuclear weapons (yield, penetration, hardening of nuclear warheads, development of multiple independently targetable reentry vehicle (MIRV), decoys, maneuverable warhead, stealth warhead) and improved the ability to deliver the weapons (precision, range, discretion of strategic bombers, Intercontinental Ballistic Missile (ICBM), Submarine-Launched Ballistic Missile (SLBM) and Anti-Ballistic Missile (ABM) systems). The risk of mutual annihilation was so high that the two super powers concluded agreements for limitation of and reduction of strategic arms (SALTs and START treaties, respectively—see Sect. 4). The number of nuclear weapons in the world had peaked at approximately 70,300 in 1986 but began to decline significantly, in particular after the end of the Cold War. The number decreased to an estimated 14,200 by early-2018. An overwhelming portion of the reduction happened in the 1990s as show in Fig. 1. Since then, the pace of reduction has slowed significantly.

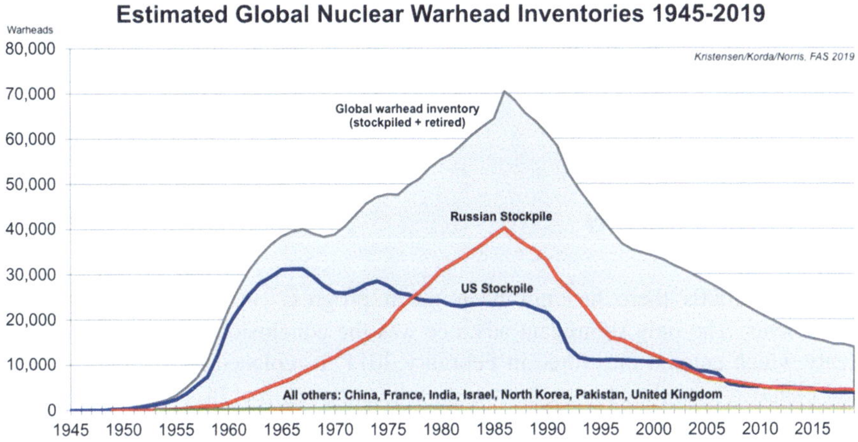

Fig. 1 The number of nuclear weapons in the world has declined significantly since the Cold War [2]. Reproduction with kind permission of FAS

In the two decades between the entry into force of the NPT and the collapse of the USSR, some important events regarding disarmament and non-proliferation occurred. In 1972 SALT I entered into force [15]. In 1974, with a nuclear test, India demonstrated its possession of nuclear weapons. In response, the Nuclear Suppliers Group (NSG) was created to limit the export of nuclear equipment, materials or technology. The 1970s and 1980s saw the strengthening of the non-proliferation regime.

2.6 *Post-Cold War Era (1990–2010)*

Progress during the first years following the collapse of USSR stalled during the late 1990s and the international context worsened. Some positive progresses were recorded but they were counterbalanced at the same time by negative steps as serious nuclear crisis just started, inter alia:

- Important progress in nuclear disarmament with START II (1992) and several agreements between USA and Russia.
- Start of the discussion for an FMCT (1992)
- Adoption of the CTBT (1996)
- Indian and Pakistan test nuclear weapons (1998)
- Proliferation crisis with the discovery of nuclear weapons program of North Korea, Iraq, Libya and Iran
- Adoption of the IAEA Additional Protocol which from the disclosure of Iraq clandestine nuclear weapons program
- Iran non-proliferation crisis (2002) and disclosure of its nuclear weapons program

- Reinforcement of export control: NSG issues the double-use items list and strengthened if after the unveiling of A.Q. Khan Proliferation's network (2004).
- North Korea nuclear tests (2006, 2009, 2013, 2016 (2 tests), 2017 (possible thermonuclear))

2.7 Current Context

Since the 2000s there has not been much progress on nuclear disarmament agreements. The only significant advance was the conclusion of the New START treaty which entered into force in February 2011. It replaced the expired START and included a modified inspection and verification regime. In 2002, the Moscow Treaty (SORT) was agreed but it did not provide for any verification measures. It was terminated[7] when New START entered-into-force in 2011. New START limits the number of deployed strategic nuclear warheads to 1550, and the number of deployed and non-deployed inter-continental ballistic missile (ICBM) launchers, submarine-launched ballistic missile (SLBM) launchers, and heavy bombers equipped for nuclear armaments to 800. The CTBT has not yet entered-into-force because a requisite number of required countries have not yet ratified. However the International Monitoring System (IMS), built to verify the treaty, is over 80% completed and operational.

Possible negotiations to ban fissile material production for nuclear weapon (Fissile Material Cut-off treaty) were attempted several times but quickly extinguished in the blocked Conference on Disarmament. Treaty text proposals of have been tabled, in particular by France, but to-date the Conference moved forward [16]. Nevertheless, important work, undertaken in the framework of the Governmental Group of Experts[8] prepares the ground for future negotiations of a cut-off treaty.

A key accomplishment was the Joint Comprehensive Plan of Action (JCPOA), agreed between Iran and the P5+1 (China, France, Germany, Russia, the UK, and the US) and the European Union, in July 2015.[9] But this achievement has been undermined by the will of the American President Donald Trump to reject the agreement and impose sanctions to Iran. This ended which means the crisis triggered

[7]New START replaced the Treaty of Moscow (SORT), which ended in December 2012. It is a follow-up to the START I treaty, which expired in December 2009. START II treaty never entered into force and the negotiations of START III treaty were never concluded.

[8]UNGA in its resolution 67/53 (2012) requested the Secretary General to establish a Group of Governmental Experts. The Group was mandated to make recommendations on possible aspects that could contribute to, but not negotiate, a treaty banning the production of fissile material for nuclear weapons or other nuclear explosive devices, 1 on the basis of document CD/1299 and the mandate contained therein. The GGE met over four two-week sessions in Geneva during 2014–2015 (CD/2023, 24 June 2015).

[9]The five permanent members of the United Nations Security Council—China, France, Russia, United Kingdom, United States—plus Germany) and the European Union.

by the discovery of an Iranian nuclear weapons program in 2002 and the inability of the IAEA to inspect Iranian suspected facilities is not over. Iran is gradually disengaging from the constraints imposed by the JCPOA.

IAEA is committed to making a contribution to nuclear disarmament. The IAEA Department of Safeguards Strategic Plan commits to: "Contribute to nuclear arms control and disarmament, by responding to requests for verification and other technical assistance associated with related agreements and arrangements" and by preparing to play an active role in the verification of a Fissile Material Cut-off Treaty when and if it enters into force.

3 Military Dimensions of Nuclear Disarmament

The military dimension in the nuclear disarmament negotiation process involves all features of the military nuclear complex: doctrines, research & development, production capacity, military nuclear fuel cycle, fissile materials and equipment, testing facilities and sites, means of delivery, nuclear strategic and non-strategic forces, conventional forces, and deployment of nuclear weapons. It must also take into account the existing nuclear disarmament commitments and the associated verification regimes, which are a very complex pattern of commitments.[10]

3.1 Existing Commitments

A state's membership in regional and/or international agreements influences its military stance with respect to nuclear disarmament verification. It adds a regional, multinational and international complexity that must be addressed.

Nuclear weapons control agreements can be organized into four interlinked categories. This non comprehensive review of regional, multinational and international agreements, that relate to nuclear weapons, include those in-force, expected to enter-in-force soon, and pending negotiation.

- **Treaties limiting nuclear weapons testing.** Historically nuclear weapon testing treaties were the first to be negotiated. The CTBT (1996) was preceded by several treaties limiting nuclear testing capacity. Partial Test Ban Treaty between USA, USSR and UK (PTBT, 1963) allowed only underground testing. Treaty on the Limitation of Underground Nuclear Weapon Tests, between USA and USSR also known as the Threshold Test Ban Treaty (TTBT, 1974) capped the yield to 150 kt. The Comprehensive Test Ban Treaty (CTBT, 1996) which forbids all nuclear weapon test explosion is not yet into force as the ratification of several

[10] For a comprehensive review of non proliferation and disarmament instruments see [17].

annex II countries required for entry into force are still missing.[11] The CTBT is equipped with a very comprehensive and efficient monitoring and verification system which is almost completed.

- **Treaties preventing the proliferation of nuclear weapons.** The keystone of the nonproliferation regime is the Non Proliferation Treaty (NPT, 1970) which limits the number of countries which have legally the right to possess nuclear weapon to the five countries which carried out a nuclear weapon test before the 1st January 1967. Countries which tested nuclear weapons after this date as India Pakistan and the Democratic People Republic of Korea (DPRK) could not accede to the NPT without renouncing their nuclear weapons. Israel is presumed to have nuclear weapons but never officially acknowledge it. Some other countries host foreign nuclear weapons and some other had in the past nuclear weapons but have renounced to them. For a comprehensive overview of countries status regarding nuclear weapons, see [18]. In coherence with the NPT countries of a same region gathered to ban nuclear weapon through *Nuclear-Weapon-Free Zone Treaties* and negotiated *Security Assurances* to prevent the use of nuclear weapons against them [19]. For the implementation of article IV of the NPT regarding peaceful uses of nuclear technology Export Control regimes (e.g. Zangger Committee, NSG (Dual use items 1992), Wassenaar) have been set by concerned group of nuclear countries.

  In accordance with Article VI of the NPT, any instrument, whether a treaty, a convention or an agreement contributing to the ultimate goal of nuclear disarmament should be "in principle" universal, non-discriminatory, multilateral, internationally and effectively verifiable. Such an instrument would have considerable impact on the international and regional strategic context. A verification protocol, along with transparency and confidence building measures, will be a key element. It must maintain a delicate balance of upholding Article I of the NPT to prevent proliferation of confidential data and technologies to non-nuclear weapon states (inter alia the limitation of access to knowledge, know-how and technologies through managed access) and allowing for the necessary transparency provide assurances of compliance and confidence in the implementation of the instrument to all parties.

- **Treaties reducing strategic forces and nuclear weapons arsenals** [15, 17]. Strategic offensive arms agreements between, USA and USSR (and then Russia) provide a great deal of negotiation and implementation experiences on mutual verification systems, technologies, inspections, managed access, and data exchange which may be useful for any future disarmament agreement negotiations Agreements related to strategic weapons systems are:

  - SALT I (Strategic Arms Limitation Talks, 1969);
  - SALT II (1972) START I (Strategic Arms Reduction Treaty 1991);

[11] There are eight Annex 2 States that have yet to ratify the Treaty—China, Egypt, Iran, Israel, and the United States of America, which have signed the Treaty, and North Korea, India, and Pakistan, which have not signed.

- START II (1992 not entered-into-force), START III Framework (1997), SORT (Strategic Offensive Reductions Treaty or Moscow Treaty, 2002);
- New START between the United States and Russia (2011);
- Nonstrategic Nuclear Arms Control: Intermediate-Range Nuclear Forces (INF Treaty; 1987).

• **Treaties to control production of fissile material for nuclear weapons.** Since the end of the cold war, important efforts to reduce available fissile material stocks for use in nuclear weapons have been completed through voluntary measures. Most of nuclear countries have declared a moratorium on production and some have irreversibly dismantled their production facilities. As of January 2017, the global stockpile of highly enriched uranium (HEU) is estimated to be about 1340 ± 125 tons. The global stockpile of separated plutonium is about 520 tons (290 in civilian custody). Most of this material (>93%) is hold by USA and Russia.[12] Fissile material disposition instruments includes (inter alia):

  - Fissile materials declared no longer needed for defense purpose[13];
  - 2000 Plutonium Management and Disposition Agreement and update;
  - Fissile Material Cut-off Treaty (FMCT negotiation not started yet. Nevertheless, some work has been done in the framework of the Group of Governmental Experts (GGE) [16]).

3.2 *A Brief Review of the Components of the Military Dimension*

• **Doctrines, posture and deterrence.** A nuclear country's policy on the use of nuclear weapons and conventional forces is very important. It indicates the role that weapons play in maintaining the status and the global image of a state[14] and reflects the state's perception of the role that nuclear weapons play in ensuring national security and defense capacity, in the context of its international policy. Most of the nuclear countries publish their nuclear doctrine. The North Atlantic Treaty Organization (NATO), an alliance relying on the nuclear capacity of three of its members, has also published its doctrine. It covers a large array of complex positions from "no first use" (China) to the deterrence (France, India) and the preemptive strike against a major conventional attack or a non-conventional

---

[12]For the actual status of fissile material disposition see [20].

[13]Verified and irreversible elimination of weapon usable nuclear material is one of the essential elements of the nuclear disarmament effort. The Action Plan adopted at the 2010 Nuclear Non-Proliferation Treaty (NPT) Review Conference calls on nuclear-weapon states *"to declare fissile material designated as no longer required for military purposes, and to place this material under safeguards to ensure that it remains permanently outside military programmes."* 2010 NPT Action Plan, action 16.

[14]For a discussion on doctrines and their roles vis-a vis nuclear disbarment see [21].

aggression (chemical, bio, cyber, etc.). Nuclear and military doctrines are major elements to take into account when considering disarmament verification.

- **Military nuclear fuel cycle.** The extend (i.e. from mine to weapon) and the development (i.e. advanced enrichment methods) of the military nuclear fuel cycle determines the capacity of a state to resume or cover clandestine activities.
- **Fissile materials.** Production and/or acquisition of weapon grade fissile materials (highly enriched uranium or plutonium) are a mandatory step in nuclear weapon development process. Putting an end to this stage of weapons manufacturing is the keystone of nuclear arms control and ultimately elimination. It would encompass the control of fissile materials production for nuclear weapons: enrichment, reprocessing, and other processes. A future cut-off treaty, equipped with an efficient verification process, would be the tool to master that stage. Absent a treaty, another way to manage the stocks of fissile material, would be to irreversibly place the disposition of fissile material no longer needed for defense purposes (fissile materials in excess of defines needs) under safeguards.
- **Testing.** Testing is the ultimate stage in the nuclear weapons research and development process before militarization, therefore prohibiting nuclear testing is important when moving towards nuclear disarmament. Except for DPRK, all NPT and non-NPT nuclear weapon states have declared or applied a de facto moratorium of testing. In the continuity of previous nuclear testing agreements (Partial Nuclear Test Ban Treaty, Threshold Test Ban Treaty) a CTBT banning any nuclear testing has been adopted and is pending ratification of the US, China, India, Pakistan, Iran and Israel. It includes a monitoring and verification system (IMS & OSI). Definite closure of nuclear test sites and associated facilities are also important signals that countries are moving towards nuclear disarmament. To-date, only France and the UK have completed this step.
- **Nuclear weapons arsenals.** Transparency nuclear arsenal composition is also a key element of progress towards nuclear disarmament. The arsenals of the US and Russia have been declared and considerably downsized within the implementation of bilateral arms control reduction treaties (see Sect. 4). France and the UK have also completed significant downsizing and declared the size and the composition of their arsenals. Published evaluations of India [7], Pakistan [8] and China [9, 22] indicate they are increasing their arsenal.
- **Nuclear strategic and non-strategic forces/Means of delivery versus conventional forces.** The size, composition and technological advance of strategic, non-strategic (tactical or theater nuclear weapons) and conventional forces, means of delivery, ground and submarine missiles and cruise missiles, military satellites, radar networks, etc. will influence the disarmament process. Some less advanced countries may maintain a small nuclear capability (i.e. Pakistan vs India or in the past NATO vs USSR) for national security purposes.

- **Former verification experiences.** Previous nuclear weapon dismantlement verification exercises provide a great deal of experience in verification process, development and implementation of ad hoc technologies, military aspects, confidentiality issues, managed access, confidence building measures and mutual verification. Some former experiences are:
  - Early days field test 34 (1967);
  - Threshold Test Ban Treaty US-USSR mutual verification (1974);
  - Black Sea US-USSR joint experiment (1989);
  - DoE studies (1996–1997);
  - Trilateral initiative (US-Russia, IAEA 1996–2002);
  - Bilateral arm reduction treaties between USA and Russia START (1991) and New START (2010);
  - UK-Norway initiative;
  - Other experiences as INF treaty (Intermediate-Range and Shorter-Range Missiles, 1987) or Conventional Armed Forces in Europe treaty (CFE 1972) has also to be taken into account.
- **The role of the IAEA in the verification process.** The IAEA is the multinational organization that carries out verification of compliance assessment for NPT commitments Nuclear Weapons Free Zone agreements and other related agreements, through a complex and efficient system of safeguards. Based on its statutes, the IAEA had not invested in the disarmament verification until proliferation crises (listed below) drove the Agency to focus on these issues:
  - DPRK (North Korea; agreed framework);
  - Iraq (IAEA): Action Team, later Iraq Nuclear Verification Office (INVO);
  - Iran—P5 +1 Joint Comprehensive Plan of Action (JCPOA);
  - Libya and A. Q. Khan nuclear black market network.

  Since the United Nations Security Council entrusted the Director General of the IAEA verification of the nuclear part of the disarmament of Iraq, the position of the Agency evolved. Building on the skills acquired on these issues and the implementation of the Additional Protocol, the IAEA is positioned to play an important role in the verification of future disarmament agreements such as the verification of a cut-off treaty.[15] The Agency is staffed by international civil servants representing the Member State from both NPT Nuclear and Non-nuclear Weapons States. This introduces both proliferation and security concerns with designing a verification regime run by the IAEA. It has not been universally agreed that the IAEA has the resources or mandate to continue to support the world community in this way but it is likely to play a pivotal role.

[15]"Contribute to nuclear arms control and disarmament, by responding to requests for verification and other technical assistance" [23].

4 Conclusions and Future Prospects

Reduction or elimination of nuclear weapons is not likely to occur in a foreseeable future. The international, regional and national security context is deteriorating and confidence is lacking. So there is no incentive for countries possessing nuclear weapons to disarm. After a long period comprising of a mix of fruitful dialog and chaotic confrontations resulting in a large decrease of arsenals, strategic forces and fissile material stocks, the relationship between the US and Russia is getting more and more difficult. Russia does not appear to be interested in the results achieved during 1990–2010 and is modernizing its strategic forces. An increasingly aggressive strategy to rebuild its lost empire is being implemented. US-China relations are also deteriorating, as China continues to develop its nuclear arsenal and rapidly modernizing its strategic forces to support its expansionist policies. Antagonism between India and Pakistan continues at a high level. Both countries continue to produce fissile materials for weapons, increasing the number of nuclear warhead and improving their means of delivery. The security situation in the South Asia remains very tense as India, Pakistan, USA, China and Russia confront unstable alliances and antagonism.

In the last decades, the UK and France [24] have made very important and unprecedented efforts towards nuclear disarmament by downsizing their arsenal and decommissioning and closing critical facilities as nuclear test sites, plutonium production reactors, reprocessing and enrichment facilities. Even though negotiations with Iran on the verification of its nuclear complex's verification resulted in the conclusion of the Joint Comprehensive Plan of Action to monitor the absence of cover nuclear program, the current situation in the Middle East does not provide an incentive for Israel to lower its nuclear posture.

There is still a perception that maintaining or developing nuclear weapons arsenals is needed to support national security.[16] Any voluntary downsizing of countries nuclear posture in the legal framework must be balanced by the assurance that its national security interests will not be jeopardized but strengthened. Nevertheless, one should be reasonably optimistic. The road to a world without nuclear weapons will be long, tortuous, with setbacks, but unavoidable. One of the great challenges to overcome will be to achieve the needed transparency, verifiability and irreversibility of disarmament measures. A systematic approach that creates a climate of confidence, while taking into account the security interests and military dimension of all stakeholders, will greatly ease the process. Achieving such confidence on a national-level followed by a regional level (European, Middle East, South Asia and Eastern Asia) to ultimately on the international level is pivotal.

[16]"Ninth, short of pretending that lions and sheep should be able to lie together before contemplating nuclear disarmament, the political conditions for a secure nonnuclear—or at least less nuclear—world are considerable. Serious political efforts should therefore accompany nuclear weapon reductions." [5, pp. 36, 38].

References

1. Finaud M (2013) Cooperative Security: A new paradigm for a world without nuclear weapons? CADMUS 2 (1):169–179
2. Arms Control Association (2019) 2019 Estimated Global Nuclear Wearhead Inventories. Sources: Arms Control Association, Federation of American Scientists, International Panel on Fissile Materials, U.S. Department of Defense, U.S. Department of State and Stockholm International Peace Research Institute (updated June 2019). https://www.armscontrol.org/factsheets/Nuclearweaponswhohaswhat
3. Norris RS, Kristensen HM (2010) Global nuclear weapons inventories 1945–2010. Bulletin of the Atomic Scientists 66(4):77–83
4. UN Security Council Resolution 1887: Non-proliferation, 24 September 2009
5. Delpech T (2012) Nuclear Deterrence in the 21st century, Lessons from the Cold War for a New Era of Strategic Piracy. Rand Corporation, Santa Monica, CA. http://www.rand.org/content/dam/rand/pubs/monographs/2012/RAND_MG1103.pdf. Accessed 30 April 2019
6. Sellal P (2010) Remarks at the Global Zero Conference, Paris, February 2nd 2010
7. Kristensen HM, Norris RS (2015) Indian nuclear forces, 2015. Bulletin of the Atomic Scientists 71(5):77–83
8. Kristensen HM, Norris RS (2016) Pakistani nuclear forces, 2016. Bulletin of the Atomic Scientists 72(6):368–376
9. Kristensen HM, Norris RS (2015) Chinese nuclear forces 2015. Bulletin of Atomic Scientists 71(4):77–84
10. Institute for Science and International Security (ISIS): Nuclear Weapons Programs Worldwide: An Historical Overview. http://isis-online.org/nuclear-weapons-programs. Accessed 30 April 2019
11. Eisenhower DD (1953) Atoms for Peace. Speech of the President of the United States of America to the 470th Plenary Meeting of the United Nations General Assembly, 8 December 1953
12. International Atomic Energy Agency (IAEA) (2016) History. https://www.iaea.org/about/overview/history. Accessed 30 April 2019
13. European Union (2012) Consolidated version of the Treaty establishing the European Atomic Energy Community. https://eur-lex.europa.eu/legal-content/EN/TXT/?uri=CELEX%3A12012A%2FTXT. Accessed 30 April 2019
14. Commission on Presidential Debates (2015) The Third Kennedy-Nixon Presidential Debate. October 13, 1960 Debate Transcript. http://www.debates.org/index.php?page=october-13-1960-debate-transcript. Accessed 30 April 2019
15. U.S.-Russian Nuclear Arms Control Agreements at a Glance. Fact Sheets & Briefs. In: Arms Control Association. https://www.armscontrol.org/factsheets/USRussiaNuclearAgreementsMarch2010. Accessed 30 April 2019
16. Draft Fissile Material Cut-off Treaty, 9 April 2015. http://www.delegfrance-cd-geneve.org/Draft-fissil-material-cut-off-treaty. Accessed 30 April 2019
17. King's College London, James Martin Center for Nonproliferation Studies (2015) NPT Briefing Book (2015 Edition). http://www.kcl.ac.uk/sspp/departments/warstudies/research/groups/csss/pubs/NPT-Briefing-Book-2015/New-sections-April-2015/NPT-BB-2015-Final-April.pdf. Accessed 30 April 2019
18. List of states with nuclear weapons. Wikipedia, the free encyclopedia. https://en.wikipedia.org/wiki/List_of_states_with_nuclear_weapons. Accessed 30 April 2019
19. Negative security assurances. In: Reaching Critical Will. http://www.reachingcriticalwill.org/images/documents/Resources/Factsheets/nsa.pdf. Accessed 30 April 2019
20. International Panel on Fissile Materials (IPFM) Home page. http://fissilematerials.org/. Accessed 30 April 2019

21. Dynkin AA (2010) Foreword at the Conference "Contemporary Nuclear Doctrines", Moscow. Institute of World Economy and International Relations, Russian Academy of Sciences and Nuclear Threat Initiative.
22. Blumenthal D, Mazza M (2011) China's Strategic Forces in the 21st Century: The PLA's Changing Nuclear Doctrine and Force Posture. http://www.npolicy.org/article_file/Chinas_Strategic_Forces.pdf. Accessed 30 April 2019
23. International Atomic Energy Agency (IAEA) (2013) IAEA Department of Safeguards Long-Term R&D Plan, 2012–2023. IAEA, STR-375
24. France Diplomatie (2016) Disarmament and non-proliferation. http://www.diplomatie.gouv.fr/en/archives-ne-pas-utiliser/archives-2016/french-foreign-policy/disarmament-and-non-proliferation. Accessed 30 April 2019

Strategic Export Control

Filippo Sevini and Willem A. Janssens

Abstract Strategic, or dual-use, goods and associated knowledge have played a key role in the development of Weapons of Mass Destruction (WMD) and their means of delivery since WWII's nuclear developments. Nuclear export controls and international safeguards have therefore developed in parallel in various phases, triggered by major international events, which showed how the insufficient scope of the controls existing at that time, as well as how legal framework's loopholes, could be exploited to acquire sensitive goods. Following a phase of 'country-to-country' support in the development of nuclear programmes, the history of proliferation has shown how in particular in the 1980s–1990s the illicit transfer of strategic goods and technology has been an additional key element and threat allowing the development of competences and capabilities. The control of strategic trade was therefore set up as a barrier against the diffusion of sensitive materials, components and technologies, which could be used for the proliferation of nuclear biological and chemical weapons of mass destruction and their means of delivery. The result has been a continuously evolving multi-layered regime which comprises Treaties, International Agreements, UN Security Council Resolutions, embargo measures and national laws. This formed a sort of system of "defence in depth" system, which has legal as well as political grounds, various procedural and technical key elements contributing to deter, delay and detect proliferation activities. The chapter reviews the background and key aspects of strategic export control, developing on the contents and challenges and its relevance to countering the proliferation of weapons of mass destruction, as well as its relevance to disarmament and arms control.

F. Sevini (✉) · W. A. Janssens
EC Directorate General Joint Research Centre, Directorate G - Nuclear Safety and Security, Ispra, Italy
e-mail: filippo.sevini@ec.europa.eu; willem.janssens@ec.europa.eu

I. Niemeyer et al. (eds.), *Nuclear Non-proliferation and Arms Control Verification*,
https://doi.org/10.1007/978-3-030-29537-0_7

1 Historical Developments

The starting point of export controls was during the Cold War in the 1950s with COCOM (Coordinating Committee for Multilateral Export Controls) [1], a multilateral export control arrangement of Western allied countries[1] established after the first Soviet nuclear test to prevent further acquisition of nuclear technology by the Warsaw Pact and China COCOM had identified as strategic those goods related to conventional weapons and weapons of mass destruction, but also those enabling economic developments as e.g. oil and gas exports.

Following the Atoms for Peace declaration of President Eisenhower, the IAEA was set up in 1957 and safeguards became the condition for nuclear supplies In the same year, the EURATOM Treaty was signed in Rome by six European States[2] setting up the European Atomic Energy Community.

The nuclear proliferation race however continued in the two geo-political blocks along the 1950s and 1960s

1.1 The Non-proliferation Treaty (NPT)

The entry into force of the Non Proliferation Treaty (NPT) in 1970 was a turning point in non-proliferation [2].

As a consequence, many steps were undertaken for the development of international nuclear safeguards and Comprehensive Safeguards Agreements (CSA), with the objectives of the prevention and detection of undeclared nuclear activities as well as of the diversion and misuse of nuclear materials from declared ones. Safeguards agreements between the IAEA and countries or regional systems of accountancy and control (EURATOM for the EU) were established.

Together with disarmament and peaceful uses, Non-proliferation (NP) is the NPT's third pillar, strongly supported by the close relationship between export controls and nuclear safeguards outlined by Article III.2's requirement for safeguards as a principal condition of the supply of nuclear items:

> Each State Party to the Treaty undertakes not to provide: (a) source or special fissionable material, or (b) equipment or material **especially designed or prepared for** the processing, use or production of special fissionable material, to any non-nuclear-weapon State for peaceful purposes, unless the source or special fissionable material shall be subject to the safeguards required by this Article.

---

[1]CoCom began operations on January 1, 1950, with a membership consisting of the United States, UK, France, Italy, the Netherlands, Belgium, and Luxembourg. In early 1950, Norway, Denmark, Canada, and West Germany joined, followed by Portugal in 1952 and Japan, Greece, and Turkey in 1953.

[2]Belgium, France, West Germany, Italy, Luxembourg, The Netherlands.

The need to interpret the term "*especially designed or prepared for*" components led to the formation of the NPT Exporters' (or Zangger) Committee, which could not come up with a definition but instead identified a list of key nuclear fuel cycle items. The resulting "Trigger List" (i.e. a list of equipment and facilities "triggering" the need for safeguards) and guidelines for the supply were communicated to Member States by the IAEA in INFCIRC/209 whose latest revision is reported in [3].

1.2 The Nuclear Suppliers Group (NSG)

The Indian "peaceful nuclear explosion", in 1974 showed that, notwithstanding the entry into force of the Non Proliferation Treaty, various countries had anyway exported nuclear technology to India, a non-signatory to the Treaty.

To address this gap, the nuclear supplier states decided to form the Nuclear Suppliers Group (NSG) [4] which, like the Zangger Committee, also issued additional Guidelines in 1978, published as INFCIRC/254/Part 1 and including an extended Trigger List [5].

Some years later in 1992, after the discovery of the covert Iraqi nuclear programme and realising the increasing role of dual use equipment, the NSG created an additional (Part II) set of guidelines for transfers of nuclear-related dual-use equipment, material and technology which could make a significant contribution to a nuclear proliferation programme.

The results are two distinct NSG guidelines, respectively the:

- "Guidelines for nuclear transfers" setting the conditions for transfers of nuclear items (i.a. nuclear safeguards and physical protection requirements) and containing two annexes, where Annex B contains the Trigger List (TL) [5]
- "Guidelines for transfers of nuclear-related dual-use equipment, materials, software and related technology", containing in annex the Dual-Use List (DUL) [6]

A broader description of the genesis of the Nuclear Suppliers Group can be found in the www.nuclearsuppliersgroup.org public web-site, developed and maintained by contribution of the European Commission Joint Research Centre Directorate for Nuclear Safety and Security in collaboration with the NSG Point of Contact, Germany, France and Spain.

The current membership of the NSG is of 48 Participating Governments (including all EU Member States), with two Observers (European Commission and Zangger Committee's chair).

2 Strategic Export Control and Nuclear Safeguards

Export control and nuclear safeguards started therefore to develop in parallel, as two intimately linked barriers against proliferation (Fig. 1). A turning point was the discovery of undeclared proliferation activities in Iraq in 1991, which led to the

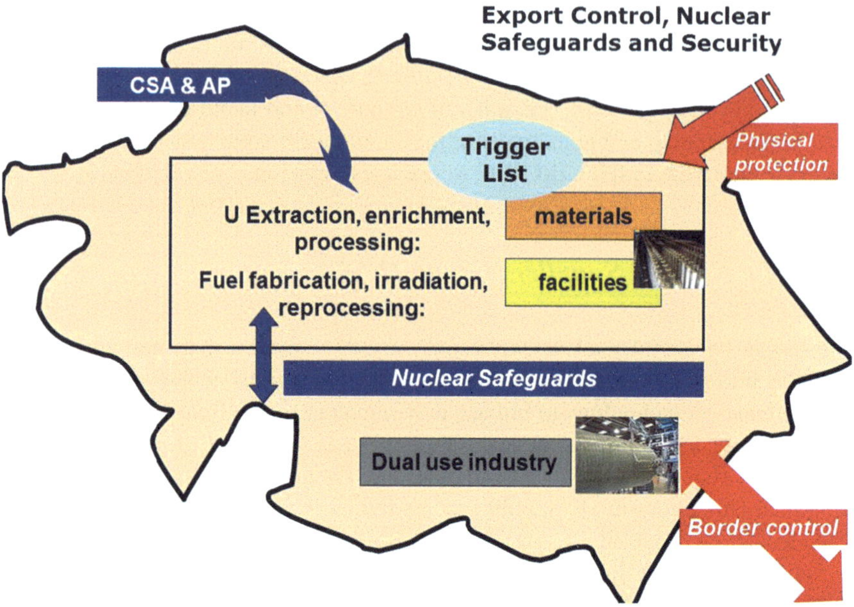

Fig. 1 Synergies among nuclear safeguards, nuclear security and export controls, including country's dual-use potential

conclusion that Comprehensive Safeguards Arrangements (CSA) were not giving enough mandate to inspections for the detection of undeclared activities.

For these reasons in 1997 the CSA where complemented by the Model Additional Protocol [7] enabling the IAEA to gain access to a much wider range of information and locations.

The extended scope of safeguards under an AP gives indeed broader IAEA access to sites and data related to nuclear fuel cycle research and development, paving the way to the so-called "integrated safeguards" system, based on classical accounting, a set of various types of inspection (short-notice, unannounced, complementary accesses) and the drawing of "broad conclusions" on State's absence of proliferation activities, based also on a series of non-classical indicators like export, import, satellite imagery, environmental analysis.

The Additional Protocol's Article 2.a. requires that States:

> ...shall provide the Agency with a declaration containing: (i) A general description of and information specifying the location of nuclear fuel cycle-related research and development activities not involving nuclear material...

and

> (iv) A description of the scale of operations for each location engaged in the activities specified in Annex I to this Protocol.

Annex I lists 15 key nuclear fuel cycle related activities:

1. The manufacture of *centrifuge rotor tubes* or the assembly of *gas centrifuges.*
2. The manufacture of *diffusion barriers.*
3. The manufacture or assembly of *laser-based systems.*
4. The manufacture or assembly of electromagnetic isotope separators.
5. The manufacture or assembly of *columns* or *extraction equipment.*
6. The manufacture of aerodynamic separation nozzles or vortex tubes.
7. The manufacture or assembly of uranium plasma generation systems.
8. The manufacture of *zirconium tubes.*
9. The manufacture or upgrading of *heavy water or deuterium.*
10. The manufacture of nuclear grade graphite.
11. The manufacture of flasks for irradiated fuel.
12. The manufacture of *reactor control rods.*
13. The manufacture of criticality safe tanks and vessels.
14. The manufacture of irradiated fuel element chopping machines.
15. The construction of *hot cells.*

Although the IAEA does not implement export controls, it benefits from their existence. The export and import of nuclear and dual-use goods are crucial indicators of possible proliferation activities. The Model Additional Protocol (AP) also requires export declarations of "Trigger list" items (see above NSG) listed in its Annex II, related to nuclear activities listed in Annex I.

Indeed, Art. 2.a.(ix) of the AP requires that States:

> ...shall provide the Agency with a declaration containing the following information regarding specified equipment and non-nuclear material listed in Annex II:
>
> For each export: the identity, quantity, location of intended use in the receiving State and date ... of export;
>
> Upon specific request, confirmation as importing State of information provided by another State concerning the export of such equipment and material

Annex II lists the items contained in the NSG Trigger List (INFCIRC 254/Part 1) available in 1995 (Rev. 2). Unfortunately, AP Annex II' list has not been amended thereafter and has therefore become obsolete with respect to the NSG's TL, amended already several times (the current version being Rev. 12). This fact creates discrepancies to exporters and authorities which is addressed in various practical ways as outlined in [8, 9].

The State Systems of Accountancy and Control (SSACs), or equivalent organizations depending on the countries' attribution of competences (e.g. EURATOM for the European Union), are responsible also for retrieving this type of information and provide it to the IAEA along with the "classical" and other required declarations. The experience of some ESARDA members with the export control provisions of the AP is summarised in [10] and sketched in Fig. 2.

All together, the facilities declarations, the export declarations, the results of inspections and other types of observations and indicators, help the IAEA to verify the absence of undeclared nuclear activities and contribute to preventing nuclear proliferation.

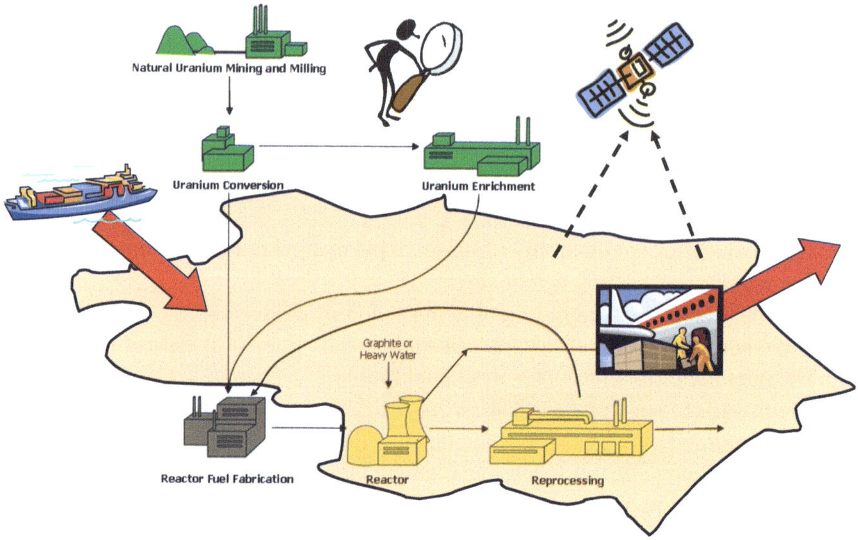

Fig. 2 State's declarations and verification activities

3 Addressing Non-nuclear Proliferation

3.1 The International Export Control Regime

Given the increasing relevance of other types of WMD, the Nuclear Suppliers Group was followed in the 1980s by the establishment of the Missile Technology Control Regime (MTCR, 1987 [11]) and the Australia Group (AG, 1985 [12]) and the later Wassenaar Arrangement (WA, 1996, [13]), with their respective control lists.

Australia Group (AG) The Australia Group (AG) was set up in 1985 and currently has a broad membership of 43 Participating Governments, including all the European Union Member States and the EU itself.

The AG defines Common Control Lists for:

- Chemical Weapons Precursors
- Dual-use chemical manufacturing facilities and equipment and related technology and software
- Dual-use biological equipment and related technology and software
- Biological agents
- Plant pathogens
- Animal pathogens

Missile Technology Control Regime (MTCR) The Missile Technology Control Regime (MTCR) established in 1987 has a more limited participation of 35 governments, with eight EU Members States and the EU itself missing. Its mission is to coordinate national export licensing efforts aimed at preventing proliferation of unmanned delivery systems capable of delivering weapons of mass destruction.

The MTCR defines the following controls:

- Category I: Complete ballistic and cruise missiles with range greater than 300 km and payload greater than 500 kg-Major subsystems such as: engines, guidance sets, space launch vehicles, sounding rockets, and unmanned aerial vehicles (UAVs)
- Category II: Items needed to construct Category 1 systems and non-Category I systems

Wassenaar Arrangement (WA) The Wassenaar Arrangement (WA) was established in 1996 after the end of COCOM, to define a comprehensive list of dual-use goods addressing the overall WMD threat (nuclear, biological, chemical and delivery means), which form the broader dual-use list available, as well as the conventional arms controls list, which is the basis of the Munition List of various countries. The WA also publishes a number of Best practice guidelines. This regime currently has 42 participating governments, while the EU and Cyprus and not are not WA members.

Comments About the International Export Control Regimes The regimes work by full consensus of its Participating Governments to define the guidelines for controlling exports. While politically binding among Participating governments, the membership of an international export control regime does not have legal implications. However it is a key element of non-proliferation commitments, as well as of the commercial policy, because the control lists decided by the regimes are included into the legislation of the Participating governments, as well as by "adherent" countries who wish to show the same level of commitments. In the EU, they constitute the integrated "EU dual-use control list" included as Annex I to Regulation 428/2009 [14], which will be described in the following sections.

3.2 *Other Treaties and Agreements*

Other international treaties are linked to non-proliferation and disarmament by targeting weapons of mass destruction:

- The Biological Toxin Weapons Convention which entered into force in 1975, with 109 Signatory States as of July 2019 [15].
- The Chemical Weapons Convention of 1992 requiring States to ban chemical weapons and allow inspections to chemical plants, includes requirements for disarmament as well as inspections of precursors' production. The CWC has been signed by 165 States so far [16].

Very relevant to WMD is also the Hague Code of Conduct of 25 November 2002 [17], an international arrangement to regulate the area of ballistic missiles capable of carrying weapons of mass destruction, currently subscribed by 138 States.

4 An International Legal Basis for Export Control: UNSCR 1540

Notwithstanding the developments of the various instruments so far described which greatly limited to diffusion of nuclear weapons, proliferation could finds other ways to take place as witnessed for example by the results obtained by Pakistan with the A. Q. Kahn's network and DPRK [18].

Proliferation in the 1980s–1990s indeed saw a change from traditional State-level proliferation, transferring "turn-key" facilities, to illicit procurement networks, with a major role played by dual-use goods and technology.

Technology holder countries suffered from gaps in export control laws (unclear rating and role of dual use items), which could be exploited by "personal contact" networks and non-State actors. Suppliers were not fully aware of the threat associated with dual use goods, or in some intentionally mislead controls.

At the same time, countries not traditionally "technology holder" which became involved in the proliferation networks as third locations, mostly lacked export control laws and therefore could not enforce controls. An international legally binding requirement was missing.

For these reasons, export controls were set at the heart of the resolution of the 2004 Security Council of the United Nations 1540 (UNSCR) of 2004 [19], which is so far the only international legally binding instrument explicitly requiring States to develop an appropriate and effective export control framework (although not clarifying how) as a response to the threat to global peace and security that could raise from the proliferation of chemical, biological, radiological and nuclear (CBRN) weapons of mass destruction (WMD).

UNSCR 1540 calls all upon all UN member states to implement a set of measures to combat illicit and criminal activities including transfer or misuse of controlled commodities by putting in place at national appropriate legislation with an efficient enforcement. It provides guidance for national implementation and it is the first international regulation for a universal control of strategic trade.

The so-called "1540 committee", which is structured on working groups and groups of expert, reports to the Security Council on the implementation of the resolution in the member states through monitoring and strengthening the different undertaken initiatives in line with the required measures. The Committee uses matrices, which are built from information provided from each country according to the operative paragraphs of the resolution, to organize information about the implementation monitoring and also as a reference tool for identifying assistance priorities that a country may require from the technical perspective or others.

The 1540 Matrix gives a quick and wide picture on the status of the global (UN member states) implementation of the resolution which would help identify countries or regions that would necessitate further assistance from other donors such as the EU.

5 The Export Control Process

After having briefly reviewed the historical and legal background of export control, summarized in Table 1, let's now look into some details what the controls include. Various countries in the world adopted national laws to rule the export of dual-use goods. These include the requirements that exporters have to comply with in order to get the authorisation to export the goods contained in the dual-use control list.

The European Union export control framework for dual-use goods was defined in the 1990s and is now part of the common Commercial Policy set by Council Regulation 428/2009 and subsequent amendments, defining the legal requirements for exporters, criteria for authorisation, procedures for consultations among authorities, information sharing and denials. A recast of this regulation is currently under discussion.

Similar export control laws exist in the US (i.e. EAR) and many other countries.

5.1 Control Lists

Taking again the example of the EU, the Dual-use Regulation includes the so-called *European Union "dual-use control list"* (Annex I), which was the result of a large

Table 1 Non-exhaustive list of key steps in non-proliferation and disarmament

Agreement	Type	Entry into force or date of establishment
Non-Proliferation Treaty (NPT)	Treaty; legally binding	1970
Nuclear Suppliers Group	International export control regime; politically binding	1974
Australia Group	International export control regime; politically binding	1985
Missile Technology Control Regime	International export control regime; politically binding	1987
Chemical Weapons Convention	Treaty; legally binding	1992
Wassenaar Arrangement	International export control regime; politically binding	1996
The Hague Code of Conduct	Guidelines; politically binding	2002
UNSCR 1540	Resolution; legally binding	2004

effort carried out in the mid 1990s, when the items listed by the WA, MTCR, NSG and AG international export control regimes mentioned above, complemented by the chemical precursors included in the Chemical Weapons Convention, were integrated into a consolidated unique list. Its latest version is published as [20].

The EU "dual-use control list" is used by various countries in the world constitutes also the basis for the US Commerce Control List (CCL)

The list is organised into ten categories, respectively as follows:

- Category 0: "Nuclear Materials, Facilities and Equipment", including:
  - Plants for the separation of isotopes of natural uranium, depleted uranium
  - Auxiliary systems for isotope separation plants
  - Plants for conversion of uranium
  - Plants for heavy water production
  - Plants for nuclear reactor fuel element fabrication
  - Plants for the reprocessing of irradiated fuel elements
  - Plants for the conversion of plutonium
- Category 1: "Special Materials and Related Equipment";
- Category 2: "Material processing";
- Category 3: "Electronics";
- Category 4: "Computers";
- Category 5 Part I: "Telecommunications";
- Category 5 Part II: "Information Security";
- Category 6: "Sensors and Lasers";
- Category 7: "Navigation and Avionics";
- Category 8: "Marine";
- Category 9: "Aerospace and Propulsion"

5.2 *Other Control Lists*

Additional to the dual-use control list, also certain United Nations Security Council's and national sanctions measures include additional dual-use items' controls or even prohibitions of exports towards specific countries; e.g. the EU measure targeting Iran, Syria and DPRK [21, 22].

Last but not least, controls apply also to items included in the Military Lists.[3] This applies to conventional arms, but also to civil goods specially designed or modified for military use.

Moreover, the authorities may impose a "catch-all clause" also on goods not specifically listed, if the exporter is informed or aware of their sensitivity which can make them anyway instrumental to a proliferation programme.

[3]Common Military List in the EU; ITAR in the US.

5.3 Key Elements of the Export Control Process

Key steps of the export control process are the authorisation process and the enforcement of controls by customs. Customs procedures in the EU are ruled by The Community Customs Code, implementing provisions and guidelines [23]. The World Customs Organisation has recently issued guidelines specifically for the enforcement of Strategic Trade Controls.

However, compliance of the exporters with the requirements is the primary ingredient to the overall success [24]. The goods appearing in the control lists need an authorisation in order to be exported, and if suppliers do not collaborate and strive to implement the system, this becomes very weak. This also applies to so-called Intangible Technology Transfers (ITT), i.e. exchanges of controlled technology (information) and software by electronic means.

The export application can be authorised based on documental checks, technical analysis and assessment of the end user and intended end use. The assessment relies on various risk-based considerations on the destination country (Fig. 3). An export can therefore also be denied and associated information transmitted to national customs responsible for enforcement [25] (to prevent unauthorized shipment), to the international regimes (if the country is a member) and to partner countries who should refrain from exporting essentially similar items to the same end user (no undercut policy).

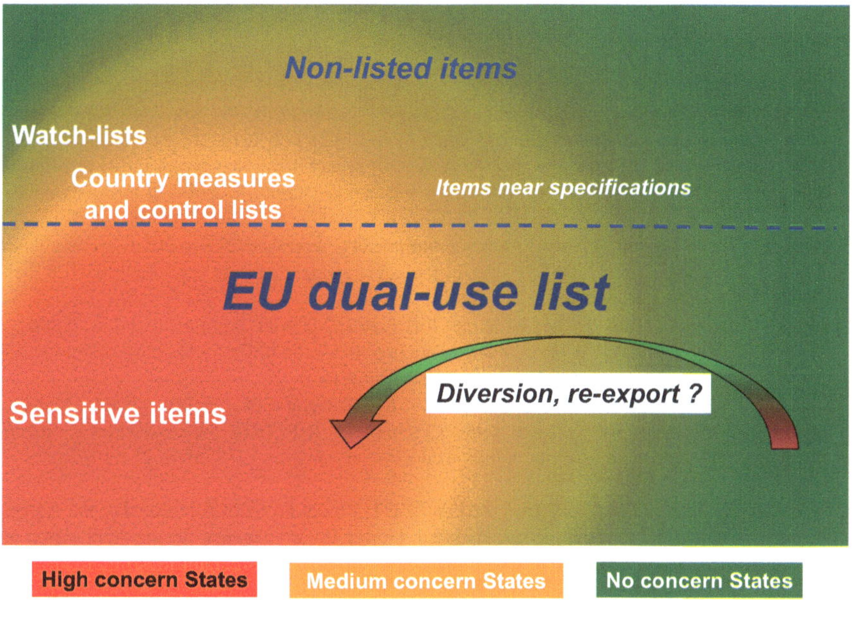

Fig. 3 Risk-based relationship between goods sensitivity and destination countries

6 Conclusions and Relevance to Disarmament and Arms Control

It would be too long to describe in more details how strategic export control is structured, but these few pages sufficiently introduced how all the various phases of the process contribute to slowing down the illicit procurement of sensitive goods and undeclared proliferation schemes.

The control steps help generating various signatures, or sets of indicators that can be used to tailor the risk assessment analyses and targeting methods. Indicators include procurement attempts; inquiries by exporters; export authorisation requests; denials; licenses; customs declarations; targeting of shipments; detection; seizures and prosecution cases.

The control of strategic trade can hence contribute both to the non-proliferation preventive efforts and the verification of the absence of undeclared activities.

Controlling and monitoring the acquisition of sensitive technology could also help to assess the extent of disarmament programs. Besides equipment of indigenous production, weaponisation programs may indeed also rely on imported items and the evidence of new acquisitions might not be compatible with announcements of declared intentions. Symmetrically, the request of technical assistance or technology as well as components needed for dismantling reprocessing and confinement of former weapons may be a sign of fulfilment of commitments.

References

1. Coordinating Committee for Multilateral Export Controls, 1949. http://www.house.gov/coxreport/body/ch9bod.html#anchor5563742. Accessed 28 Feb 2019
2. International Atomic Energy Agency (IAEA) (1970) Treaty in the Non-proliferation of Nuclear Weapons. IAEA INFCIRC/140. https://www.iaea.org/default/files/publications/documents/infcircs/1970/infcirc140.pdf. Accessed 28 Feb 2019
3. Zangger Committee. http://www.zanggercommittee.org. Accessed 28 Feb 2019
4. Nuclear Suppliers Group (NSG). http://www.nuclearsuppliersgroup.org. Accessed 28 Feb 2019
5. Nuclear Suppliers Group (2016) Guidelines for Nuclear Material, Equipment and Technology, INFCIRC/254/Rev.13/Part 1. https://www.iaea.org/sites/default/files/publications/documents/infcircs/1978/infcirc254r13p1.pdf. Accessed 28 Feb 2019
6. Nuclear Suppliers Group (2016) Guidelines for Transfers of Nuclear-related Dual-use Equipment, Materials, Software and Related Technology, INFCIRC/254/Rev.10/Part 2. https://www.iaea.org/sites/default/files/publications/documents/infcircs/1978/infcirc254r10p2.pdf. Accessed 28 Feb 2019
7. International Atomic Energy Agency (IAEA) (1997) Model Protocol Additional to the Agreement(s) between State(s) and the International Atomic Energy Agency for the Application of Safeguards, INFCIRC/540 (Corrected). https://www.iaea.org/sites/default/files/infcirc540.pdf. Accessed 28 Feb 2019
8. Sevini F, Chatelus R, Ardhammar M, Idinger J, Heine P (2014) States' reporting of Annex II exports (AP) and the significance for safeguards evaluation. In: Proc. IAEA Symposium on International Safeguards, Vienna, 2014

9. Sevini F (2014) Nuclear export controls update paper. In: Proc. 55th INMM Annual Meeting, Atlanta, 2014
10. Sevini F, et al. (2011) Some ESARDA Parties' experience with Additional Protocol export control declarations. In: Proc. 33th ESARDA Symposium, Budapest, 2011. https://esarda.jrc.ec.europa.eu/. Accessed 28 Feb 2019
11. Missile Technology Control Regime. http://mtcr.info. Accessed 28 Feb 2019
12. Australia Group. http://australiagroup.net. Accessed 28 Feb 2019
13. Wassenaar Arrangement. http://www.wassenaar.org. Accessed 28 Feb 2019
14. Council of the European Union (2009) Council Regulation (EC) No 428/2009 of 5 May 2009 setting up a Community regime for the control of exports, transfer, brokering and transit of dual-use items (Recast)
15. The Biological and Toxin Weapons Convention. https://www.opbw.org. Accessed 28 Feb 2019
16. Organisation for the Prohibition of Chemical Weapons (OPCW). http://www.opcw.org. Accessed 28 Feb 2019
17. The Hague Code of Conduct against Ballistic Missile Proliferation (HCoC). http://www.hcoc.at. Accessed 28 Feb 2019
18. Albright D (2010) Peddling peril. How the Secret Nuclear Trade Arms America's Enemie. Free press, New York
19. United Nations Security Council (2004) Resolution 1540 Adopted by the Security Council at its 4956th Meeting, on 28 April 2004, S/RES/1540 (2004)
20. Commission Delegated Regulation (EU) 2018/1922 of 10 October 2018
21. Council of the European Union (2012) Council Regulation (EU) No 267/2012 of 23 March 2012 concerning restrictive measures against Iran and repealing Regulation (EU) No 961/2010, later amended as 1375/2016
22. Council of the European Union (2009) Council Regulation (EU) No 1283/2009 of 22 December 2009 amending Council Regulation (EC) No 329/2007 concerning restrictive measures against the Democratic People's Republic of Korea, and later amendments
23. European Parliament and Council of the European Union (2013) Regulation (EU) No 952/2013 of the European Parliament and of the Council of 9 October 2013 laying down the Union Customs Code
24. Sevini F. et al. (2014) Strengthening Strategic Export Controls by Internal Compliance Programs, second Revision, JRC Technical note JRC92964, 2014, EUR 27059. https://ec.europa.eu/jrc. Accessed 30 April 2018
25. WCO Strategic Trade Control Enforcement Implementation Guide. http://www.wcoomd.org. Accessed 28 Feb 2019

Part III
Methods and Models

The more quantitative nature of a systems approach will require integrating a variety of methods and models in a transparent and objective manner so that both treaty partners and the broader global community will have confidence in the results of the verification regime. Some are based on mathematical or statistical concepts like game theory and information management. Others strive to understand the complete system, such as formal pathway analysis—that can also be used to quantify effectiveness and efficiency of the regime. Broader concepts integrate open source analysis and employ a more qualitative method of county profiling. The emerging concept of societal verification may augment the more traditional state-controlled technical verification methods. The inherent uncertainties across such a system drive the need for metrics to allow for a clear evaluation of any proposed verification regime.

NTI Nuclear Security Index: A Model for State-Level Approaches to Nonproliferation and Arms Control

Samantha Neakrase and Michelle Nalabandian

Abstract The NTI Nuclear Security Index (NTI Index) is a first-of-its-kind public assessment of nuclear security conditions on a country-by-country basis in 176 countries. Initially launched in 2012 (a fifth edition is planned for release in June 2020), the NTI Index helps spark international discussions about priorities required to strengthen security and most important, encourages governments to provide assurances and take actions to reduce risks. Developed with the Economist Intelligence Unit (EIU) and with input from a respected international panel of nuclear security experts, the NTI Index draws on NTI's nuclear expertise and the EIU's experience in constructing indices, and the reach of the EIU's global network analysts and contributors. This chapter examines whether the NTI Index methodology could apply to the state-level concept (SLC) in the nonproliferation and arms control verification context, including a discussion of the benefits and constraints of developing an index tool for assessing a country's compliance with nonproliferation and arms control commitments. For further details on the NTI Index, please see the NTI Index website: www.ntiindex.org.

1 Introduction

A state-level approach as described in the introductory chapter of this book *Systems Concept—Structuring a New Verification Approach* by Mona Dreicer et al. is an approach designed to build confidence in a country's ability to comply with certain commitments by examining a broad range of information at the national level to provide a coherent and comprehensive picture of the state's actions [1]. A state-level approach to verifying behavior must be objective, reproducible, transparent, standardized, and clearly documented [1]. The concept derives from an approach called the state-level concept (SLC) that the International Atomic Energy Agency

S. Neakrase (✉) · M. Nalabandian
Nuclear Threat Initiative (NTI), Washington, DC, USA
e-mail: neakrase@nti.org; nalabandian@nti.org

I. Niemeyer et al. (eds.), *Nuclear Non-proliferation and Arms Control Verification*,
https://doi.org/10.1007/978-3-030-29537-0_8

(IAEA) has begun using in the safeguards context to transition from a facility-by-facility approach to examining safeguards compliance to an examination of the country as a whole (see the chapter *The Evolution of Safeguards* by Jill Cooley in this book). The IAEA only applies the SLC in cases where the state's declarations under its comprehensive safeguards agreement have been found to be correct and complete [2].

Further, this book examines various methods and concepts that may be used to apply the SLC to nonproliferation and arms control verification, and this chapter will discuss specifically how the methodological principles used in the development of the NTI Nuclear Security Index (NTI Index) could apply in the context of nonproliferation and arms control verification. Section 2 of the chapter describes the NTI Index and the methodology used in its development. Section 3 discusses how the NTI Index methodology could apply to the SLC in the nonproliferation and arms control verification context, including benefits, factors, and constraints. The chapter concludes that an index could be a useful tool to assess a country's compliance with nonproliferation and arms control commitments, but has significant challenges and therefore should be considered only as one among many tools in the toolbox.

2 About the NTI Index: Goals and Methodology

The NTI Index [3], developed in collaboration with the Economist Intelligence Unit (EIU), is a unique public assessment of nuclear security conditions around the world. Initially launched in January 2012, with a fifth edition due to be released in 2020, the NTI Index is designed to encourage governments to take actions and build confidence in the security of their materials. A first-of-its-kind resource, the NTI Index also provides a framework for establishing priorities for nuclear security and tracking progress over time.

2.1 Goals of the NTI Index

NTI developed the NTI Index to address several critical limitations in nuclear security. First, despite high-level attention given to nuclear materials security at the Nuclear Security Summits, there was no consensus on priorities for securing all nuclear materials. No one had answered the important question of 'what actions a country can take to improve its nuclear security - what matters most?' By developing a broad framework for analysis, in consultation with an international panel of experts, NTI sought to catalyze a discussion on priorities for nuclear materials security.

Second, despite progress made internationally to improve global nuclear security and the commitments made by governments at the Nuclear Security Summits, there was no way for countries to track progress or hold each other accountable. An index that is released periodically is a useful tool to track progress, and with the release of a fourth edition in September 2018, the NTI Index has been able to highlight key achievements in nuclear security as well as identify challenges where improvements are needed. In fact, at the 2014 Nuclear Security Summit (NSS), six countries cited their NTI Index scores as a measure of progress in their official national statements and progress reports. At the 2016 NSS, several additional countries mentioned the NTI Index either in their national statements or progress reports. The 2020 NTI Index will further indicate whether progress has continued since the 2016 NSS or if progress has stalled.

Finally, NTI hoped to promote action by countries to improve their own nuclear security. By providing a public assessment, highlighting progress, and identifying challenges, countries can use the NTI Index as a tool to ensure that their attention and resources are being allocated to the most critical areas that require improvements. Consultations held between NTI and several governments have confirmed that the NTI Index has indeed become a valuable resource. As a result of the NTI Index, a number of countries reached out to NTI to gain a greater understanding of and advice in setting priorities in nuclear security. Some governments have also used the NTI Index to guide actions to enhance nuclear materials security.

2.2 NTI Index Methodology

The following summarizes four key components of the NTI Index methodology:

Broad Framework The NTI Index provides a holistic approach for assessing nuclear security, covering a broad range of factors that could affect a country's ability to secure its nuclear materials and facilities. The framework is made up of five categories of indicators that are weighted to reflect their relative importance.

Robust Data and Rigorous Analysis The research process is led by the EIU, leveraging its global network of analysts who are experienced in researching country laws and regulations in original languages. EIU analysts rely on public and open-source information, including national laws and regulations, government reports and public statements, and reports from non-governmental organizations and international organizations such as the IAEA. The EIU does not review the effectiveness of implementation at a country's nuclear facilities.

International Perspective NTI works with an international panel of experts from around the world to provide input on the NTI Index framework. The panel advises NTI and the EIU on the selection of indicators and their relative importance, and provides input on options for strengthening each subsequent edition of the NTI Index. Panel members do not represent their country's interests or score individual countries; rather they play an advisory role in their personal, not professional, capacities. The international panel ensures that the NTI Index reflects diverse viewpoints and ongoing international discussions on nuclear security priorities.

Transparent Process To ensure that the NTI Index process is as transparent as possible, NTI and the EIU engage with governments in two ways. First, NTI offers briefings to governments to familiarize them with the NTI Index process. Second, the EIU provides governments with the opportunity to review and comment on the preliminary results to ensure that the NTI Index data is as accurate and up-to-date as possible.

For further details on the NTI Index methodology as well as all data collected, please see the NTI Index website: www.ntiindex.org.

3 Applying the NTI Index Methodology to the State-Level Concept

Indices have several characteristics that make them extremely useful tools to measure performance that cannot be directly observed, such as a country's nuclear security conditions or its compliance with nonproliferation and arms control commitments. An index can aggregate a wide range of related data and evaluate it in a consistent manner, track progress over time, and motivate countries to improve performance. It can also provide an objective way to measure a complex topic as long as the indicators are framed in such a way to limit the risk of subjectivity and to increase the likelihood that the same scores could be obtained by another set of researchers, a key indicator of objectivity. Finally, an index can provide a broad framework for analysis that encompasses a wide range of factors, thus providing a more holistic and multidimensional approach to assessing a country's actions.

However, before using an index methodology in the context of the SLC, particularly applied to nonproliferation and arms control verification, a number of factors and constraints should be considered. The NTI Index methodology and the lessons learned from developing the NTI Index provide useful context to consider those factors. Notably, several of the constraints and challenges of the NTI Index may be magnified and difficult to overcome in the nonproliferation and arms control context. Finally, the NTI Index is different from the SLC in some important ways, which are described below.

3.1 Developing a Framework

An index provides a broad framework for analyzing a given topic using a holistic approach and taking into account a variety of relevant variables. The first step in developing an index is to define what the index is measuring, assess what categories and indicators reflect policies, actions, and other factors that are relevant to the topic, and to then prioritize those factors. This process is a useful analytic exercise in and of itself to frame future debates and policy considerations. This type of exercise is similar to the process that is required to develop and apply an SLC, because the SLC, by definition, examines a broad set of factors and takes a holistic approach. Therefore, this element of the methodology would be appropriate when applying the SLC to nonproliferation and arms control.

As was the case with the NTI Index, to ensure that a nonproliferation or arms control index is credible and represents an international perspective, it will be important for its framework to be developed in a transparent manner, gathering input from international experts, governments, and other key stakeholders. Identifying what matters in assessing a subject and choosing what should be in an index can be extremely challenging, and experts may ultimately disagree with the final framework. However, if the process is transparent and inclusive, disagreement with the final assessment can be a healthy part of ongoing discussions. In the context of nonproliferation and arms control, the debate over what matters is likely to be thornier than for nuclear security. This is particularly so given the interests at stake (i.e., the more serious consequence of non-compliance with international obligations) that could result from a low score. Seeking input from a diverse group of credible international experts and engaging with a broad set of stakeholders will be necessary to ensure a robust index framework that avoids serious political opposition.

3.2 *Research and Data Gathering*

Research and data gathering make up the bulk of the work of an index. In the case of the NTI Index, NTI works with the EIU, which leads the research and data gathering process. It is important for NTI to work with a credible, experienced, and objective organization like the EIU to ensure that the index results will be reliable and seriously considered. A nonproliferation and arms control index would similarly benefit from partnership with an independent group of analysts to ensure there is no appearance of subjectivity or a political agenda, allowing the data to stand on its own.

As is the case with most public indices, research for the NTI Index relies on public and open-source information. This approach ensures that results are transparent, objective, and repeatable. There are a number of constraints associated with this approach, however. First, not all countries make their laws and regulations publicly available, and as a result data can be difficult to attain. In the case of the NTI Index, some assumptions and proxy measures are used to fill gaps in the publicly available data. In the nonproliferation and arms control context, data access may be even more difficult given that much of the information is deemed sensitive.

Second, relying on publicly available information also means that an index framework cannot always include all factors deemed important or relevant to an issue and must be limited to those that can be assessed using public information. In the case of the NTI Index, although facility-level assessments would provide important "ground-truth" information, this level of data gathering is not possible because of the sensitive nature of specific security arrangements. The NTI Index relies instead on the assumption that a country with the appropriate laws and regulations in place is more likely to have sound security procedures at each facility than a country without appropriate laws and regulations. This of course means that the NTI Index includes only "indicators" of security conditions, and not the complete set of best security practices that nuclear facilities should apply. Once again, this will likely be the case, if not more so, in a nonproliferation and arms control context. Note that an index in the nonproliferation and arms control context that is compiled by a government or international organization may be able to use non-public information gathered directly from governments or through other means.

Third, data constraints mean that results must be interpreted carefully and not assigned more meaning than is appropriate based on the specific category or type of data gathered. From the perspective of the NTI Index, NTI chose to measure "nuclear security conditions" rather than "nuclear security performance" or "good or bad security." The phrase "nuclear security conditions" acknowledges the fact that the NTI Index is not a facility-by-facility assessment of security practices and instead built on the assumption that a country with appropriate laws and regulations, and participates in global nuclear security instruments or agreements, is more likely to have sound security practices on the ground. The results therefore do not necessarily provide a complete picture of on-the-ground implementation and can be misinterpreted if not read with a full understanding of the methodology. Therefore, one must be careful to properly interpret and characterize results, asking questions such as: What does a high or low score mean? What is average and what does an average score mean? In a nonproliferation or arms control context, would a high score mean compliance with a treaty? Would a low score mean non-compliance? Would a low score trigger non-compliance consequences based on the terms of international treaties? Considerations such as these will necessitate extreme care in deciding how to interpret a high or low score and how to frame what the index is measuring, so as not to derail important international nonproliferation and arms control efforts.

3.3 Government Engagement and Outreach

If governments perceive a product like the NTI Index as being nontransparent and are taken by surprise when results are published, this can negatively impact the quality of follow-on discussions and impede the achievement of project goals, such as catalyzing a dialogue and promoting action (it is often difficult to predict how governments will react to a product like the NTI Index and ensure a constructive discussion). In the case of the NTI Index, some governments are wary of providing certain information to the EIU in the data confirmation process, and several governments decline to engage in this process entirely, due either to internal policies prohibiting engagement with nongovernmental organizations or other secrecy concerns. For this reason, extensive outreach, particularly to governments who might have misgivings about the project or might be sensitive to the results, is important throughout the index development process.

As previously noted, these sensitivities will likely be exacerbated in the context of nonproliferation and arms control and there is a risk that results can be misunderstood or misinterpreted. Given the sensitivities involved, engagement with governments and a high level of transparency with regard to the results and the data will be vital to ensure a constructive response.

3.4 Operational Capacity

Operational constraints will limit whether it is practical to undertake the development of an index for nonproliferation and arms control. The NTI Index separately assesses three groups of countries: In the 2020 edition, 22 countries with 1 kg or more of weapons-usable nuclear materials will be assessed across five categories of indicators; 154 countries with less than 1 kg of or no weapons-usable nuclear materials will be assessed across only three of the categories; and 48 countries either with or without materials but with certain types of nuclear facilities are assessed across a modified set of the five categories and indicators. Only the groups of 22 countries and 48 countries are assessed by researching national laws and regulations in native languages. Had research been undertaken of all 176 countries' laws and regulations, the project would have been far more intensive. The operational difficulty of the project depends on how many countries are included in the assessment and whether the research requires an in-depth review of primary documents in native languages or can rely on secondary sources or other previously collated sources of data.

3.5 Key Differences Between the NTI Index and the State-Level Concept

The NTI Index methodology has much in common with the state-level concept. The goal of the state-level approach is to provide a broad picture of a country's actions from a variety of sources. The NTI Index approaches nuclear security conditions from this very perspective—assessing a country's national-level actions to draw conclusions about its nuclear security conditions, taking into account a broad set of factors and drawing from a variety of publicly available sources. In addition, one stated goal of the state-level approach is to have an objective, reproducible, and standardized approach to assessing a country's actions. An index does exactly this, and is particularly useful for measuring performance that cannot directly be observed, such as nuclear security conditions and nonproliferation and disarmament actions. However, there are important differences between the NTI Index and the

SLC. What the NTI Index is attempting to accomplish is substantially different from the goals of a nonproliferation and arms control index.

First, the SLC, as articulated by the IAEA, provides a customized approach to assessing a state's compliance with safeguards commitments, taking into account state-specific factors based on different cheating pathways. However, the NTI Index uses a common set of criteria to assess all countries and does not provide a customized approach for each country. Therefore, while an index may provide some set of common data that is indicative of a country's compliance with certain nonproliferation and arms control obligations, it would not be able to take into account the differences in how a country might be able to circumvent its obligations or take into account country-specific factors.

Second, the goals of the NTI Index and a nonproliferation and arms control verification index will likely differ substantially, with much higher stakes associated with the results of the latter. The goals of the NTI Index are to catalyze a debate on priorities, track progress over time and provide accountability, and promote action. Some countries may dislike the results or disagree with the methodology, but there are few consequences beyond scoring poorly and the appearance of not adhering to international norms. A nonproliferation and arms control verification index could have similar goals to the NTI Index, but if it is explicitly, or even implicitly, assessing a country's compliance with a treaty or other set of international commitments, the consequences could be far more serious. For instance, a country that scores poorly might be subject to some type of international sanction or enforcement action. With such high stakes, a country scoring poorly might disengage entirely from important nonproliferation and arms control negotiations or processes.

4 Conclusions

Although indices can be useful analytic tools, they also have several important constraints, particularly with regard to data availability and interpretation, which must be considered and addressed in the index development process. The benefits and challenges of using an index must be carefully managed and the goals must be clearly articulated. Using an index in the context of nonproliferation and arms control would meet some of the goals of using a state-level approach to verifying compliance with treaties or other commitments. However, there are enough differences between the NTI Index and using the SLC in the context of nonproliferation and arms control to warrant a cautious approach. Carefully developed, an index could be a supplemental tool in the verification toolbox, but not the only tool.

References

1. Allen K, Dreicer M, Chen C et al (2015) Systems Approach to Arms Control Verification. ESARDA Bulletin 53:83–91
2. Rockwood L (2014) The IAEA's state-level concept and the law of unintended consequences. Arms Control Today. https://www.armscontrol.org. Accessed 28 Feb 2019
3. Nuclear Threat Initiative (2016) NTI Nuclear Security Index. http://www.ntiindex.org. Accessed 28 Feb 2019

Formalizing Acquisition Path Analysis

Morton Canty and Clemens Listner

Abstract The theory of directed graphs and non-cooperative games is applied to the problem of verification of State compliance to international treaties on arms control, disarmament and non-proliferation of weapons of mass destruction. Hypothetical treaty violations are formulated in terms of illegal *acquisition paths* for the accumulation of clandestine weapons, weapons-grade materials or some other military capability. The paths constitute the illegal strategies of a sovereign State in a two-person *inspection game* played against a multi- or international Inspectorate charged with compliance verification. The effectiveness of existing or postulated verification measures is quantified in terms of the Inspectorate's expected utility at Nash equilibrium. A case study involving a State with a moderately large nuclear fuel cycle is presented.

1 Introduction

Inspection games [1, 2] deal with the problem faced by an inspector who is required to control the compliance of an inspectee to some legal or otherwise formal undertaking. One of the best examples of an inspector, in the institutional sense, is the International Atomic Energy Agency (IAEA) which, under a United Nations mandate, verifies the activities of all States (inspectees) signatory to the Nuclear Weapons Non-proliferation Treaty. An inspection regime of this kind is a true conflict situation, even if the State voluntarily submits to inspection, because the Inspectorate must assume that the State has a real incentive to violate its commitments. The primary objective of the Inspectorate is to *deter* the State from illegal behavior or, barring this, to detect the illegal activity. Both protagonists will act strategically, taking into account the strategic alternatives of the opponent. In this

M. Canty (✉)
Consultant, Jülich, Germany

C. Listner
Consultant, Berlin, Germany

I. Niemeyer et al. (eds.), *Nuclear Non-proliferation and Arms Control Verification*,
https://doi.org/10.1007/978-3-030-29537-0_9

chapter the theory of inspection games is applied to the problem of quantification of acquisition path analysis based on the IAEA Physical Model of Member States' nuclear capabilities. Hypothetical treaty violations are formulated in terms of an inspection game in normal form and the effectiveness of existing verification measures is quantified in terms of the Inspectorate's payoff, i.e., its expected utility at game-theoretical equilibrium.

A two-person inspection game in *normal* or *strategic form* is a list of actions, called *pure strategies*, for each of the protagonists (inspector, inspectee), together with a rule for specifying each player's payoff (utility) when both have chosen a specific action. Each player seeks to maximize his or her own payoff. A *Nash equilibrium* of the game is a specification of strategies for both players with the property that neither player has an incentive to deviate unilaterally from his or her specified strategy [3]. A *solution* of the game is a Nash equilibrium which is either unique, or which, for some reason, has been selected among alternative equilibria.

2 The Game-Theoretical Model

In order to quantify the effectiveness of IAEA safeguards measures within a controlled State relative to the Physical Model and its associated acquisition paths, we define a two-person, non-cooperative game in strategic form with the Inspectorate as player 1 and the State as player 2. We assume initially that the State, if it violates its commitments, will concentrate all of its effort on one acquisition path and that Inspectorate will control a single illegal activity within one selected path only. The latter assumption will be relaxed later. The State then has $n + 1$ strategic alternatives or pure strategies, namely to acquire clandestine weapons capability along one of the n illegal acquisition paths, or to behave legally. The Inspectorate's pure strategies are to choose which activity to control.

We define the utilities for (Inspectorate, State) initially as follows:

$$\begin{aligned} (-a, -b) \quad & \text{for detected illegal behavior,} \\ (-c, d_i) \quad & \text{for undetected illegal behavior along the } i\text{th acquisition path,} \\ & i = 1 \dots n, \\ (0, 0) \quad & \text{for legal behavior and no false accusation,} \\ (-e, -f) \quad & \text{in case of a false accusation,} \end{aligned} \tag{1}$$

where it is assumed that

$$c > a > e > 0, \quad b > f > 0, \quad d_1 > d_2 > \cdots > d_n > 0. \tag{2}$$

The above ordering implies that the best result for the Inspectorate is payoff nil, i.e. legal behavior on the part of the State, then to have raised a false alarm $(-e)$, to have detected a violation $(-a)$, and finally *not* to have detected one $(-c)$. On the other hand the State would prefer to acquire clandestine Weapons capability along some path without being detected (d_i), whereby it tends to favor path i over path j when $i < j$.[1] Otherwise the State is inclined to behave honestly (0). Its next best outcome is to have to accept a false accusation $(-f)$ and least of all does the State want to be caught red-handed $(-b)$.

Non-detection probability and false alarm probability for path i, $i = 1 \ldots n$, are β_i and α, respectively. The false alarm probability may be taken to be path-independent and chosen at some conventional value such as 5%. If illegal activity coincides with inspection along path i, the expected utilities (Inspectorate, State) are obtained by summing the utilities for detected and undetected violation, weighted by the detection or non-detection probability:

$$\Big((-a)(1-\beta_i)+(-c)\beta_i,\ (-b)(1-\beta_i)+(d)\beta_i\Big). \tag{3}$$

2.1 Extensive Form

Figure 1 illustrates the simplest nontrivial situation in which the two paths $P_1 = \{1, 2\}$ and $P_2 = \{1, 3\}$ involve a common illegal activity $\{1\}$. If the Inspectorate controls only one location/activity, we obtain the so-called *extensive form* two-person game illustrated in Fig. 2. At its first decision point the State decides whether or not to behave legally. For the legal behavior decision, each player pays the expected false alarm costs, and the inspection strategy is irrelevant. If the State decides on an illegal acquisition path, the outcome then depends upon the Inspectorate's decision to control or not to control $\{1\}$. In the former case, since the Inspectorate only controls once, the outcomes are determined by the non-detection probability β_1 as shown. In the latter case the game proceeds, the State choosing either location/activity $\{2\}$ or $\{3\}$in order to complete its acquisition. The Inspectorate, in ignorance of this choice, likewise chooses either $\{2\}$ or $\{3\}$. The Inspectorate's incomplete information is represented by the oval *information set* in Fig. 2. The four possible outcomes are then easily seen to be as shown at the bottom of the figure.

[1] Activities without loss of generality, of course, since we can always number the paths according to decreasing attractiveness.

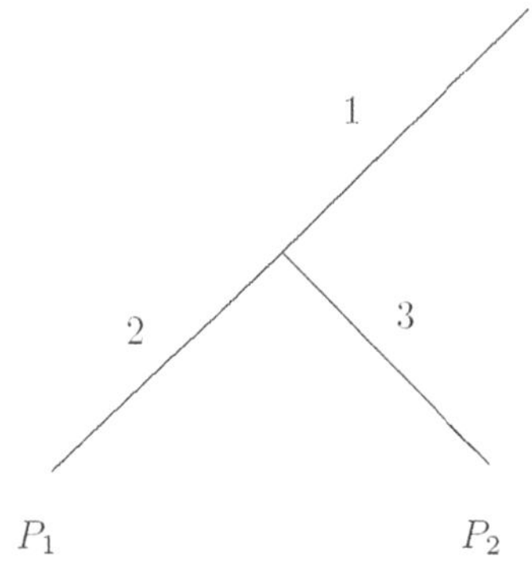

Fig. 1 Two acquisition paths with a common location/activity

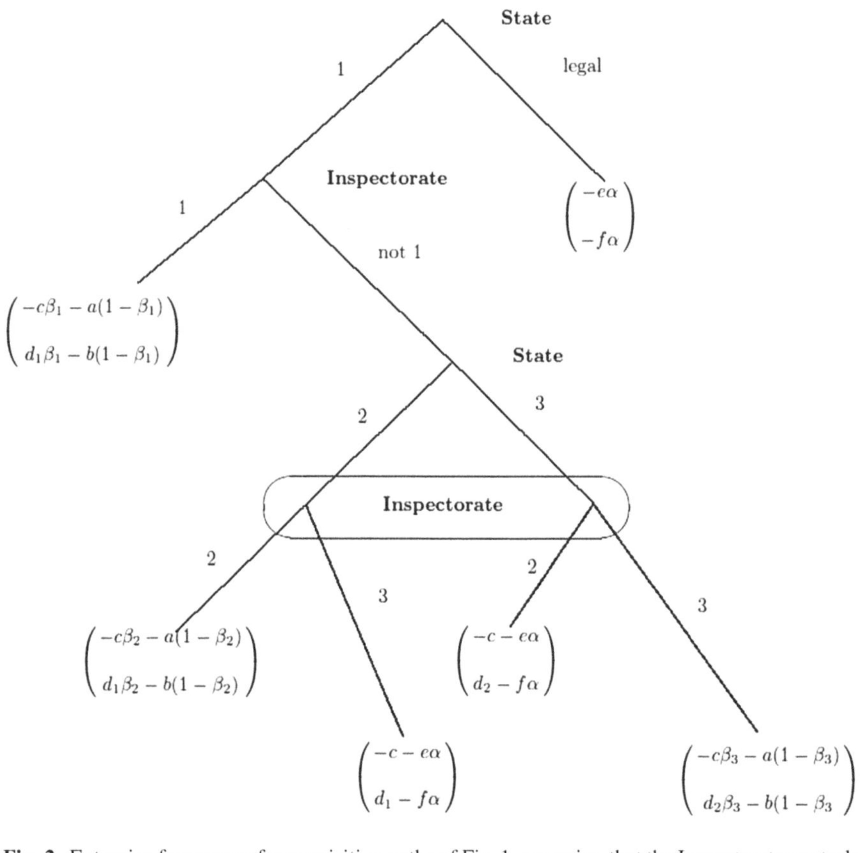

Fig. 2 Extensive form game for acquisition paths of Fig. 1, assuming that the Inspectorate controls only one location/activity and that $d_1 > d_2$. The oval is the Inspectorate's information set if it does not control location/activity 1

	12	13	legal
1	$-b(1-\beta_1)+d_1\beta_1$ $-c\beta_1-a(1-\beta_1)$	$-b(1-\beta_1)-d_2\beta_1$ $-c\beta_1-a(1-\beta_1)$	$-f\alpha$ $-e\alpha$
2	$-b(1-\beta_2)+d_1\beta_2$ $-c\beta_2-a(1-\beta_2)$	$-f\alpha+d_2$ $-c-e\alpha$	$-f\alpha$ $-e\alpha$
3	$-f\alpha+d_1$ $-c-e\alpha$	$-b(1-\beta_3)+d_2\beta_3$ $-c\beta_3-a(1-\beta_3)$	$-f\alpha$ $-e\alpha$

Fig. 3 Normal or bimatrix form of a simultaneous version of the game of Fig. 2

2.2 *Normal Form*

While the game shown in Fig. 2 can be solved by backward induction, a closed form solution is not possible.[2] However if we ignore the sequential or causal aspect, e.g., that {1} is a prerequisite for {2} or {3}, the strategies can be chosen simultaneously, and the game can be reduced to the normal or bimatrix form shown in Fig. 3.[3] The elements in the lower left-hand corners of each cell comprise the 3×3 matrix A of payoffs to the Inspectorate for each of the 9 pure strategy combinations, the upper right-hand elements comprise similarly the State's payoff matrix B.

More generally, let P_j, $j = 1 \ldots n$, represent the acquisition paths available to the State, each consisting of a set of one or more of m possible locations/activities. Let K_i, $i = 1 \ldots k$ represent the pure strategies of the Inspectorate, each consisting of a subset of the m locations/activities. The size of these subsets will in general restricted by the available inspection manpower. Thus if the jth location/activity entails w_j inspection person-days of effort and W person-days are available in all, we require

$$\sum_{j \in K_i} w_j \leq W, \quad i = 1 \ldots k. \tag{4}$$

Having enumerated the list of pure strategies satisfying the above condition, we can eliminate all those which are proper subsets of other pure strategies on the list as these will be strictly dominated.[4] Hypothetical clandestine locations/activities which cannot be detected by agreed upon inspection activities might be assigned $\beta_j = 1$ and $w_j = 0$ so as to include them in the analysis. This is discussed in more detail in Sect. 2.5 below.

[2]For an example of an analytic solution of a sequential inspection game in extensive form, see [1].

[3]It might reasonably be assumed that a State intent upon treaty violation would have to plan its strategy well in advance. Likewise, inspection plans must be formulated and agreed upon on a long-term basis.

[4]This means essentially that they will never be part of the Inspectorate's equilibrium. For the formal definition of strategy domination, see [4], Definition 1.4.10.

If the State chooses path P_j and the Inspectorate chooses strategy K_i, then the probabilities of non-detection and no false alarm are given respectively by

$$\begin{aligned} \beta_{i,j} &= \prod_{\ell \in K_i \cap P_j} \beta_\ell \\ 1 - \alpha_{i,j} &= \prod_{\ell \in (K_i - K_i \cap P_j)} (1 - \alpha_\ell), \end{aligned} \tag{5}$$

where we now allow that both non-detection probabilities and false alarm probabilities be location/activity-specific. The State's legal strategy is $P_{n+1} = \{\ \}$, the empty set, for which $\beta_{i,n+1}$ is not defined. Note that

$$|K_i - K_i \cap P_{n+1}| = |K_i|. \tag{6}$$

We now have a general normal form bimatrix game $\langle P, Q, A, B \rangle$ where $P = (p_1, p_2 \ldots p_k)^\top$ and $Q = (q_1, q_2 \ldots q_{n+1})^\top$ denote the mixed strategies of Inspectorate and State, respectively, and where the payoff matrix A for the Inspectorate and B for the State have elements

$$\begin{aligned} A_{i,j} &= \begin{cases} -c \cdot \beta_{i,j} - a \cdot (1 - \beta_{i,j}) - e \cdot \alpha_{i,j} & for\ \mathrm{j = 1 \ldots n} \\ -e \cdot \alpha_{i,j} & for\ \mathrm{j = n+1} \end{cases} \\ B_{i,j} &= \begin{cases} d_j \cdot \beta_{i,j} - b \cdot (1 - \beta_{i,j}) - f \cdot \alpha_{i,j} & for\ \mathrm{j = 1 \ldots n} \\ -f \cdot \alpha_{i,j} & for\ \mathrm{j = n+1} \end{cases} \end{aligned} \tag{7}$$

for $i = 1 \ldots k$. Although as already mentioned a closed form solution cannot be given, the game can be solved numerically with standard methods [4].

2.3 Graph Representation

An acquisition path model, such as the simple one shown in Fig. 1, can be considered as a directed graph. Each edge or location/activity and can be assigned a scalar number representing its length, e.g., some assessment of feasibility, cost, etc. A sequence of edges beginning at the root and ending at a node (an acquisition path) is characterized by its overall length, for example by the simple sum of the edge weights, and defines the attractiveness parameter d_i introduced in Eq. (1). One can apply standard algorithms in order to determine all feasible paths and sort them according to attractiveness. The shortest path from the root to a given end node can be found using, e.g., the Dijkstra algorithm [5]. However it may be strategically inopportune for a State to choose the most attractive path, and therefore for a comprehensive analysis of the State's options, all paths have to be assessed. This can be easily accomplished using enumeration techniques that prevent cycles.

2.4 Equilibrium Multiplicity

The normal form game described in Sect. 2.2 will in general be degenerate, i.e., possess infinitely many equilibria [4, 6]. In order to investigate its equilibrium properties, acquisition paths drawn from a small pool of activities (of the order 10, with randomly chosen detection and false alarm probabilities, inspection costs and attractivities) were used to simulate matrix games of the form given by Eq. (7). These games were then solved both by complete vertex enumeration [7] and by the Lemke Howson method [8]. In all cases the equilibrium was either unique, or the set of multiple equilibria was such that the State's equilibrium strategy was the same for all equilibria. If the State's equilibrium strategy is unique, all of the extreme equilibria are necessarily payoff equivalent, otherwise the other player (the Inspectorate) would have an incentive to deviate unilaterally from its equilibrium strategy, contradicting the definition of Nash equilibrium.

In order to solve the larger matrix games that arise in realistic contexts, the Lemke-Howson algorithm was used, an algorithm which is more efficient than vertex enumeration, but which is guaranteed to find one equilibrium only. While the algorithm is efficient in practice, in the worst case the number of pivot operations may be exponential in the number of pure strategies in the game. In addition, it has been shown that it is PSPACE-complete to find any of the solutions that can be obtained with the Lemke-Howson algorithm [9]. In view of the simulation results, we consider it safe to assume that any single equilibrium found by the Lemke Howson method is sufficient for the strategic analysis.

2.5 Parameters

In performing acquisition path analysis with this approach, the following quantities are required:

1. For each acquisition path $i = 1 \ldots n$ identified from the physical model:
   - P_i: the subset of locations/activities comprising the path
   - d_i: the operator's incentive to choose that path
2. For each location/activity $j = 1 \ldots m$ identified in 1:
   - β_j: non-detection probability if inspected
   - α_j: false alarm probability if inspected
   - w_j: inspection effort required if inspected

3. The global parameters:

   - b: State's perceived sanctions in the event of detection
   - f: State's false alarm cost for false accusation
   - a: Inspectorate's loss in the event of detection (failed deterrence)
   - e: Inspectorate's false alarm cost (credibility loss)
   - c: Inspectorate's loss in the event of non-detection
   - W: The total amount of inspection effort planned for the State

We can identify the *technical quantities* in the above list as P, β, α, w, and W. They can be estimated, at least approximately, given the negotiated inspection regime within the State. For example, for a bulk handling facility in which the activity "protracted diversion of one SQ of material" is part of an acquisition path, the existing material accountancy system will determine β for a given value of α. For the activity "falsification of r sealed items from bulk storage", the seal verification sampling plan will determine β for $\alpha = 0$.

The detection probability for activities taking place at clandestine locations are obviously difficult to quantify. The false alarm probability, on the other hand, can confidently be taken to be nil: if a clandestine facility is discovered, there will be little ambiguity as to its purpose. Similar considerations apply for misuse of declared facilities. For example cascade manipulation should be easily detectable at enrichment plants in States which allow limited frequency unannounced access to cascade halls, again with zero false alarm probability.

The simplest assumptions for clandestine facilities/activities would be to set the detection probability to 100% for States which have implemented the Additional Protocol and are in all respects cooperative, and to zero otherwise. Unfortunately, the latter assumption begs the question of Nuclear Safeguards altogether, since it says that a fully clandestine acquisition path is by definition undetectable. Clearly a more pragmatic and differentiated scheme is needed. Several such approaches are discussed in some detail in [10, 11].

The *non-technical quantities* are the utilities d, b, and f, for the State and a, e and c for the Inspectorate. We can apply the graph-theoretical methods suggested in [12] to determine *all* acquisition paths and order them according to increasing length ℓ_i (meaning difficulty or "unattractivity"):

$$\ell_1 \le \ell_2 \le \cdots \le \ell_n,$$

where the ordering is determined by a shortest path algorithm. The utilities d_i in (2.1) for a successful acquisition along path i are taken to be inversely proportional to the path lengths and normalized to some convenient value d for the easiest path:

$$d \ge d_i = d\ell_1/\ell_i > 0, \quad i = 1 \ldots n. \tag{8}$$

The utility $b > 0$ then measures the State's perceived sanctions for a detected illegal acquisition attempt relative to $d_1 = d$, the utility for successful acquisition along the shortest or most attractive path. It will vary from State to State and is a purely "political" assessment. A "neutral" assessment might choose $b = d$.

A less arbitrary assignment of b can be derived from IAEA practice. Generally it is assumed that nuclear material diversion should be detectable within the IAEA Safeguards system with "a high degree of assurance", meaning on the order of 90%. Let us assume that such a degree of assurance just suffices to deter a State from illegal activity. Then in terms of the utilities b and d,

$$d \cdot \beta - b \cdot (1 - \beta) = 0,$$

or, with $\beta = 0.1$,

$$\frac{d}{b} = 9. \tag{9}$$

So, implicit in the Safeguards requirement for high assurance is the assumption that the incentive to violate the Treaty is about an order of magnitude larger that the perceived sanctions that would accompany detection of that violation. Although this could be interpreted as being grossly unfair to fully cooperative member States, nevertheless it is consistent with IAEA policy and with member States' perception of that policy. Therefore in the default parameterization of the game-theoretical model, Eq. (8) is taken to be, with $d = 9$,

$$9 \geq d_i = 9\ell_1/\ell_i > 0, \quad i = 1 \ldots n \tag{10}$$

and $b = 1$. If and when the IAEA is in a position to attribute realistic incentives on a state-by-state basis, then b might be treated as an adjustable parameter reflecting a realistic internal assessment of the motives of the inspected State.

Similarly, the false alarm cost $0 < f < b$ is also measured relative to d. It is also "political" and potentially as controversial as b.[5] A pragmatic choice might be $f = 1$.

The Inspectorate's utilities essentially define, via the Inspectorate's Nash equilibrium payoff, a relative scale upon which different acquisition path configurations can be evaluated and compared. Suppose we use the values

$$c = 100, \quad a = 10, \quad e = 0.$$

[5] A false accusation cost Iraq dearly, although it was not the IAEA that raised it.

Then the Inspectorate's equilibrium payoff H_1^* is confined to the interval $[-100, 0]$, with the lower limit being achieved for successful clandestine weapons acquisition on the part of the State, and the upper limit for deterrence of illegal activity. Intermediate values will arise when the Nash equilibrium involves illegal activity with finite detection probability. Therefore on a scale 0 to 100, we might measure the effectiveness E of a given inspection regime on the basis of the acquisition path analysis as

$$E = H_1^* + 100. \tag{11}$$

More generally, for any value of c,

$$E = (H^* + c) \cdot \frac{100}{c}. \tag{12}$$

3 A Case Study

Figure 4 shows a directed graph representation of a generic fuel cycle. The nodes correspond to different material processing states and the edges to the processes themselves. In the context of illegal weapons acquisition, in addition to legitimate activities each processing edge can involve undeclared (diverted/illegally imported) material and be carried out in a clandestine location or by means of misuse of a declared facility. Any such illegal path terminating at a direct use (DU) node my be considered to be a weapons acquisition path.

A brute force path enumeration, allowing for all possible combinations of diversion/clandestine/misuse activities, leads to a combinatorial explosion which prohibits the use of the analysis methods discussed above. The number of pure inspection strategies as defined in Sect. 2.2 is 2^n, where n is the number of distinct edges in the graph. The elimination of dominated strategies reduces this somewhat, but for n larger than about 20, strategy enumeration and subsequent Nash equilibrium calculation becomes generally infeasible. Fortunately, for any given State, simple plausibility arguments can be brought to bear which restrict the number of possible paths/edges considerably. Thus the misuse of a declared reprocessing plant obviously does not apply for States without commercial reprocessing so that only clandestine reprocessing need be considered. Similarly, diversion of declared, highly enriched Uranium will usually only apply to States having research reactors, and so on.

Figure 5 represents an assessment of the current situation for a Sate with a moderately large fuel cycle. The plausibility arguments have reduced the edge count to 16 and the number of distinct acquisition paths to 22. To illustrate the evaluation

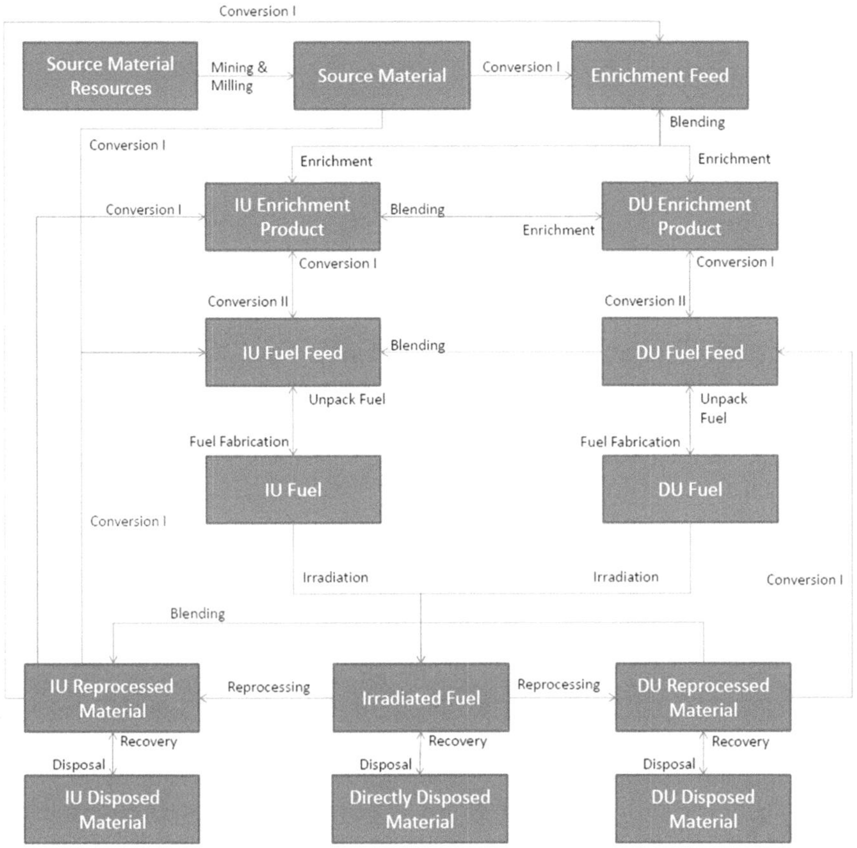

Fig. 4 Generic fuel cycle physical model. IU = indirect use, DU = direct use nuclear material

process, sample calculations were carried out with highly simplified assumptions with respect to the parameters:

- All false alarm costs were set to nil.
- The perceived sanction utility b was set to 9, that is, equal in magnitude to the incentive d to acquire a weapon along the most attractive (least costly) path, see Sect. 2.5.
- Edge costs were assessed subjectively on a scale 0 to 2, and the overall acquisition path cost was taken to be the sum of the costs of the contributing edges.
- Inspection efforts (in person-days) were estimated from the most recent Safeguards Implementation Report provided by the IAEA.
- Since the State was assumed to have implemented the Additional Protocol, detection probabilities for all clandestine and misuse activities were set to one.
- Detection probabilities for diversion from bulk and item facilities were set to 0.6.

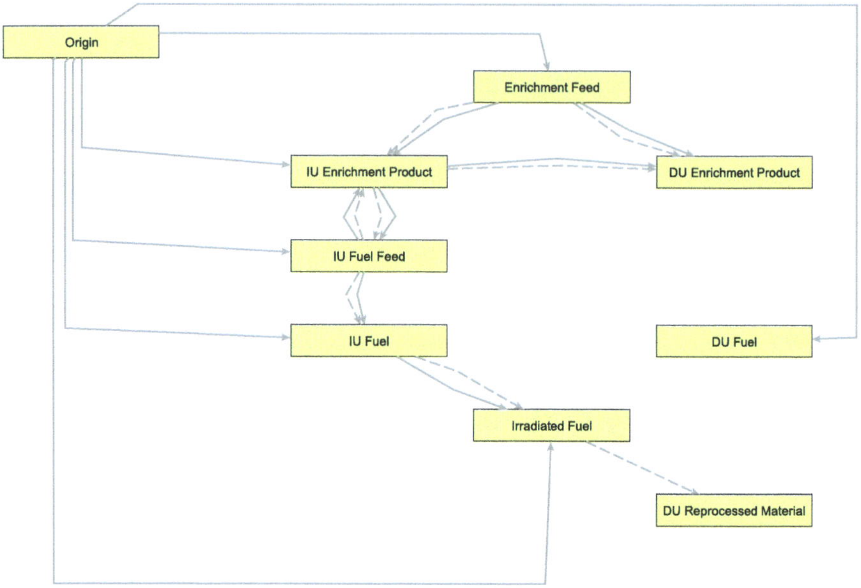

Fig. 5 Acquisition paths

The edges which contribute to the 22 identified acquisition paths are listed in Table 1 along with their associated parameters. Calculations were carried out for inspection plans involving different overall inspection effort expenditures.

Results are shown in Table 2. For efforts up to 100 person-days, the State will use path [0] (diversion of direct use fuel). The Inspectorate's recommended inspection activities [3] or [3, 5, 7] are Nash equilibrium strategies, but of course can't detect or deter the State's intensions. Therefore the effectiveness score is 0%. With total effort 150 person-days, the Inspectorate should play a mixed strategy, inspecting activity [0] with 21% probability, otherwise inspecting activities [3, 5, 7, 8, 11]. The State's counter strategy is now to choose path [1, 2] (diversion of enrichment product followed my facility misuse to obtain direct use enrichment product). This is again not detectable so the effectiveness is still 0%. For 200 person-days the situation remains essentially unchanged, but with 250 person-days, all 22 paths become detectable with finite probability. Now the Inspectorate should mix the pure strategies [0, 3, 5, 7], [0, 3, 5, 8] and [1, 3] in the proportions shown. The State will mix two illegal strategies, namely [0] and [1, 2] and accept the associated risk of detection. The equilibrium payoff to the Inspectorate H_1^* now leads to an improved effectiveness score of 33%. The situation is similar for 300 person-days. But for 350 person days the Sate's Nash equilibrium strategy is legal behaviour: the risk of detection with the associated consequences outweighs the incentive to acquire a clandestine weapon. The effectiveness score is 100%.

Table 1 Diversion path activities (edges)

Edge	Activity	Semantic	Effort	Cost	$1-\beta$
0	→ DU Fuel	Diversion	143	1	0.60
1	→ IU Enrichment product	Diversion	200	2/3	0.60
2	IU Enrichment product → DU Enrichment product	Misuse	200	2/3	1.00
3	IU Enrichment product → DU Enrichment product	Clandestine	38	5/3	1.00
4	→ IU Fuel	Diversion	85	1	0.60
5	IU Fuel Feed → IU Enrichment product	Misuse	28	1	1.00
6	→ Irradiated Fuel	Diversion	860	1	0.60
7	Irradiated Fuel → DU Reprocessed material	Clandestine	28	2	1.00
8	IU Fuel Feed → IU Enrichment product	Clandestine	28	2	1.00
9	→ IU Fuel	Diversion	143	1	0.60
10	IU Fuel → Irradiated Fuel	Misuse	645	1	1.00
11	IU Fuel → Irradiated Fuel	Clandestine	28	2	1.00
12	IU Fuel Feed → IU Fuel	Misuse	215	1	1.00
13	IU Enrichment product → IU Fuel Feed	Misuse	645	1	1.00
14	IU Fuel Feed → IU Fuel	Clandestine	28	2	1.00
13	IU Enrichment product → IU Fuel Feed	Clandestine	28	2	1.00

Note that this score is achieved (under the assumption that $b = d = 9$) with an effort which is considerably less than that which would be required to inspect all 16 edges completely. A higher effort would of course have been required for $b = 1$, see Sect. 2.5. The game-theoretical analysis automatically optimizes the Inspectorate's behaviour, ensuring maximum efficiency.

More extensive applications of the methodology, including a differentiation between member States that have signed the Additional Protocol and those that have not, are given in [10].

Table 2 Safeguards effectiveness from the game-theoretical acquisition path analysis

Total effort (p-d)	Inspectorate strategy Pr. [edge set]	State strategy Pr. [edge set]	Effectiveness (%)
50.0	1.00 [3]	1.0 [0]	0
100.0	1.00 [3,5,7]	1.0 [0]	0
150.0	0.79 [3,5,7,8,11]	1.0 [1,2]	0
	0.21 [0]		
200.0	0.79 [3,5,7,8,11,13,14]	1.0 [1,2]	0
	0.21 [0,3,5]		
250.0	0.61 [0,3,5,7]	0.62 [0]	33
	0.07 [0,3,5,8]	0.28 [1,2]	
	0.32 [1,3]		
300.0	0.68 [0,3,5,7,8,11]	0.62 [0]	33
	0.32 [1,3,5,7]	0.28 [1,2]	
350.0	0.57 [0,3,5,7,8,11,13,14]	1.00 [legal]	100
	0.43 [0,1]		

4 Discussion and Outlook

We have presented a structured methodology and a software implementation for acquisition path analysis which is as objective as possible within the highly political context of international treaty verification. The methodology clearly isolates purely technical aspects (such as inspection costs w_i, detection capabilities β_i) from judgemental aspects (such as proliferation incentives d_i, sanctionability b). It is reproducible, transparent, standardized and easily documented. Selection of model parameters on the part of an Inspectorate clearly requires expert judgement, but, since the parameter values can be easily varied within the model, the effect of different assumptions on the outcome can be easily investigated. The analyst receives immediate feedback on the effect of his or her assumptions.

Finally it should be emphasized that the procedure is modular. The directed graph approach on its own, without the strategic analysis, can be very valuable as a tool to structure hypothetical control regimes which do not as yet exist, e.g., a biological weapons treaty, bilateral nuclear weapons cutoff agreements, agreements on ballistic delivery systems, or the application of comprehensive IAEA safeguards in nuclear weapons states. Once a viable verification framework is developed for such regimes, the strategic aspect can be included. Investigations of this kind are in progress.

Acknowledgements This report was prepared in close cooperation with Irmgard Niemeyer (Forschungszentrum Jülich), Arnold Rezniczek (UBA Unternehmensberatung GmbH, Herzogenrath) and Gotthard Stein (Bonn).

References

1. Avenhaus R, Canty MJ (1996) Compliance Quantified, an Introduction to Data Verification. Cambridge University Press, Cambridge
2. Avenhaus R, von Stengel B, Zamir S (1999) Inspection Games. In: Aumann RJ, Hart S (eds) Handbook of Game Theory, Vol 3. Handbooks in Economics 11. North Holland, Amsterdam
3. Nash JF (1951) Non-cooperative games. Annals of Maths 54:286–295
4. Canty MJ (2004) Resolving Conflicts with Mathematica: Algorithms for Two-person Games. Academic Press, Amsterdam
5. Dijkstra, EW (1959) A note on two problems in connexion with graphs. Numerische Mathematik 1.1:269–271
6. Nisan N, Roughgarden T, Tardos E, Vazirani VV (eds) (2007) Algorithmic Game Theory. Cambridge University Press, Cambridge
7. Avis D, Rosenberg G, Savani R, von Stengel B (2009) Enumeration of Nash equilibria for two-player games. Economic Theory 42:9–37
8. Lemke CE, Howson JT (1964) Equilibrium points of bimatrix games. SIAM Journal on Applied Mathematics 12:413–423
9. Goldberg PW, Papadimitiou CH, Savani R (2011) The complexity of the homotopy method, equilibrium selection, and Lemke - Howson solutions. In: Proc. 52nd FOCS, pp 67–76
10. Listner C, Niemeyer I, Canty MJ, Murphy CL, Stein G, Rezniczek A (2015) Acquisition path analysis quantified – shaping the success of the IAEA's State-level Concept. Journal of Nuclear Materials Management XLIII(4):49–59
11. Listner C, Niemeyer I, Canty MJ, Stein G (2016) A strategic model for state compliance verification. Naval Research Logistics 63(3):260–271
12. Huszti J, Németh A, Vincze A (2012) Applicability of the directed graph methodology for acquisition path analysis. ESARDA Bulletin 47:72–79

Metrics for Detecting Undeclared Materials and Activities

Nicholas Kyriakopoulos

Abstract The State-level approach to safeguards aims to draw safeguards conclusions about the State as a whole in contrast to facility-specific classical safeguards. To do so, the integrated safeguards approach aims to integrate all safeguards-relevant information about each State. To achieve this goal procedures and metrics are needed for generating standardized and measurable results on the basis of the available data. While the information pertaining to the declared nuclear fuel cycle is quantitative and measurable, activities outside the declared nuclear fuel cycle may either not be well defined or quantifiable. Some of that information may be quantitative while other may be descriptive and qualitative. Also information about possible clandestine undeclared activities, by its nature, includes uncertainties about its validity and accuracy. The uncertainties inherent in the information available for evaluating the State as a whole present a challenge to the development of a safeguards system that is objective and transparent. This chapter examines the challenges associated with the development of such a system and specifies the necessary attributes and metrics of the information used in the development of such a system. It associates a set of metrics for the attributes of correctness, completeness, transparency and timeliness that can be used to measure the relevance of the information to the evaluation of the State as a whole.

1 Introduction

Classical safeguards seek to verify the accuracy of the declarations by the States. The existence of clandestine nuclear activities in Iraq in the presence of classical safeguards, made it necessary to re-evaluate the application of safeguards. As a result, it became apparent that in order to ascertain compliance with the NPT, safeguards need to be applied to the State as a whole. The mechanism for doing

N. Kyriakopoulos (✉)
The George Washington University, Washington, DC, USA
e-mail: kyriak@gwu.edu

I. Niemeyer et al. (eds.), *Nuclear Non-proliferation and Arms Control Verification*, https://doi.org/10.1007/978-3-030-29537-0_10

so is the Additional Protocol. Under it the IAEA collects information which, in conjunction with the application of classical safeguards, is to be used to ascertain whether or not a State complies with its obligations under the NPT.

Unlike classical safeguards that require the establishment of a National System of Accounting and Control for all declared materials in the State, the Additional Protocol specifically excludes material accountancy and "mechanistic verification" of the declarations by the State. Thus, for a State under safeguards the part of the State that consists of declared nuclear fuel cycle is monitored using material balance accounting, while the remaining part is not subject to such monitoring. In order to achieve the objective to evaluate the State as a whole, the IAEA has developed the concept of integrated safeguards that aims to utilize the totality of the available information to draw its conclusions concerning compliance. In order for the safeguards to be non-discriminatory, they must be objective and transparent [1].

This concept for evaluating the State as a whole involves conducting state-specific critical path analysis of all possible paths leading to the acquisition of nuclear materials and using this framework to formulate a State-specific safeguards system. The challenge for the IAEA is to translate the concept into a set of procedures and metrics that would generate standardized and measurable results. It is not an easy task. In this chapter, we discuss the sources of the difficulties for creating an objective and transparent system for evaluating a State as a whole. We also specify the necessary attributes of the metrics of the information available to the IAEA for determining compliance of a State with its obligations under the NPT.

The first section identifies potential proliferation paths a State could undertake and classifies them into distinct categories. It is followed by a critical analysis and characterization of the data available to the IAEA for evaluating the State as a whole. The final section identifies specific attributes that the data must have in order to generate objective and transparent results when used in evaluating a clandestine proliferation detection system.

2 Potential Clandestine Proliferation Paths

The effort to develop a systematic and objective approach for evaluating a State as a whole has led to the concept of Acquisition Path Analysis that considers an exhaustive list of strategies a State could pursue to acquire weapons grade nuclear material. The potential acquisition strategies available to a State would be a function of the nuclear fuel-cycle profile of that State and identifies plausible paths with signatures that could be detected. Under such a concept, the State-specific safeguards approach can develop a monitoring and evaluation strategy that optimizes detection of undeclared nuclear material or activities [2, 3].

One approach towards the development of a State-specific safeguards regime is to categorize all potential acquisition paths in a given State on the basis of the "attractiveness" of use. Attractiveness can be evaluated in terms of the ability of the State to pursue a particular acquisition path and the ability of the

IAEA to deter the State from pursuing it. The major parameters associated with attractiveness are technical difficulty, proliferation time, proliferation cost, and detection probability. The first three parameters are associated with the conditions in a potential proliferator State while the forth one is associated with each State-specific safeguards regime. Theoretically, when a State decides to embark on a clandestine acquisition path it would be logical to conclude that the State would pursue an acquisition strategy that minimizes the technical difficulties, proliferation time, and proliferation cost. Such an approach, however, may not be optimal from the perspective of the State, if the State also takes into account the cost associated with the potential detection of the chosen strategy. At an early stage of the clandestine proliferation activities, a State would aim to minimize the cost of detection at the expense of a non-optimal technical approach and with little regard to the monetary cost of pursuing the clandestine activity. Conversely, at a later stage, it might wish to minimize the proliferation time with little regard to the cost of detection. How a State takes into account those factors becomes a policy decision which is impossible to quantify. Thus, models of possible acquisition paths need not necessarily be based on minimization of the technical difficulties, proliferation time and proliferation cost.

The scope of the acquisition strategy as defined by the IAEA is limited to the acquisition of nuclear material. Although the nuclear material is the essential component of a nuclear weapons program, there are additional components that would need to be in place in order to assemble an explosive device. In other words, a plausible strategy for a clandestine nuclear weapons program could be to establish a number of non-nuclear development paths, such as acquisition of highly accurate timing circuitry, acquisition of capabilities for precision tooling, etc., in preparation for or in addition to the clandestine acquisition of nuclear material. It would not be unusual for a State to acquire the necessary technologies for producing weapons-grade nuclear material before initiating the clandestine diversion process. In this broader context, the term acquisition is used to denote capability, indigenous or imported, to build a nuclear explosive device.

A State has a number of options for initiating a clandestine weapons program and pursuing an acquisition strategy that takes into account a number of internal and external conditions unique to that State. For example, in a State with indigenous advanced electronics manufacturing, the capability to acquire the trigger mechanism would already be present in the form of advanced circuitry and engineering personnel. However if there were none or minimal electronics manufacturing capabilities in the State, many steps would be needed for attainment of the capability to manufacture a working trigger mechanism. Similar reasoning applies to the paths for the clandestine acquisition of highly enriched uranium or plutonium. Thus, although the various acquisition paths could have been well documented in theory, the number of steps each path would contain is not well-defined and it is correlated to the technological capabilities of the specific State.

Any attempt to develop metrics for evaluating detection probabilities needs to be based on the definition of the processes to be detected and the associated parameters to be measured. One can identify the following distinct categories of acquisition strategies: Spin-off a declared nuclear fuel cycle, parallel acquisition path, or ab initio development. In the spin-off model, the State initiates a clandestine acquisition path from some stage of the declared fuel cycle and keeps it separate afterwards. Such a separation involves facilities in locations different from the declared ones and personnel not connected from the declared activities. In a parallel acquisition model, the processes of the declared fuel cycle could be also used for undeclared purposes such as using personnel, equipment and facilities to engage in a clandestine program. A State adopting an ab initio strategy undertakes the initiation of a clandestine process totally separate from any declared nuclear activities. Under such strategy the facilities and personnel for the clandestine activities are totally separate from those associated with the declared fuel cycle.

There are different costs to the State associated with each acquisition strategy, one associated directly with each path and one with the detection of the clandestine activity. For example, if a particular State considers the cost of detection to be the deciding factor in undertaking clandestine activities, it could chose the ab initio strategy and incur the higher cost compared to that of a parallel path strategy that would have lower development costs but higher probability of detection, due to the proximity with the declared and safeguarded activities.

Absent of any hard data, it would be reasonable to assume that the probabilities of detecting diversion are conditioned upon the particular acquisition path a proliferator has chosen. An objective and transparent safeguards system needs to develop a mechanism for detecting proliferation by a State under the three possible scenarios. In other words,

$$P_{detection} = P_{ab\ initio} \cdot P_{detection|ab\ initio} + P_{spin-off} \cdot P_{detection|spin-off} + P_{parallel} \cdot P_{detection|sparallel}$$

Quantifying those probabilities is a problem awaiting solution. One, however, can set up a framework for addressing the problem. The first step for evaluating these probabilities is the development of realistic models for each of the three potential clandestine processes. The considerations that a State takes into account in order to decide which of the three paths to pursue are primarily internal and depend on the political, economic, scientific and technological profile of the State. Consequently, in order to calculate these probabilities, there is a need to develop metrics for measuring the state of political, economic, scientific and technological development for each State.

Calculation of the conditional probabilities of detection for each clandestine path requires the existence of process models for the paths and sufficient data for estimating the states of the underlying processes. Until fairly realistic process models are constructed, there is no reliable mechanism for establishing metrics for detecting the existence of such activities. The IAEA faces the challenge of detecting undeclared activities that are not well defined or understood using whatever data are available. Under the circumstances, an alternative and more productive approach is to maximize the extraction of information from the available data, i.e., do the best one can with what is available [4].

3 Characteristics of the Data Available to the IAEA

Detection of proliferation activities is based on the detailed analysis of the information entered into the safeguards system. To set up an objective mechanism for maximizing the extraction of relevant information from all the data available to the IAEA, it is necessary to develop a scheme for classifying the data on the basis of some unique and measurable characteristics, and on their contribution to the detection of non-compliance. For consistency of notation, all inputs to the safeguards systems are data that can be stored in a safeguards database and may be referred to as safeguards data. A partition of these data by source creates three subsets, IAEA-generated data, State-generated data, and open-source data. State-generated data are further partitioned into agreed safeguards data, i.e., data given by the States to the IAEA as required by the respective safeguards agreements, and voluntary data, i.e., information offered to the IAEA on a voluntary basis. Similarly, the open-source data are partitioned into public domain data, i.e., published data that the IAEA can seek and obtain, and private source data, i.e., data made available to the IAEA from sources other than State Parties. Another partition on the basis of unambiguous and measurable characteristics is into quantitative (numerical) and descriptive (text, voice, images) data. Quantitative data may further be partitioned into three subsets on the basis of the mechanism providing the data, sensing (measurement), observations (inspector) and file (data records). The partitions vary in their contribution to the evaluation of compliance [5, 6].

The wide range of data entered into the safeguards database requires the identification of attributes that associate the data with the verification function of the safeguards system. At the highest level the safeguards data may be partitioned into verifiable and unverifiable subsets. For example, materials accounting can be used by the IAEA to verify the flow of nuclear material for a declared nuclear facility, while such a mechanism is not available for similar data from sites declared under the Additional Protocol. An inspection at a site can only verify the conditions there at the time of the inspection but not at the time of the declaration. Absent verification there is great uncertainty whether or not the data represent what they purport to represent. The degree of contribution of the data to the determination of non-compliance is measured by the attribute of relevance. Relevant data have a direct and definitive connection to the process under investigation, while no such connection exists for irrelevant data. The interval between the two extremes represents the range of uncertainty regarding the contribution of the data to the decision about non-compliance. One should make the distinction between the attribute of relevance and the noise characteristics of data. One can have noise-free data that have no relation to a particular process or very noisy data that are directly related to a process under investigation. Although in reality the space between relevance and irrelevance is continuous, for practical purposes the range may be quantized into discrete steps. Determining the size of the quantization steps is another challenge for the safeguards system; in other words, how does one decide into how many intervals to divide the relevance range?

3.1 Safeguards Data

There is a major distinction between the characteristics of the data collected under classical safeguards (INFCIR/153) and those collected under the Additional Protocol (INFCIRC/540). Under classical safeguards, the declared nuclear fuel-cycle is a completely specified and bounded physical system. Within the declared fuel cycle, all processes are described in terms of physical location, function and their relation to the other declared processes. The measurement points, variables and frequency of measurements, as well as the measuring tools (instruments, inspectors), are selected in order to verify the accuracy of a State's declaration. The input data are quantifiable, reproducible, and therefore verifiable. In addition, they can be ranked as relevant because they have a direct connection to the process, i.e., the declared nuclear fuel cycle which is completely specified. As such they satisfy the requirement of objectivity and transparency. The long history of verification to a high degree of accuracy of the declarations by the States is evidence that the system is reliable, objective and transparent.

In contrast, the integrated safeguards system is still work in progress. The domain of the system extends beyond the declared facilities and processes within those facilities. All activities taking place within a State are considered and a subset is identified as having potential connection with clandestine activities. In turn, the subset is partitioned into the declared fuel cycle activities and, for lack of a better word, "potentially suspect" activities. While for the first set the processes and variables are completely specified, the second set consists of processes that have some connection to the declared nuclear fuel-cycle but the connection is imprecise and measurements incomplete. For example, Article 2 of the Additional Protocol requires the declaration of locations for "nuclear fuel cycle-related research and development activities not involving nuclear materials". It could be possible that a State is engaged in extensive research and development activities for non-nuclear fuel cycle applications while maintaining contingency plans for a clandestine program if the need arises. Similarly, although "location, operational status and estimated annual production of uranium mines" must be declared, material accountancy for verifying the declared estimates is not required. Similar ambiguities exist for the declarations involving technologies generally referred to as dual use such as, e.g., "zirconium tubes", "critically safe tanks and vessels", "hot cells outfitted with equipment for remote use", etc. Thus, the domain of the Additional Protocol encompasses locations and processes (activities) that may have some relevance to a nuclear program but they are neither precisely defined nor unambiguously measurable.

Location of facilities is quantitative and measurable. However, expressions such as "operational activities of safeguards relevance", "description of the scale of operations", "information regarding quantities, uses..." are imprecise and subject to interpretation. There is no clear connection between this type of data and the mechanism for detecting non-compliance. Even quantitative information such as quantities of imported or exported items is highly ambiguous, because there is no monitoring mechanism to verify declared quantities. The descriptive information falls into the category of qualitative data for which objective evaluation is difficult or practically impossible to undertake. The quantitative information, such as quantities produced or items imported or exported, could represent the true state of the process in a State that does not intend to engage in clandestine activities. A State planning a clandestine program could submit misleading data for which there is no systematic mechanism to test their validity. The integrated safeguards system needs to develop mechanisms for differentiating the two conditions. Thus, it would be more appropriate to consider the quantitative data provided under the Additional Protocol as noisy or fuzzy.

The data collected through environmental sampling can generate indicators of the presence of nuclear materials in undeclared locations. The data may be relevant to the extent they may contain information about a clandestine activity; they may also be fuzzy or noisy depending on the extent of possible concealment between a clandestine nuclear source and the location of the sampling instruments. Thus, while in terms of their validity as measurements performed by the IAEA may be classified as relevant, lack of knowledge of the degree of concealment between the source and the location where the measurements are taken introduces uncertainties that have an impact on their contribution to the detection system.

3.2 *Non-safeguards Data*

The data collected by the IAEA from sources outside the safeguards agreements may range from irrelevant to relevant; all are considered to be corrupted to some degree at their source. There are two classes of corrupted data, inherently noisy and intentionally corrupted. An obvious example of intentionally corrupted data would be if a State provides the IAEA with data about an allied State with the intent to shield it from intrusive scrutiny of incriminating information. Conversely, information offered about an enemy State could be misinformation for the purpose of putting that State on the defensive. In both cases the data would be intentionally highly corrupted for the purpose of generating misleading outputs from the safeguards system.

A similar partition exists for the information collected from the open literature. Data from refereed scientific publications reflect the actual state of scientific and technological progress. The information contained in such sources includes low levels of noise because the refereeing process is a filter; thus, the data may be deemed as relevant. On the other hand, data obtained from the public media may or may not be relevant and are subject to the same types of corruption as that inherent to the data offered by a State.

One then can partition the data entering the safeguards system into two categories, one associated with a robust verification regime and a second generating, at best, ambiguous results regarding non-compliance. This partition is illustrated in Fig. 1 where the main conditions for deterministic verification, i.e., objective,

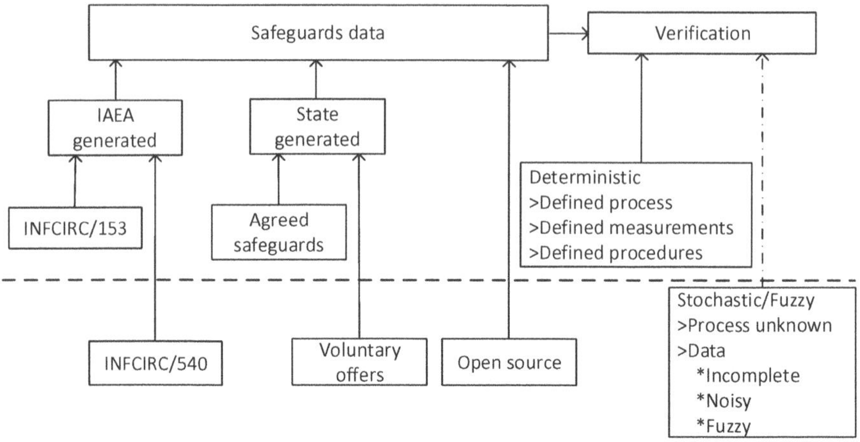

Fig. 1 Safeguards data and their contribution to the verification regime

transparent and reproducible, are detailed knowledge of the underlying process, with sufficient data and procedures to monitor it. The absence of these three requirements results in uncertain or fuzzy conclusions concerning compliance.

4 Development of Metrics for Detecting Undeclared Activities

The development of an objective and transparent safeguards system implies the use of quantitative analysis using independently verifiable measurements and well-documented algorithms. It also requires the establishment of a set of metrics for assessing the performance of the tools used in detecting clandestine activities. One can identify two categories of metrics. Detector metrics measure the design of the detector, i.e., the set of algorithms used to process the data and data metrics for measuring the quality of the input data, i.e., the value of the information contained in the data. Measuring the quality of the input data is prerequisite to the development of metrics for evaluating the performance of the detector.

Detector design presupposes knowledge of the processes that generate the data upon which the detector operates. In other words, in order to design a proliferation detector one needs to have a model of the underlying proliferation process, identify the measurements needed to detect the process and develop appropriate monitoring procedures. The procedure for developing performance metrics compares the required data with those available to the IAEA and defines metrics for such a comparison. This approach has the potential of producing more reliable information about the existence of clandestine activities. However, the effort is time-consuming, because it requires the construction of an exhaustive list of possible State-specific proliferation scenarios that encompass a range of scientific, technological, economic, political and geographic factors [8–10].

Even in the absence of process models for clandestine activities, analysis of the information content in the data available to the IAEA can provide a measure of how much information is provided about such potential activities. Such an analysis can generate indicators of some, but not all, possible clandestine activities. In other words, the evaluation of the State as a whole would yield information only about activities to which the available data have some connection. This approach has the potential of adaptive improvement by incorporating learning schemes based on the results of iterative evaluations of each State [11].

For a system designed to detect clandestine proliferation the following attributes for the input data are essential:

- *Correctness*: denotes the accuracy with which the elements of a data record reflect the true values of the variables the record purports to represent;
- *Completeness*: denotes the degree to which a data record contains all the data needed for a given task;
- *Transparency*: denotes the degree to which a data record can be verified;
- *Timeliness*: denotes the degree to which the time a data record is received by the time when the data in the record were generated.

In addition, in order to prevent nuclear proliferation, it is desirable that the IAEA is able to detect specific clandestine activities as close to the time the activities take place as possible. In Sect. 3 the attribute of relevance was defined as the degree of contribution a data record makes to the detection of non-compliance. One can then identify correctness, completeness, transparency and timeliness as the constituent components of relevance.

In order for these attributes to be used in an objective and transparent manner they must be quantified. Table 1 relates the information quality attributes with the categories of data collected under the Additional Protocol. Table 2 relates those attributes with all categories of data available to the IAEA. For consistency, the normalized continuous range of values for each attribute is with 0 indicating absence of any connection between the data and some clandestine activity and 1 a definite (perfect) one. As an example, a State declares that at a given time it has X number of zirconium tubes. The attribute of correctness might be assigned a value less than but close to 1 that may reflect the uncertainty about the quality of the tubes. If the State has declared some but not all the stock the attribute of completeness would have a value somewhere in the range. Similarly, if the IAEA cannot undertake an exhaustive auditing of the declaration, the attribute of transparency would merit a value at the low end of the range, even close to zero. Similar values can be associated with the attribute of timeliness. If the number of tubes is declared once a year, but the time when the inventory of the stock took place is not specified, the timeliness attribute would also have a value in the range.

Table 1 Quality attributes and potential ranges of values of additional protocol data

Categories	Attributes		Correctness	Completeness	Transparency	Timeliness	Relevance
Quantifiable	Facilities: location	Storage	0↔1	0↔1	0↔1	0↔1	0↔1
		Manufacturing	0↔1	0↔1	0↔1	0↔1	0↔1
		Mining	0↔1	0↔1	0↔1	0↔1	0↔1
		Concentration plants	0↔1	0↔1	0↔1	0↔1	0↔1
	Operations: scale	Inventory	0↔1	0↔1	0↔1	0↔1	0↔1
		Production capacity	0↔1	0↔1	0↔1	0↔1	0↔1
		Quantities produced	0↔1	0↔1	0↔1	0↔1	0↔1
	Exports	Quantities	0↔1	0↔1	0↔1	0↔1	0↔1
		Destinations	0↔1	0↔1	0↔1	0↔1	0↔1
	Imports	Quantities	0↔1	0↔1	0↔1	0↔1	0↔1
		Sources	0↔1	0↔1	0↔1	0↔1	0↔1
Frequency of declarations	Initial		0↔1	0↔1	0↔1	0↔1	0↔1
	Annual past activities		0↔1	0↔1	0↔1	0↔1	0↔1
	Advanced planned activities		0↔1	0↔1	0↔1	0↔1	0↔1

Table 2 Quality attributes and potential ranges of values of all data available to the IAEA

Categories		Attributes	Correctness	Completeness	Transparency	Timeliness	Relevance
Type	Descriptive	Verbal	$0 \leftrightarrow 1$	$0 \leftrightarrow 1$	$0 \leftrightarrow 1$	$0 \leftrightarrow 1$	$0 \leftrightarrow 1$
		Written	$0 \leftrightarrow 1$	$0 \leftrightarrow 1$	$0 \leftrightarrow 1$	$0 \leftrightarrow 1$	$0 \leftrightarrow 1$
	Quantitative	Measurements	$0 \leftrightarrow 1$	$0 \leftrightarrow 1$	$0 \leftrightarrow 1$	$0 \leftrightarrow 1$	$0 \leftrightarrow 1$
		Observations	$0 \leftrightarrow 1$	$0 \leftrightarrow 1$	$0 \leftrightarrow 1$	$0 \leftrightarrow 1$	$0 \leftrightarrow 1$
		Files	$0 \leftrightarrow 1$	$0 \leftrightarrow 1$	$0 \leftrightarrow 1$	$0 \leftrightarrow 1$	$0 \leftrightarrow 1$
Source	IAEA-generated		$0 \leftrightarrow 1$	$0 \leftrightarrow 1$	$0 \leftrightarrow 1$	$0 \leftrightarrow 1$	$0 \leftrightarrow 1$
	State-generated	Agreed safeguards	$0 \leftrightarrow 1$	$0 \leftrightarrow 1$	$0 \leftrightarrow 1$	$0 \leftrightarrow 1$	$0 \leftrightarrow 1$
		Voluntary	$0 \leftrightarrow 1$	$0 \leftrightarrow 1$	$0 \leftrightarrow 1$	$0 \leftrightarrow 1$	$0 \leftrightarrow 1$
	Open-source	Public domain	$0 \leftrightarrow 1$	$0 \leftrightarrow 1$	$0 \leftrightarrow 1$	$0 \leftrightarrow 1$	$0 \leftrightarrow 1$
		Private source	$0 \leftrightarrow 1$	$0 \leftrightarrow 1$	$0 \leftrightarrow 1$	$0 \leftrightarrow 1$	$0 \leftrightarrow 1$
Frequency	Periodic		$0 \leftrightarrow 1$	$0 \leftrightarrow 1$	$0 \leftrightarrow 1$	$0 \leftrightarrow 1$	$0 \leftrightarrow 1$
	Ad hoc		$0 \leftrightarrow 1$	$0 \leftrightarrow 1$	$0 \leftrightarrow 1$	$0 \leftrightarrow 1$	$0 \leftrightarrow 1$

5 Conclusions

In this chapter we have investigated the problem of the development of a set of metrics for the detection of clandestine nuclear activities. Development of reliable models for all possible clandestine nuclear proliferation paths is a problem awaiting solution. Considering the uncertainties inherent in such clandestine activities the models are, by necessity, stochastic or fuzzy. In other words the model parameters are either random or fuzzy variables. The integrated safeguards concept by the IAEA sets out the requirement to be objective and non-discriminatory and that the evaluation of the detector must be quantitative, using clearly defined algorithms and quantitative data. The design of an objective detector requires fairly detailed information about the process and sufficient measurements for minimizing the probability of a false positive resulting in falsely accusing a State of illegal activities.

In seeking to develop measures for detecting clandestine activities, a more promising approach would be to develop procedures for maximizing the extraction of information from all the data available to the IAEA. To this end, we have identified a minimum set of essential attributes that are necessary for measuring the quality of the available data. These are correctness, completeness, transparency and timeliness. The utility of a particular data set in the context of the safeguards system is measured with the attribute of relevance, which has the four essential attributes as constituent parts. For a given set of data, a mechanism remains to be developed for assigning values to each of the attributes and combing them to draw inferences about the information content in the particular set.

References

1. Cooley JN (2011) Progress in Evolving the State-level Concept. Paper presented at the Seventh INMM/ESARDA Joint Workshop Future Directions for Nuclear Safeguards and Verification, Aix-en-Provence, France, 17–20 October 2011
2. Listner C, Canty MJ, Rezniczek A, Stein G, Niemeyer I (2013) Approaching acquisition path analysis formally - a comparison between AP and non-AP States. In: Proceedings of the 35rd ESARDA Annual Meeting, Bruges, Belgium, 27–30 May 2013
3. Listner C, Canty MJ, Stein G, Rezniczek A (2014) Quantifying Detection Probabilities for Proliferation Activities in Undeclared Facilities. In: Proceedings IAEA Symposium on International Safeguards, vol. IAEA-CN-220/303. https://www.iaea.org/safeguards/symposium/2014/home/eproceedings/sg2014-papers/000303.pdf. Accessed 29 Feb 2019
4. Kyriakopoulos N, Zhou X (2015) Clandestine Proliferation: A Stochastic Process Model. In: Proceedings 37th ESARDA Symposium on Safeguards and Nuclear Non-Proliferation, Manchester, UK, 19–21 May 2015, pp 358–371
5. Kim LK, Renda G, Cojazzi GGM (2015) Methodological Aspects on the IAEA State Level Concept and Related Acquisition Path Analysis. In: Proceedings of 37th ESARDA Symposium on Safeguards and Nuclear Non-Proliferation, Manchester, UK, 19–21 May 2015, pp 383–396
6. Wilson B, El Gebaly A, Schuler R, Schot PM, Ferguson M (2014) Information on the Nuclear Fuel Cycle From Disparate Sources as a Verification Tool. In: Proceedings IAEA Symposium on International Safeguards, (Available on line) https://www.iaea.org/safeguards/symposium/2014/home/eproceedings/sg2014-papers/000058.pdf. Accessed 29 Feb 2019

7. Meadow CT, Yuan W (1997) Measuring the Impact of Information: Defining the Concepts. Information Processing & Management 33(6):697–714
8. Alique O, Vaccaro J, Svedkauskaite J (2015) The Use of Measurement Uncertainty in Nuclear Materials Accountancy and Verification. In: Proceedings of 37th ESARDA Symposium on Safeguards and Nuclear Non-Proliferation, Manchester, UK, 19–21 May 2015, pp 1010–1017
9. Pipino LL, Lee YW, Wang RY (2002) Data Quality Assessment. Communications of the ACM 45(4ve): 211–218
10. Kaiser M, Klier M, Heinrich M (2007) How to Measure Data Quality? - A Metric-Based Approach. In: ICIS 2007Proceedings, Paper 108. http://aisel.aisnet.org/icis2007/108. Accessed 29 Feb 2019
11. Roberts PS, How Well Will the IAEA Be Able To Safeguard More Nuclear Materials in More States? http://www.npolicy.org/article.php?aid=1186&tid=4. Accessed 29 Feb 2019

Game Theoretical Models for Arms Control and Disarmament Verification

Dedicated to the Memory of Carl A. Bennett (1922–2014)

Rudolf Avenhaus and Thomas Krieger

Abstract In the context of arms control and disarmament, inspections are procedures designed to provide data with which a State's compliance or non-compliance to a treaty, an agreement or other set of rules can be assessed. There is always, potentially at least, a conflict between the inspection authority (in the following Inspectorate) and the State (in the following Inspectee) who is required to comply, but who may have an interest not to do so. Non-cooperative game theory provides the appropriate tools to analyse these kind of conflicts. The solution concept is hereby the so-called Nash equilibrium. Since inspections, beyond their assessment of compliance or non-compliance, should induce the State to legal behaviour, one may say that effective inspections should be designed such that the Inspectee's equilibrium strategy of the game theoretical model describing these inspections is legal behaviour of the Inspectee. Game theoretical analyses of arms control and disarmament verification have been performed since the early sixties of the last century. It was, however, the Treaty on the Non-Proliferation of Nuclear Weapons which stimulated more detailed analytical work. During the last 10 years two kinds of topics have been addressed in greater detail, unannounced interim inspections and acquisition path analysis. In this paper, assumptions necessary for a quantitative modelling of unannounced interim inspections are discussed, and one example is presented in major detail. Some thoughts on the art of modelling conflicts of both technical and political nature complement the previous thoughts.

R. Avenhaus (✉)
Universität der Bundeswehr München, Neubiberg, Germany
e-mail: rudolf.avenhaus@unibw.de

T. Krieger
Forschungszentrum Jülich GmbH, Jülich, Germany
e-mail: t.krieger@fz-juelich.de

I. Niemeyer et al. (eds.), *Nuclear Non-proliferation and Arms Control Verification*,
https://doi.org/10.1007/978-3-030-29537-0_11

1 Introduction

In the context of arms control and disarmament, inspections are procedures designed to provide data with which a State's compliance or non-compliance to a treaty, an agreement or other set of rules can be assessed. There is always, potentially at least, a conflict between the inspection authority (in the following Inspectorate) and the State (in the following Inspectee) who is required to comply, but who may have an interest not to do so. Of course, if the Inspectee were not tempted to violate the agreement, i.e., to behave illegally, then inspections would be unnecessary, but experience shows that this cannot be assumed in general.

Thus, with the data at hand, it is natural that quantitative models of the conflict posed by inspections are described in terms of non-cooperative games with two players, Inspectee and Inspectorate. The solution concept of non-cooperative games is the so-called Nash equilibrium, see, e.g., [1] or [2], which says that an equilibrium pair of strategies of both players has the property that any unilateral deviation from this pair of strategies does not improve the payoff of the deviator. Since inspections, beyond their assessment of compliance or non-compliance, should induce the Inspectee to legal behaviour, in other words deter him from illegal behaviour, one may say that *effective* inspections should be designed such that the equilibrium strategy of the Inspectee in the inspection game describing these inspections is compliance to the agreement, i.e., legal behaviour. *Efficient* inspections consequently are effective inspections achieved at minimum cost. Note that in chapter *Formalizing Acquisition Path Analysis* by Mort Canty and Clemens Listner in this book, another definition of effectiveness is considered.

Game theoretical analyses of arms control and disarmament verification have been performed since the early sixties of the last century, see, e.g., [3]. It was, however, the Treaty on the Non-Proliferation of Nuclear Weapons (NPT), inaugurated in 1968, which stimulated more detailed and practically oriented analytical work. The NPT was intended to freeze the status quo of nuclear weapons and non-nuclear weapons states, the former pledging themselves to a long-term reduction and ultimate elimination of their nuclear arsenals, while the latter agreed not to acquire such arsenals. The International Atomic Energy Agency (IAEA) in Vienna was given responsibility to verify compliance to the NPT, in particular to inspect the peaceful nuclear activities of the member states [4].

In the course of subsequent negotiations of the practical implementation of IAEA inspections, material accountancy emerged as the basic verification principle: Through periodic comparisons of book and physical inventories at nuclear installations, a quantitative statement regarding the continued presence of nuclear material was to be made. This system requires that plant operators, via their national control authorities, report all relevant data to the IAEA, while the international inspectors verify those data by making independent measurements on a random sampling basis. Extensive use of automatic surveillance equipment and seals is also made in order to minimize the number of measurements.

The details of the game theoretical models with the help of which these problems are analysed can be found e.g., in [5]. Also, in [6] the state of the art of about 10 years ago was described. There, three types of models were named, and examples were presented: Conceptual models treat the problem of arms control and disarmament in a very general way, providing insight into the nature and necessity of verification. Structural models demonstrate how sensible optimal inspection strategies depend on specific assumptions. Operational models provide the inspector with concrete advice as to how to spend his allotted inspection effort and time effectively and efficiently.

During the last 10 years two kinds of topics have especially been addressed due to the extended safeguards system laid down into the so-called Additional Protocol [7]. It foresees that, contrary to the original design, also undeclared facilities and activities have to be taken into account when safeguards measures are taken. Therefore, unannounced interim inspections have become much more important, and furthermore, acquisition path analyses have been performed in major detail in order to study systematically undeclared activities on one hand, and inspections strategies which might detect them, on the other.

For these reasons, in the following a game theoretical model is presented which deals in a simplified way and under well-specified assumptions with this topic, and it is shown what effective inspections means in quantitative terms. In the concluding chapter, some observations are offered on quantitative modelling of conflicts which are both technical and political by nature. Acquisition path analysis will be the subject of the chapter *Formalizing Acquisition Path Analysis* by Morton Canty and Clemens Listner in this book.

2 Unannounced Interim Inspections

Short notice and unannounced interim inspections (in the following shortly interim inspections) are a useful safeguards tool because they may shorten the detection time, i.e., the time between the start of an illegal activity and its detection. More than that because of their unpredictability they may help to deter any person or organization from illegal behaviour. Therefore, the problem of interim inspections to be carried out both by European Atomic Energy Community (EURATOM) and the IAEA has been discussed for many years and it has been settled in the so-called IAEA/EURATOM Partnership Approach, see [8].

On the other hand unpredictability cannot mean arbitrariness, on the contrary, international safeguards have to be performed in line with agreed rules which mean that these inspections have to be planned and carried through accordingly; they have to be arranged on the basis of probabilistic objectives, for example expected detection times.

Interim inspections pose specific problems different from standard safeguards measures like material accountancy and data verification. While the latter measures provide very detailed information and while they are characterized by the use of advanced statistical techniques for the compilation and evaluation of measurement data, interim inspections are aiming at the immediate detection of illegal activities or, positively formulated, confirmation of legal behaviour. Therefore, primarily simple techniques for the checking of seals, or comparing installations in facilities with the design information provided by the facility attachments, are used, and above all, time is important: Time available for the inspector in the facility, and time elapsed between the start of an illegal activity and its detection.

In order to be able to perform substantive analyses of interim inspections many assumptions have to be made which may be disputed since they are either not explicitly formulated in available nuclear safeguards documents or deal with the behaviour of facility operators, organizations or States in case they might behave illegally. In [9] these assumptions are classified and it is shown how many different quantitative models would have to be studied if one would try to analyse the consequences of all these assumptions. Figure 1 represents a modified version of this classification.

In this figure a hierarchy of assumptions is shown. First, the objective of the Inspectorate may be to detect any illegal activity as soon as possible (Playing for time) or within a given time interval (Critical time). Second, interim inspections and the start of an illegal activity may take place at any point of time (Continuous time) or only at discrete ones (Discrete time). Third, both the Inspectorate and the Inspectee may plan their activities at the beginning of the reference time interval (No) or sequentially in its course (Se). And fourth, inspections may be performed with ($\alpha \geq 0, \beta > 0$) or without committing statistical errors ($\alpha = \beta = 0$). Note that due to different analytical techniques, $\alpha = \beta = 0$ and $\alpha = 0, \beta > 0$, should not be considered special cases of $\alpha \geq 0, \beta \geq 0$. This way one arrives at 36 different models, but of course, even more assumptions are required in order to fully specify a mathematical model of interim inspections.

2.1 *Game Theoretical Model*

For the purpose of illustration we consider now one inspection problem which is taken from practice, see [9], and analysed in detail in [12], and which is described by one out of the 36 models classified in Fig. 1.

In Europe, adjacent to nuclear power reactors there are storages which contain spent fuel elements of the reactors in sealed casks and which are safeguarded both by EURATOM and IAEA. Every 3 months interim inspections may be performed, and at the end of a calendar year a so-called Physical Inventory Verification (PIV) is carried through which is assumed to provide exact knowledge about the amount of spent fuel in the storage. We designate the time points 0 and 4 for the PIVs and 1,2,3 for the interim inspections where periods 0 to 1 or 1, 1 to 2, 2 to 3, and 3 to 4 are a quarter of a year each.

The Inspectee is assumed to start one illegal activity at time points 0,1,2 or 3 by, e.g., replacing in one or more casks spent fuel elements by dummies. The

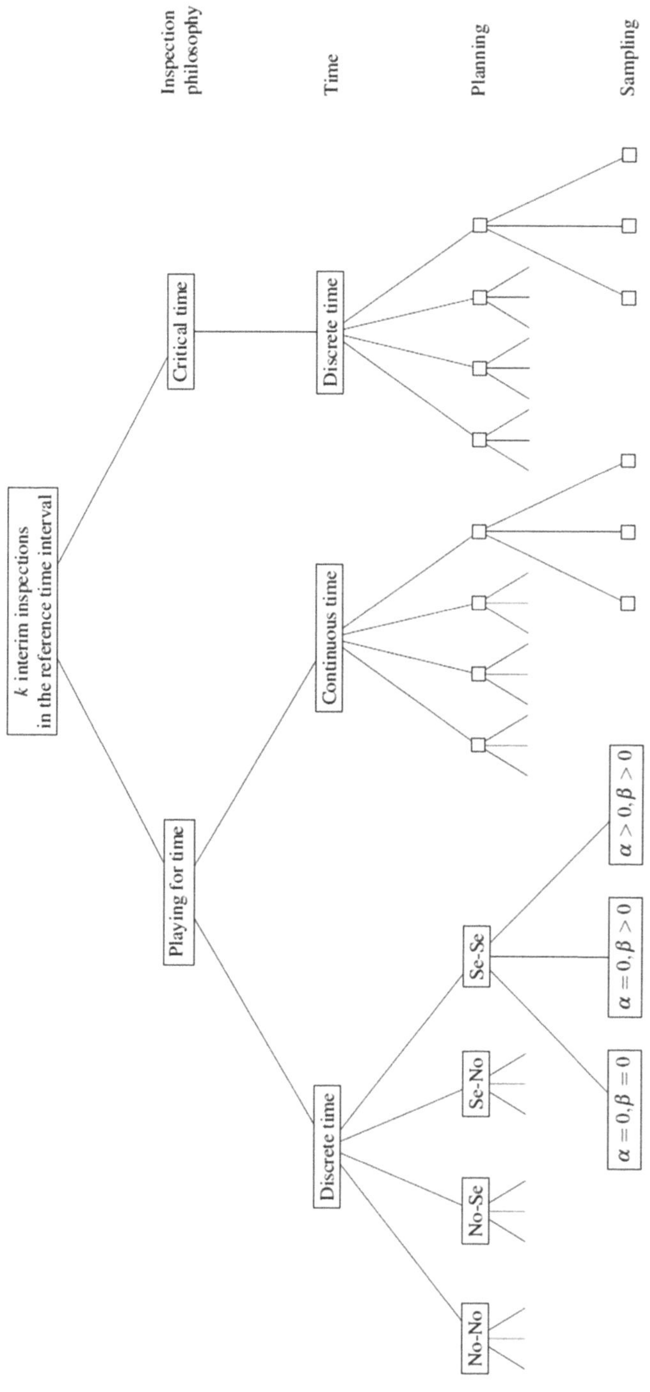

Fig. 1 Classification of models for inspections over time

Inspectorate is assumed to perform two interim inspections at time points 1,2 or 3. An interim inspection means checking some of the seals. Since errors of the first kind can be excluded, we deal with an attribute sampling procedure, where an illegal activity will be detected with probability $1 - \beta$ by the next inspection or, if not earlier, with certainty at the end of the year (the PIV). A detailed discussion of these assumptions can be found in [9] and [11].

Since the Inspectee is assumed to want to maximize the time elapsed between the start of the illegal activity and its detection, whereas the Inspectorate wants to minimize it, we consider a zero sum game with the expected detection time (counted in quarters of years) as payoff to the Inspectee.

Note that the assumption that the Inspectee will behave illegally sounds strange at first sight. This so-called diversion hypothesis[1] has caused considerable discussion. From a quantitative point of view it is a necessary first step: If we want to determine the equilibrium which implies legal behaviour of the Inspectee, we have to consider deviations from this equilibrium, i.e., illegal behaviour of the Inspectee.

There exist four variants of this general model, see Fig. 1: The Inspectee may decide at the beginning, i.e., at time point 0, when to start his illegal activity (No-...), or he may only decide whether to start the illegal activity immediately or to postpone it; in the latter case he may decide again after the first interim inspection, and so on (Se-...). The Inspectorate may decide at time point 0 when to perform both interim inspections (...-No), or he may decide only when to perform the first interim inspection, and—at the first interim inspection—when to perform the second one (...-Se). As an example here only the fully sequential variant, i.e., the Se-Se-variant, is analysed, the other ones can be found in [12].

The extensive form of this Se-Se inspection game is represented in Fig. 2.

The Inspectee decides at time point 0 to start the illegal activity immediately (ill) with probability $1 - g_0$ or to postpone it (l) with probability g_0. In the latter case he decides after the first interim inspection to start the illegal activity immediately (ill) with probability $1 - g_1$ or to postpone it (l) again with probability g_1. In the latter case he starts it after the second interim inspection by assumption. The Inspectorate decides at time point 0 only if to perform the first interim inspection at time points 1 or 2 with probabilities $1-h_0$ and h_0. In case the first interim inspection is carried through at 1, it decides to perform the second one at time points 2 or 3 with probabilities $1 - h_1$ and h_1. In case the first interim inspection takes place at time point 2, the second one has to be performed at time point 3. Thus, the sets of strategies of both players are given by

$$G = \{\mathbf{g} := (g_0, g_1) : g_0, g_1 \in [0, 1]\} \quad \text{and} \quad H = \{\mathbf{h} := (h_0, h_1) : h_0, h_1 \in [0, 1]\}.$$

[1]In [10] it is written: "This diversion hypothesis should not been understood – and in general is not understood – as an expression of distrust directed against States in general or any State in particular. Any misunderstanding might be dispelled by comparing diversion hypothesis with the philosophy of airport control. In order to be effective, airport control has to assume a priori and without any suspicion against a particular passenger that each handbag might contain prohibited goods."

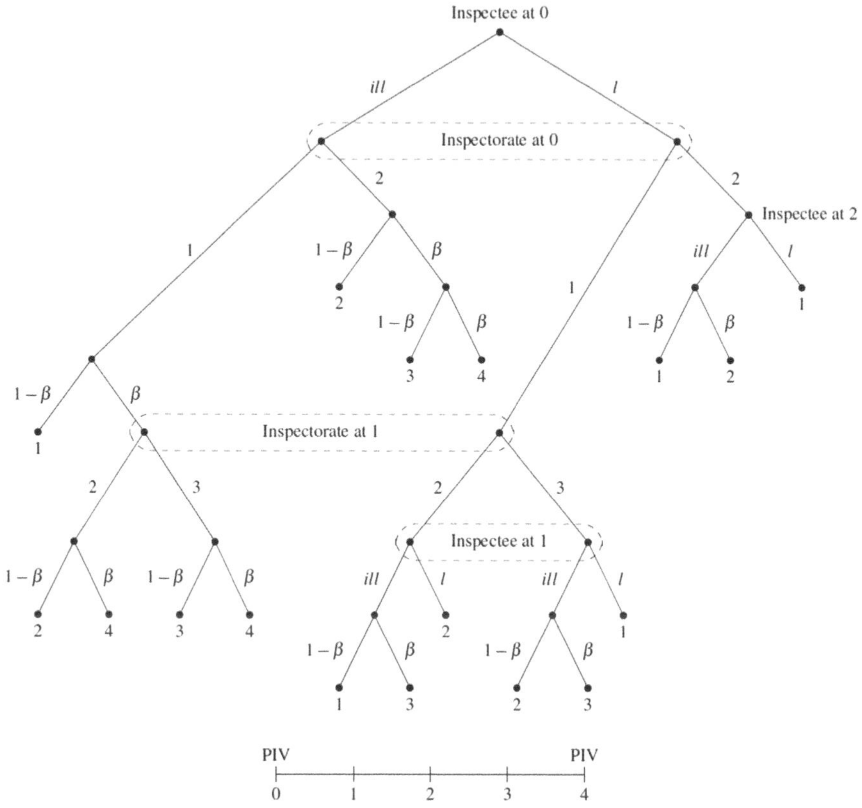

Fig. 2 Graphical representation of the extensive form of the Se-Se variant of the inspection game. *ill*: immediate start of the *ill*egal activity. *l*: postponing the start of the illegal activity, i.e., behaving *l*egally at this stage. The three encircled areas denote the information sets of both players

The ovals indicate the information sets of the players: At an information set the player, who has to make a decision, does not know at which node in this set he stands. For example: The information set "Inspectorate at 0" indicates that it does not know the Inspectee's decision at time point 0. At the end nodes the detection times are given.

We see that, if the Inspectorate decides at time point 0 to perform his first interim inspection at 2 and the Inspectee decides at time point 0 to postpone the illegal activity, then the Inspectorate's choice for 3 is dominated thus, the Inspectee's payoff is $1 - \beta + 2\beta = 1 + \beta$.

The payoff to the Inspectee, i.e., the expected detection time, as a function of the strategies $\mathbf{g} = (g_0, g_1) \in G$ and $\mathbf{h} = (h_0, h_1) \in H$ of the two players is given by

$$Op(\mathbf{g}, \mathbf{h}) = (1 - g_0) \Big\{ (1 - h_0) \Big(1 - \beta + \beta [(1 - h_1) (2 + 2\beta) + h_1 (3 + \beta)] \Big) + h_0 (2 + \beta + \beta^2) \Big\}$$

$$+ g_0 \Big\{ (1 - h_0) \Big((1 - h_1) [(1 - g_1) (1 + 2\beta) + g_1 2] + h_1 [(1 - g_1) (2 + \beta) + g_1] \Big) + h_0 (1 + \beta) \Big\} . \qquad (1)$$

It summarizes, together with the sets of strategies G and H of both players the formal description of the inspection game.

2.2 Analysis

As mentioned in Sect. 1 the solution concept of a non-cooperative game is the Nash equilibrium, which—since we deal with a zero sum game—simplifies to the saddle point concept: Saddle point strategies $\mathbf{g}^*$ and $\mathbf{h}^*$, also called optimal strategies, are described by the so-called saddle point criterion which in our case is given by

$$Op(\mathbf{g}, \mathbf{h}^*) \le Op(\mathbf{g}^*, \mathbf{h}^*) \le Op(\mathbf{g}^*, \mathbf{h}) \qquad (2)$$

for all $\mathbf{g} \in G$ and all $\mathbf{h} \in H$. It formalizes the idea that the Inspectee wants to maximize the expected detection time while the Inspectorate wants to minimize it. We formulate the game theoretical solution in

Theorem 1 *Given the Se-Se inspection game the extensive form of which is represented in Fig. 1. Then optimal strategies* $\mathbf{g}^* = (g_0^*, g_1^*)$ *and* $\mathbf{h}^* = (h_0^*, h_1^*)$, *and the optimal expected detection time* $Op^* := Op(\mathbf{g}^*, \mathbf{h}^*)$ *are given as follows:*

- *For* $0 \le \beta \le 1/2$ *an optimal strategy for the Inspectee is given by*

$$g_0^* = \frac{2 - 2\beta + \beta^2 - \beta^3}{3 - 3\beta + 2\beta^2 - \beta^3} \quad \text{and} \quad g_1^* = \frac{1}{2 + \beta^2}$$

and for the Inspectorate by

$$h_0^* = \frac{1 - 2\beta}{3 - 3\beta + 2\beta^2 - \beta^3} \quad \text{and} \quad h_1^* = \frac{1 - 2\beta}{2 - \beta}$$

with the optimal expected detection time

$$Op^* = \frac{4 - \beta + \beta^2}{3 - 3\beta + 2\beta^2 - \beta^3} .$$

- *For* $1/2 \leq \beta \leq 1$ *an optimal strategy for the Inspectee and for the Inspectorate are given by*

$$\mathbf{g}^* = (0, 0) \qquad \textit{and} \qquad \mathbf{h}^* = (0, 0)$$

with the optimal expected detection time

$$Op^* = 1 + \beta + 2\,\beta^2 .$$

Proof For $0 \leq \beta \leq 1/2$ a lengthy calculation shows that $Op(\mathbf{g}, \mathbf{h}^*) = Op^*$ for all $\mathbf{g} \in G$ and $Op(\mathbf{g}^*, \mathbf{h}) = Op^*$ for all $\mathbf{h} \in H$, i.e., the strategies $\mathbf{g}^*$ and $\mathbf{h}^*$ are determined such that they render the adversary indifferent with respect to his strategy choice. Thus, the saddle point conditions (2) are fulfilled as equalities. For $1/2 \leq \beta \leq 1$ one sees that with (1)

$$Op(\mathbf{g}, \mathbf{h}^*) = 1 + \beta + 2\,\beta^2 - g_0\,(g_1 + \beta)\,(2\,\beta - 1)\,,$$

which—because of $2\,\beta - 1 \geq 0$—is maximized for $\mathbf{g}^* = (0, 0)$, i.e., the left hand inequality in (2) is fulfilled. Furthermore, we have with (1)

$$Op(\mathbf{g}^*, \mathbf{h}) = (1 - h_0)\Big(1 - \beta + \beta\,[\,(1 - h_1)\,(2 + 2\,\beta) + h_1\,(3 + \beta)\,]\Big) + h_0\,(2 + \beta + \beta^2)\,,$$

which is—in a first step—minimized by $h_1^* = 0$ leading to

$$Op(\mathbf{g}^*, (h_0, 0)) = 1 + \beta + 2\,\beta^2 + h_0\,(1 - \beta^2)\,,$$

which is minimized by $h_0^* = 0$, because $1 - \beta^2 \geq 0$, i.e., the right hand inequality of (2) is also fulfilled. □

Note that at $\beta = 1/2$ both the optimal detection time and the Inspectorate's optimal strategy are continuous, but not the Inspectee's one. Thus, any convex combination of both Inspectee's optimal strategies at $\beta = 1/2$ is also optimal. We did not mention this explicitly in the theorem, since in practice this precise value of $\beta = 1/2$ will hardly ever be realized.

2.3 *Effective and Efficient Inspections*

In Sect. 1 we mentioned that an inspection system is considered effective if the Inspectee's equilibrium strategy of the game describing this system is legal behaviour. So far we have presented a purely technical analysis in which illegal behaviour of the Inspectee was assumed and which, therefore, cannot deal with this

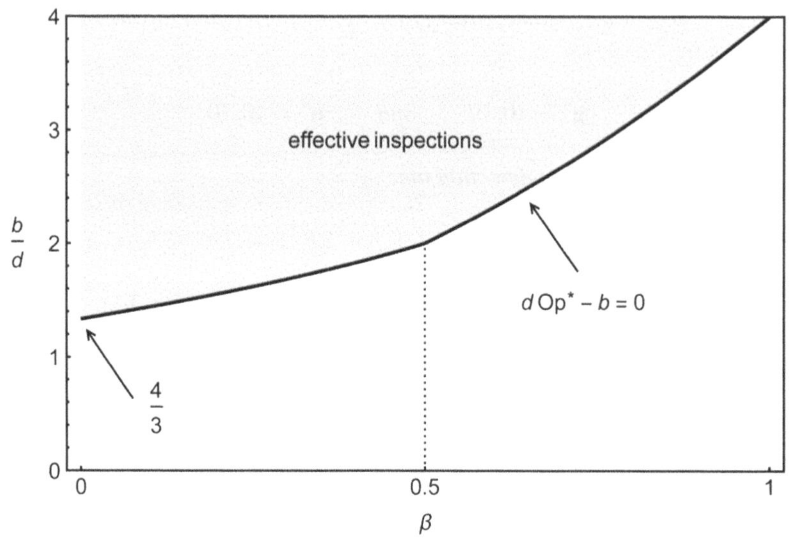

Fig. 3 Area of effective inspections. Op^* is given by the theorem

issue. In order to be able to do this, payoff parameters have to be introduced: Let the payoff to the Inspectee be $d\,Op(\mathbf{g}, \mathbf{h}) - b$ for illegal behaviour and 0 for legal behaviour, where $d > 0$ is the gain of the Inspectee per unit detection time and $b > 0$ the sanctions in case of detected illegal behaviour. Since, in the sense of a conservative estimate, the longest possible detection time (for $\beta = 1$) is 4, for $4\,d - b < 0$ the Inspectee will always behave legally. For smaller b the Inspectee will behave legally if $d\,Op^* - b < 0$. With Op^* given by the theorem, this is a condition between b/d and β, see Fig. 3.

The estimation of the payoff parameters b and d may pose considerable practical, even political problems; we will return to this issue in Sect. 3.

We see that in the framework of our model effective inspections require that the larger β is, the larger has to be the ratio of sanctions to gains of the Inspectee which is reasonable. Furthermore, we mentioned already that efficient inspections are effective inspections at minimum cost. If we assume that β is a monotonously decreasing function of increasing cost, then efficient inspections are given by $d\,Op^* - b = 0$, see Fig. 3.

Let us mention, finally, that in order to determine the equilibrium strategy of the Inspectorate in case the Inspectee behaves legally also the Inspectorate's payoff parameters have to be introduced and then, the saddle point conditions (2) have to be replaced by the Nash equilibrium condition, see, e.g., [1], since we deal no longer with a zero sum game. It turns out that there are many equilibrium strategies of the Inspectorate and—even more important—that the Inspectorate's optimal strategy in case of illegal behaviour of the Inspectee is part of the Inspectorate's set of equilibrium strategies in case of legal behaviour of the Inspectee, see, e.g., [9] and

[11]. This is very helpful for practical applications since in practice in most cases the values of the payoff parameters can be estimated only very roughly.

2.4 *Extensions*

As mentioned before, in [12] four variants have been analysed for $k = 2$ interim inspections and $N = 3$ possible time points for these inspections. It has been shown that it depends only on the Inspectee's choice (in the sense of the variants mentioned above) which optimal expected detection time can be achieved, and that for $\beta = 0$ all four variants lead to the same one, $4/3$. Plausible interpretations for these results can be offered only in special cases.

The theorem indicates that it is very difficult to analyse models for larger N and k and $\beta > 0$. For $\beta = 0$, however, solutions have been obtained for any k in case the Inspectee decides sequentially [11] and [12]: Both for sequential (Se-...) and non-sequential behaviour (No-...) of the Inspectorate the optimal expected detection time is $(N + 1)/(k + 1)$. This might be guessed, but the details of these results are far from being trivial. For example, for larger N and k only the optimal expected time points for the interim inspections, but not their distributions are uniquely determined.

Furthermore, under the following assumptions classified in Fig. 1, namely (1) Playing for time, (2) interim inspections can be performed at any point of time (Continuous time), (3) both players decide sequentially (Se-Se) and (4) errors of the first and second kind are taken into account ($\alpha > 0$, $\beta > 0$), equilibria are presented in [13] for any number k of interim inspections.

Note that, even though the model in [13] is very general and has been solved completely, in this contribution we chose as example a model which uses on a first level only technical parameters. Thus, its solution can be immediately applied in practice. Of course, in order to analyse its effectiveness and efficiency, on a second level payoff parameters are used.

Also, as indicated in Fig. 1, there are models where the objective of the Inspectorate is the detection of an illegal activity within a so-called critical time which lead to quite different optimal strategies of both players, see, e.g., [11].

3 Conclusions: The Art of Modeling

In Sect. 2.3 we have introduced so-called political parameters which describe the Inspectee's incentives in case of legal and illegal behaviour. This issue has caused considerable disputes over the years. In [6] the pros and cons are discussed in some detail, but before referring shortly to them a few more general remarks about the art of modelling, in particular in the context of arms control and disarmament verification shall be made.

When analysts who have a reasonable background in Statistics, Game Theory and Operations Research in general, and who have some experience with applied work, are asked for answers by practitioners who know their problems but not the difficulties of quantitative modelling, the following frequently happens: Either analysts arrive at formal models and their solutions which are satisfying from a mathematical, not to say aesthetical point of view, but not so much from a practical one. Or the problem and its solution is so special, that it is interesting only for that single case. Or the analysts have to work with approximations and present second best solutions which may answer the practitioners' questions but are not satisfying analysts' standards of rigorousness, originality and importance.

Only at rare occasions—lucky moments in their lives—analysts come across problems and their solutions which are satisfying both from the practical and theoretical points of view and beyond, which are of more general interest, in other words, these analysts have found some interesting general structure.

In the case of the analysis of interim inspections all these situations are happening: the theorem and also the solutions of the other variants are interesting both from the practitioner's and theoretician's points of view, and properties of the solutions can be found in those of other inspection problems. Considering more complicated cases like $N > 3$ and $\beta > 0$, solutions get less attractive. Here, the needs of the practitioner may be satisfied with approximations, in the case of interim inspections, e.g., with the help of continuous time models, see, e.g., [11].

If the practitioner, or more generally the person or organization which proposes to analysts a study of a problem, formulates the problem in form of very precise assumptions, then it is just good luck or bad luck whether or not the analysis will become interesting from the analysts' point of view. The real situation, however, is different in general. The practitioner may not know so precisely which assumptions are required for the quantitative analysis, or he may not be so certain about all assumptions required, or he may change his opinions in the course of the analysts' work. This, in turn gives room to the latter one to formulate assumptions himself and ask for their confirmation. This way both the practitioner and the analyst can guide the work in directions which renders the analysis feasible and convincing for both sides: Sometimes the model itself may be justified by an interesting solution which, by the way, requires that the analyst is able to interpret this solution in a way which convinces the practitioner. Of course, all this requires that both sides listen carefully to each other, even in cases where they do not fully understand the other side. In other words, all this requires patience and time.

This shows that modelling is a kind of art which cannot be taught and learned with the help of textbooks and courses alone. It has to be practised for long times under favourable circumstances; a key word for those who have or want to encourage studies of the kind an example for which is presented here, is capacity building on both sides, the practitioner and the analyst. Our experience shows, that it is difficult to get diplomats, politicians and administrators interested in the type of thinking outlined her. Howard Raiffa [14] has described the general problem many times and very convincingly. We can only repeat his plea for our case, namely that practitioners and analysts approach each other in the interest of really solving the vital problems of verification:

> ... Not enough research on the processes of international negotiations is being done. What is being done is not adequately coordinated and disseminated. Present research efforts are not cross-fertilized: across disciplines, between practitioners and researchers, and across national boundaries. Regrettably, a lot of profound theorizing by economists mathematicians, philosophers, and game theorists on topics related to negotiation analysis has had little or no impact on practice. An important reason is clearly the lack of effective communication and dissemination of theoretical research results. Such communication could be improved if there were more intermediaries who are comfortable in both worlds and who could act as inventive go-betweens to facilitate the transfer of information that shows how theory can influence practice and how practice can influence the research agendas of theorists. The information must flow in both directions: many practitioners have developed valid, extremely useful, and often profound insights and analyses, which should help to guide the agenda of researchers in this field.

Let us illustrate these wise words with the problem of the estimation of payoff parameters which we mentioned in Sect. 2.3. In IAEA safeguards it was argued by many that an international verification organization has no business doing this sort of political analysis. Instead, the point was taken that the detection capability is an external variable that must be determined bureaucratically. Typical, some ad hoc measure, such as x percent probability for detecting a violation strategy y within time z, was specified which determined the overall inspection effort required for its achievement. This of course begs all questions regarding efficiency and effectiveness. C. A. Bennett, to whom this contribution is dedicated, commented these attempts many years ago as follows: *They are looking for a technical solution of a political problem.*

There are situations where even peaceful theoreticians have to fight with as peaceful administrators, if necessary passionately, for the better cause.

References

1. Nash JF (1951) Non-Cooperative Games. Annals of Mathematics 54(2):286–295
2. Myerson RB (1991) Game Theory: Analysis of Conflict. Harvard University Press, Cambridge
3. Avenhaus R, Canty MJ, Kilgour DM, von Stengel B, Zamir S (1996) Inspection games in arms control (Invited review). European Journal of Operational Research 90:383–394
4. International Atomic Energy Agency (IAEA) (1972) The Structure and Content of Agreements Between the Agency and States in Connection with the Treaty on the Non-Proliferation of Nuclear Weapons, INFCIRC/153 (Corrected)
5. Avenhaus R, Canty MJ (1996) Compliance Quantified - An Introduction to Data Verification. Cambridge University Press, Cambridge,
6. Avenhaus R, Canty MJ (2006) Formal Models of Verification. In: Avenhaus R, Kyriakopoulos N, Richard M, Stein G (eds) Verifying Treaty Compliance. Limiting Weapons of Mass Destruction and Monitoring Kyoto Protocol Provisions. Springer, Berlin-Heidelberg, pp 295–319
7. International Atomic Energy Agency (IAEA) (1997) Model Protocol Additional to the Agreement(s) between State(s) and the International Atomic Energy Agency for the Application of Safeguards, INFCIRC/540 (Corrected)

8. EURATOM (2005) Commission Regulation (EURATOM) No. 302/2005, vol. L54. Official Journal of the European Union, Brussels
9. Avenhaus R, Krieger Th, Cojazzi GGM (eds) (2010) Unannouced Interim Inspections. JRC Scientific and Technical Reports JRC 58623, Publication Office of the European Union, EUR 24512 EN
10. Grümm H (1983) Safeguards Verification – Its Credibility and the Diversion Hypothesis. IAEA Bulletin 25(4):27–28
11. Avenhaus R, Krieger T (2020) Inspection Games over Time – Fundamental Models and Approaches. Forschungszentrum Jülich GmbH, Jülich, Germany
12. Avenhaus R, Krieger T (2013) Planning Interim Inspections: The Role of Information. In: Proc. 35th ESARDA Annual Meeting, Bruges, Belgium
13. Avenhaus R, Canty MJ (2005) Playing for time: A sequential inspection game. European Journal for Operations Research 167(2):475–492
14. Raiffa H (2002) Contributions of Applied Systems Analysis to International Negotiation. In: Kremenyuk VA (ed) International Negotiation: Analysis, Approaches, Issues, 2nd Edition. Jossey-Bass Publishers, San Francisco, pp 5–21

Societal Verification for Nuclear Nonproliferation and Arms Control

Zoe N. Gastelum

Abstract Societal verification—the use of data produced by the public to support confirmation that a state is in compliance with its nonproliferation or arms control obligations—is a concept as old as nonproliferation and arms control proposals themselves. With the tremendous growth in access to the Internet, and its accompanying public generation of and access to data, the concept of societal verification has undergone a recent resurgence in popularity. This chapter explores societal verification through two mechanisms of collecting and analyzing societally-produced data: mobilization and observation. It describes current applications and research in each area before providing an overview of challenges and considerations that must be addressed in order to bring societally-produced data into an official verification regime. The chapter concludes by emphasizing that the role of societal verification, if any, in nonproliferation and arms control will supplement rather than supplant traditional verification means.

1 Introduction

Societal verification, despite its recent surge in visibility, is not a new concept. As early as the 1950s, concepts for public involvement in treaty monitoring and verification were proposed in which members of the public would have the ability, if not the obligation, to report treaty violations to an international body. Across the numerous models of societal verification that have emerged since then, two societal verification concepts for nonproliferation and arms control treaties have remained relevant: first, that "society has a role in generating and analyzing information", and second, that "it can contribute to the verification of a specific set of commitments" [1]. For the purposes of this chapter, societal verification refers to a process in which information generated by the public is used to support confirmation that state

Z. N. Gastelum (✉)
Sandia National Laboratories, Albuquerque, NM, USA
e-mail: zgastel@sandia.gov

I. Niemeyer et al. (eds.), *Nuclear Non-proliferation and Arms Control Verification*,
https://doi.org/10.1007/978-3-030-29537-0_12

parties are in compliance with their nonproliferation or arms control obligations, with emphasis placed on the role of societally-produced data to potentially enhance, rather than replace, traditional verification mechanisms.

The Nuclear Threat Initiative, in their study "Redefining Societal Verification", aptly points out that societally-produced information might come from individuals or a select group of experts, and the use of that information could support verification efforts by states or international organizations [1]. This chapter also accounts for the use of societally-produced information by unofficial monitoring bodies such as non-governmental organizations or special interest groups. It is likewise significant to mention the various roles or purposes that societal verification may hold within a broader treaty monitoring regime, including building confidence among treaty parties, providing transparency to the public regarding treaty compliance, or to queue additional investigations by treaty monitoring parties.

The most recent interest in societal verification has been spurred, in part, by the terrific growth of publicly available data, pervasive sensing from social media and personal wearables such as fitness trackers and body cameras, computing capabilities, and the interconnected nature of the Internet as part of the "information age." Recent studies from the Pew Research Center have shown that, despite disparities in Internet access between developing and advanced economies, developing nations appear to be closing the divide. Between 2013 and 2015, developing countries increased their levels of adult access to the Internet from 45 to 54% of adults, compared to a stagnant 84% of adult users in the United States during the same time period [2, 3]. Both developing and advanced economies saw significant use of social media, with a global median of 76% of Internet users reporting use of social media platforms in 2016 [3]. While social media platform engagement can differ significantly by age, gender, socioeconomic status, and ethnicity, a Pew Research Center study from early 2018 showed stability or growth among American adults for platforms such as Facebook (used by 68% of American adults), Instagram (35%), Pinterest (29%) and Twitter (24%), with the most significant use across all the surveyed social media platforms from 18- to 24-year-olds [4]. With more users across a wider geographic area accessing the Internet and engaging in social media, the allure of using the resulting body of publicly available, societally-produced data to support verification is immense.

This chapter begins by defining societal verification through an examination of two distinct methods of information collection—societal mobilization, and societal observation. Each collection mechanism is explained outside of the realm of nonproliferation and arms control, to illustrate the general potential of societally-produced information. Then, societal mobilization and societal observation are explored specifically related to nonproliferation and arms control verification, including an overview of recent research in this area that relies on broad access to the Internet and social networking platforms. Most research in the field of societal verification for nuclear nonproliferation and arms control includes acknowledgement of the significant technical and policy challenges to implementing societal verification as part of nonproliferation or arms control monitoring regime, and the next section attempts to summarize some of those challenges. The final section of this chapter concludes that while societal verification may support broader nonproliferation and arms control verification efforts, the potential benefits of societally-produced data must be weighed against its numerous challenges.

2 Societal Mobilization

Societal mobilization refers to an appeal that is broadcast to the public (which may include a group of individuals or a sub-set of experts) requesting information, analysis, opinion or a technical solution. In most cases, societal mobilization does not refer to an entire analysis or activity being outsourced to the public (crowd-sourced). Rather, required information or analysis is decomposed into smaller pieces that non-experts can support with minimal time, training, or other resources, and then is recompiled for an expert analyst to evaluate. When societal mobilization is used to request information or analysis from a set of experts rather than the public-at-large, even that request represents a piece of a larger research question or goal. Technology competitions are the exception, in which complete algorithms, equipment, or other solution are requested in full and may be evaluated against a set of objective criteria or performance tests.

One example of societal mobilization in which micro-problems are posed to the public is Amazon's Mechanical Turk, in which users complete small Human Intelligence Tasks that include questions such as "is this website suitable for a general audience?" or "are these two products the same?" [5]. Responses are collected to provide insight into a broader question or problem set posed by a researcher or corporation. In another domain, Foldit uses humans' unique pattern recognition and puzzle-solving skills to identify likely protein folding patterns, providing insight to scientists for how to treat certain diseases [6].

Another example of societal mobilization is the United Nations Institute for Training and Research's Operational Satellite Applications Programme (UNITAR-UNOSAT), which hosts a research project called Geo Tag-X that focuses on photograph analysis to support disaster relief efforts. Geo Tag-X asks users to answer a series of questions about an image, such as "is the shelter raised off the ground?" and "does the shelter have a second cover?" to prioritize images for specialist analysis and deployment of resources [7]. Similarly, Tomnod asks users to assess objects of interest within satellite imagery, to support disaster relief, population mapping, and environmental awareness among other topics [8].

Some cases of societal mobilization involve expert groups self-mobilizing on a set of problems they want to solve. For example, Stack Overflow [9] is a question and answer website for computer programmers in which users can ask or respond to programming questions, and rank the usefulness of the responses. The social curating mechanism of ranking the most relevant responses results in a better experience for future users, who are guided to the top-ranked responses to their questions.

Self-organizing online groups have also mobilized within the true crime domain, offering new perspectives, ideas, or even evidence for historical criminal cases that were otherwise stagnant. Some prominent examples have resulted from the podcasts "Serial" and "Up and Vanished" which spurred the public into revisiting leads, tips, evidence collection, or legal processes without explicit requests to do so [10, 11]. The seeming success of those self-organizing crime solving communities has led to active requests for analysis of evidence from online communities. A current example at the time of this writing is the Bellingcat Investigation Team and Forensic Architecture project, which is requesting citizen analysis of evidence surrounding the death of seven Venezuelan anti-government activists who were killed in what the Venezuelan government described as a shoot-out but others have questioned as a possible retribution execution [12].

So-called "citizen science" activities also contribute to societal mobilization, in which the public may be provided with sampling equipment, monitoring tools, or information on how to use homemade or personally purchased equipment to collect data for public distribution and subsequent analysis by a group of experts. After the disaster at the Fukushima Daiichi nuclear power plant, for example, the group Safecast was formed to monitor and share radiation measurements world-wide over the Internet [13] and in 2017 expanded their scope to air quality monitoring [14]. In another example, a project in Amsterdam distributed low-cost sensors to measure carbon monoxide, nitrous oxide, temperature, humidity, light intensity, and noise levels around the city [15]. Similar citizen engagement has been used to monitor elections, such as in the Sudan Vote Monitor project that used a text messaging platform to support observation efforts for Sudan's 2011 referendum on the independence of South Sudan [16].

Examples of societal mobilization in search of a complete technical solution include the 2006–2009 Netflix Prize seeking to improve the DVD and streaming service's recommendation engine, the similarly named Zillow Prize seeking to improve the accuracy of the Zillow real estate site's housing value estimates, and the UAV (unmanned aerial vehicle) Challenge Medical Express to develop a UAV to retrieve medical samples from remote locations [17].

3 Societal Observation

In contrast to societal mobilization, in which the public is called into action for a specific problem-solving or data-collection task, societal observation is the collection and analysis of publicly available information that is not dependent on direct societal engagement. Rather, societal observation collects data produced by the public that would be available regardless of the intended end-use, and exploits it for internal expert analysis. In the current environment of social connectedness via the Internet, societal observation increasingly includes the analysis of publicly available social media data consisting of text, images, videos or audio published to social platforms with the explicit intent to share that data with the public.

One example of societal observation is mining popular search terms to support event detection. Google Flu Trends was a web service operated by the Google search engine that attempted to predict influenza outbreaks based on search terms used on its site. Ultimately Google Flu Trends was not reliable for predicting current influenza cases across multiple years, and the service was discontinued after it was discovered that the relevant terms for predicting influenza were more closely related to winter than illness symptoms, making it less applicable for off-season influenza cases [18, 19]. Research continues in other areas of applying search term frequency as a means of societal observation.

Publicly available photographs offer another potential data source for societal observation. Image recognition software is becoming increasingly advanced, allowing analysts to better understand a large body of photographs without manual observation. For example, Ditto Labs is a start-up company that offers image recognition software for companies seeking to better understand the presence of their brand on social media platforms. The software uses a logo recognition algorithm to identify clothing, food, gadgets, or other brands in photos posted to social media. The software analyzes the sentiment of users based on facial expression, detects co-occurrence of logos with other products, and looks at background scenes in attempt to analyze how and where people use certain products [20]. In February 2016, Google released a Beta version of its Cloud Vision API open for users to analyze either their own or publicly available images. The machine learning algorithms used in Cloud Vision allow users to identify and search for objects and logos, assess sentiment based on facial expressions, extract text in multiple languages, and detect potentially offensive content [21].

Societal observation can also benefit from analysis of publicly available video content. In February 2013, for example, Russian drivers captured astonishing video of the meteor that impacted Earth near Chelyabinsk, Russia on the dashboard cameras that are used widely in that country to document traffic accidents [22]. Scientists used the videos, in combination with security camera recordings and more traditional asteroid observation sensors, to estimate the trajectory, size, and energy of the event [23, 24].

4 Mobilization for Nuclear Nonproliferation and Arms Control Verification

Societal mobilization may be implemented in a number of ways to support nuclear nonproliferation and arms control verification. Recall that social mobilization requires the public to be mobilized to answer specific questions, collect a certain type of data using sensors or monitoring equipment, or develop complete technical solutions such as algorithms or equipment. For nonproliferation and arms control verification, these questions might include: What is the function or purpose of a given piece of technology? Where is a specific building located? or, How is the building being used? Societal mobilization might also include the deployment of sensors, equipment, or other data collection mechanisms that allow the public to collect information that can be used by expert analysts to support nonproliferation or arms control verification. As with societal mobilization for non-nuclear activities, mobilization for verification generally focuses on small pieces of a larger analytical effort that are difficult to answer within the mobilizing body's own purview. While no official mobilization activities have been initiated by a government or verification body for nuclear nonproliferation and arms control, there are several examples of closely related activities that illustrate how such societal mobilization could take place to support nonproliferation and arms control verification activities.

The United States government has experimented with several societal verification "challenges" that, while not explicitly for nonproliferation or arms control verification, can be seen as proxies to develop a stronger understanding of how societal mobilization challenges might be used for these areas. In 2009, the Defense Advanced Research Projects Agency (DARPA) conducted the Network Challenge (commonly referred to as the Red Balloon Challenge), in which participants were asked to locate ten 8-ft weather balloons moored in various locations around the United States. A team from the Massachusetts Institute of Technology (MIT) was the first to correctly locate and photograph all ten balloons, using an incentive structure over social media to entice a geographically distributed group of participants to not only locate the balloons, but also recruit their friends to search [25].

In 2012, the U.S. State Department conducted the TAG Challenge to expand the scope tested by the Network Challenge in two ways: first, the "targets" (five individuals wearing challenge-branded t-shirts) were internationally located across the United States and Europe; second, the "targets" were mobile, walking around their respective cities [26, 27]. The winning team in the TAG Challenge (consisting of some members of the MIT Network Challenge winning team) did not ultimately locate all five "targets", thus illustrating important considerations for societal mobilization, including the use of social media platforms, language, and time barriers.

In the field of image analysis for nuclear verification efforts, in 2016 researchers from the James Martin Center for Nonproliferation Studies (CNS) launched the Project on Crowdsourced Imagery Analysis, or Geo4Nonpro [32]. The project aims to create an open repository of satellite images and expert analysis of known or suspected proliferation-related sites to support verification activities, identify best practices for crowdsourcing in the field of satellite imagery analysis, and study potential paradigms for incorporating public verification into arms control regimes [33].

Experts at CNS have also been involved in some of the most prominent examples of expert self-mobilization for nonproliferation analysis, in which they regularly assess potential proliferation activities at sites around the world and have undertaken multiple analyses in which they utilized social media platforms to reach out to technology, regional, or policy experts to support their verification research. Many of their findings have been published in topical research papers on the CNS website [34] or on the Arms Control Wonk blog [35] founded by Jeffery Lewis,

Director of the East Asia Nonproliferation Program at CNS. Researchers at the Institute of Science and International Security are another group of self-mobilizing nonproliferation analysts who regularly publish their reports online [36].

As with the societal mobilization campaigns described in the previous section for environmental monitoring, societal mobilization for nonproliferation can also include the distribution of sensors to the public for expert collection and analysis of the resulting data. Claire Sullivan of the University of Illinois at Urbana-Champaign has proposed intercommunicating networks of sensors, such as cell phones, to collect radiation measurements [37]. While Dr. Sullivan's research focuses on radiation measurements for nuclear forensic activities, similar sensor networks could be deployed to willing participants to capture measurements or samples as they move throughout their livelihoods for nonproliferation verification purposes. Communicating sensor networks would allow expert analysts to cross-reference potentially suspicious readings or samples with others from the same area to determine if an anomalous measurement is precisely that, if there are location-specific contributors such as weather patterns or higher concentrations of naturally occurring radioactive materials, or if additional measurements or sampling activities should be undertaken at a specific location.

A recent Underhanded C Contest provides an example of societal mobilization towards a technical solution for nuclear nonproliferation. The annual Underhanded C Contest challenges participants to write a program in the C computer programming language that appears to be benign and straight-forward, but that contains malicious code. In the 2015 contest, Underhanded C partnered with the Nuclear Threat Initiative to develop a nuclear disarmament scenario, in which two state parties have agreed to an inspection regime in which measurements of objects representing nuclear weapons would be compared to a reference measurement of a trusted object [28]. An information barrier would be implemented to distill the results from a potentially sensitive measurement to a simple "yes" if the measured object was sufficiently similar to be deemed a nuclear weapon; otherwise "no". Contestants were asked to design a code that would make it appear that the detectors and were performing as expected during testing, but would then proceed to confirm as "yes" objects that were not sufficiently similar, so that the inspected party would could get credit for dismantling objects that were not actually nuclear weapons (thus allowing them to maintain larger nuclear weapons reserves than the other treaty party). The winner of the contest developed a program that would change the number of bytes being analyzed in a verification scenario, allowing an object with a very small amount of fissile material to trigger a false positive result [29].

The International Atomic Energy Agency (IAEA) has also joined the movement to solicit technical solutions via crowdsourcing, in some cases seeking solutions that could be directly applied to safeguards verification activities. In 2016, the IAEA launched its Technology Challenge for Digital Image processing for the Improved Cerenkov Viewing Device (ICVD), which sought software to "process the videos captured through the eyepiece of an ICVD, and turn them into a single composite image with a better resolving power [30]. That challenge was followed in 2017 by the IAEA Robotics Challenge to develop unmanned ground and floating vehicles for repetitive verification activities such as positioning an ICVD in a spent fuel pond, counting items, recording identification tags. The top three unmanned floating vehicle teams from the Robotics Challenge were invited to participate in proof-of-concept deployments, working with nuclear facility operators and IAEA safeguards inspectors [31].

5 Observation for Nuclear Nonproliferation and Arms Control Verification

Societal observation for nuclear nonproliferation and arms control refers to a review of societally-produced data that would exist regardless of the research objective. Recall that the collection and analysis of publicly-produced and available data for societal observation in nuclear nonproliferation and arms control verification activities does not involve any public requests for information, external expert groups collaborating on an analysis mission, or dissemination of relevant sensor or measurement equipment by the entities interested in verifying nonproliferation and arms control.

As one of the foremost international verification bodies for nuclear nonproliferation, and with an active contingent of open source analysts in the Department of Safeguards, the IAEA has explored how to integrate societal observation as an additional form of open source information that it can use to support its safeguards verification activities. In 2014, IAEA safeguards analysts Thomas Lorenz and Yana Feldman published a review of social media for IAEA safeguards, including several examples in which information found on social media was relevant to the assessment of a state's safeguards declaration [38]. In one example, the authors describe a case in which a government ministry distributed press releases about energy projects that could be relevant for assessing the state's Additional Protocol declarations via its social media profile, while the ministry's website was not updated. In another example, Lorenz and Feldman found a social media profile of a newly established medical center that did not yet have a website, which included photographs of equipment that could aid in design information verification and in-field activities planning. The analysts also found a social media profile for an environmental group that opposed uranium mining in their neighboring state, which provided up-to-date information on the neighbor's mining plans and progress.

In 2016, IAEA safeguards analysts Marcy Fowler, Jennah Khaled, and Thomas Sköld (née Lorenz) presented follow-on analysis in which they describe how the Agency can use data from social media platforms for real-time monitoring of safeguards-relevant events, to conduct topic and event analysis to provide context for conferences, negotiations, or other relevant activities, and to understand networks related to nuclear activities and materials [39]. The authors note several challenges specific to using social media data including the short-lived, high-volume nature of the data and the continuously evolving platforms and information types that are available. Lorenz and Feldman suggest that while not appropriate for all safeguards mission areas, societally produced and disseminated data available from social media platforms may, once verified, be integrated with state declarations and results of in-field activities for a broader analysis of all safeguards-relevant information [38].

In a different approach to the analysis of social media data, Lee and Zolotov [40] presented in their 2013 paper *New Media for Nonproliferation and Arms Control* a case study in which researchers at the CNS examined the reach of their own social media presence related to the 2013 Preparatory Committee (PrepCom) meeting in advance of the 2015 Treaty on the Non-Proliferation of Nuclear Weapons Review Committee meeting. In the study, the research team examined their re-tweets, mentions, topic frequency, and also a social network of their Twitter followers. The examination of the 2013 PrepCom social media presence is significant because of the event's high visibility within the nonproliferation community. The team's targeted analysis could indicate potential for monitoring other nonproliferation or arms-control relevant meetings such as treaty negotiations for a follow-on to the New START arms control agreement, IAEA Board of Governors meetings, or topical conferences and workshops.

New media analytics has also been proposed to look for nonproliferation or arms control events-of-interest using targeted search terms. For example, in 2013 a research team from Pacific Northwest National Laboratory (PNNL) reported on the development of proxy search strings that indicated initial promise for monitoring the lead-up and response to largely visible events. In one case, the research team searched for meteor events as a proxy for an atmospheric nuclear test, using the search string "loud boom." While the search string did result in a spike in Twitter data for days associated with meteor events, it also resulted in spikes associated with thunder storms, fireworks on Independence Day, and a loud noise during a televised Presidential debate, "which apparently caused President Barack Obama to turn and look behind him during the discourse" [41, 42]. The following year PNNL researchers analyzed social media indicators of sudden increases in population and industrial activities such as those associated with the shale oil extraction boom in North Dakota as a proxy for the development of a clandestine nuclear program in remote areas [45]. While neither research topic was intended to be definitive, they could indicate a potential for using societal observation for monitoring largely visible or noticeable events related to nonproliferation and arms control.

6 Challenges and Considerations

Numerous studies mentioned in this chapter have previously described the technical, analytical, and policy challenges related to societal verification for nonproliferation and arms control verification. More detailed discussions of societal verification challenges can be found in publications from the Nuclear Threat Initiative [1], Lee and Zolotov [40], Gastelum el al. [42], and Henry et al. [46]. Further consideration regarding the increasing volume of multimedia data among societally produced information is described in Barletta et al. [43] and Dupuy et al. [44].

One of the most important challenges underlying the use of societal verification is the definition of the verification question. The verification question influences how societally produced information is collected (through mobilization or observation), the sensors that might be deployed, the platform on which to collect and analyze the data, and the types of validation and analysis that might be done to the resulting data. Lee refers to this as the "intent of a project" that must be defined early in the project planning process [47]. Nonproliferation and arms control questions supported via societal verification should be:

- Specific in their aim, across technology and geography
- Observable by either humans or their deployed sensors and equipment
- Verifiable to some degree, and
- Have the potential for relevant follow-up traditional verification activities.

The verification question to be supported by societally-produced data would also benefit from a definition of how the data will be incorporated into the nonproliferation or arms control monitoring regime that it intends to address. Current nonproliferation and arms control treaties do not include explicit guidance on how to use societally-produced information for verification, or even if such data may be included in official verification activities as part of the broader National Technical Means that some treaties allow.

A challenge that arises anytime a new data stream is introduced for nonproliferation and arms control is information validation. This has been emphasized for the use of open source data, and especially social media data. Data resulting from societal verification campaigns must be recognized for what it is: information coming from people or other sensors outside of the control of the verification body who may "spoof" or alter data, misreport, mislead, or misuse the equipment in an accidental or malicious manner that causes false output. Of the numerous examples of the intentional or unintentional spread of misinformation, some exemplar cases include false entries to the Wikipedia online collaborative encyclopedia [48], a violent attack at a Washington pizzeria spurred by a conspiracy theory alleging ties to a child abuse ring [49], and the misidentification of participants in a white supremacist rally resulting in the harassment of innocent look-alikes [50].

This type of wide distribution of misinformation with limited correction is the focus of a study by Kate Starbird and her team at the University of Washington and Northwest University, who found that misinformation is propagated through Twitter at a significantly higher rate than corrections to that misinformation. Starbird et al, analyzed a set of false rumors spread on Twitter following the April 15, 2013 Boston Marathon bombings. Using data from the Twitter Streaming API from between about 3 h after the bombing to several days after the arrest of the second suspect in the case, the researchers collected 92,785 tweets related to a false rumor that an 8 year-old girl was killed while running the race (runners must be at least 18). Of those 92,785 tweets, 90,668 were identified as misinformation and only 2046 were corrections of the rumor. Though the researchers showed that the frequency of both the misinformation and correction tweets occurred within close time proximity of each other, the misinformation continued to propagate at low levels even after the corrections had subsided [51]. Data from societal verification, therefore, should be subject to the same scrutiny of other open source data in which there may be errors or misinformation, either intentionally or not.

Societal mobilization applications provide unique challenges and opportunities for data validation, and the validation of crowdsourced data is an active area of research. Some of the methods currently being studied in multiple domains include the use of "gold standard" data that has been labeled by an expert, majority voting schemes in which the input from multiple users is compared and the most common result is taken as "true," the use of a control group to judge the suitability of crowdsourced responses (essentially crowdsourcing the question 'which answer is most appropriate?'), and assessments of user-specific factors such as time to complete a task, answers across multiple tasks, etc. [52, 53].

Though societal verification does not explicitly require the use of social media platforms, many modern societal mobilization and observation efforts for nonproliferation and arms control verification include social media as either the mechanism for users to engage in a verification mission, or the platform by which to collect and share data from such an effort. Yet, with so many social media platforms available, it can be difficult to determine which to choose to launch a societal verification effort from, or from which to collect data. The choice of social media platform will have significant impact on the data type (text, photograph, video, radiation sensor, etc.), the languages present in text data, and the total volume of data available. Cultural differences among users will also determine how and what social media users may report on. Data fusion across multiple social media platforms, incorporating multimedia content, analyzing content in multiple languages, and storing, processing, and analyzing massive data sets all pose unique technical challenges for using social media or other publicly available data sources [42].

It is also noteworthy that the use of social media for societal verification imposes a selection bias, as the choice of social media platforms involved in a societal verification activity will directly influence the countries represented in the user population, the information and communication technology used to access the platform (for example, through the Internet or via text messaging), and government oversight of the platform. Thus, it must be recognized that social media platforms cannot be assumed to be representative of larger populations; rather they represent the online behavior of their sub-set of users. For any social media platform, differences in demographics, location, access to resources, and other factors can lead to significant biases of the information available from that source.

Frank Pabian and a team at the European Commission's Joint Research Centre discuss the "serendipity factor" of finding previously unknown or undeclared nuclear activities within societally-produced data. Especially for societal observation, Pabian et al. note that "unknowns might just as well be stumbled upon as found through the application of any ... search and discovery methodology" [46]. Thus, even with clear and focused research questions, leveraging appropriate data collection methods and platforms, using the right analytical tools and techniques, and having the required analyst expertise, it may be pure accident that a verification-relevant discovery is made [45, 46].

Legal, ethical, and moral concerns are brought up repeatedly in conversations about societal verification. It is a consensus of the nonproliferation and arms control authors writing on this topic that any societal verification has to ensure that the participants are not subject to physical, emotional, economic or political danger based on their involvement. As such, many experts agree that successful mechanisms for societal mobilization would require buy-in or endorsement from the states in which verification activities would occur. For societal observation, risks are considered smaller since the data is not specifically elicited for verification purposes, but government concurrence is still important, especially in states with higher levels of government control of the press.

This sentiment among the nonproliferation analysis community to "do no harm" to the society that it mobilizes or observes was especially apparent in an October 2009 blog post by Lewis on Arms Control Wonk. After learning that one of his readers had visited and photographed an alleged nuclear reactor in Syria, Lewis "[discouraged] in the strongest possible terms, readers from doing anything illegal or that might otherwise endanger yourself, your host or people around you. Many governments have no sense of proportion when it comes to the line between what is innocent behavior in a free society—taking pictures of public events; using your intellect to draw conclusions—and espionage" [54].

7 Conclusions

Societal verification has unique strengths and benefits. It can be used to bring together large, diverse groups of multidisciplinary experts to solve a problem. It can mobilize large populations into data collection machines which would have previously been too costly or time consuming to collect. And it has the potential to build transparency and confidence among treaty parties, expert groups, and the public at large regarding implementation of nonproliferation and arms control agreements. The preceding sections demonstrate a strong potential for societal verification for nuclear nonproliferation and arms control, with successful case studies spanning non-nuclear, nuclear-proxy, and even directly relevant nonproliferation and arms control applications. At the same time, it is necessary to consider how, and to what end, to incorporate societal verification techniques into broader nonproliferation and arms control verification efforts. In that vein, societal verification is not a replacement for traditional verification activities. Information from societal verification requires the same, if not more, verification and validation rigor applied to other open source data. As the U.S. Under Secretary for Arms Control and International Security Rose Gottemoeller explained, societal verification offers a unique opportunity to supplement, rather than to supplant, traditional verification mechanisms, such as the queueing of "sensors and satellites to make better use of our scarce and expensive National Technical Means" [55].

Sandia National Laboratories is a multimission laboratory managed and operated by National Technology and Engineering Solutions of Sandia, LLC, a wholly owned subsidiary of Honeywell International, Inc., for the U.S. Department of Energy National Nuclear Security Administration under contract DE-NA0003525.

References

1. Redefining Societal Verification (2014) In: Nuclear Threat Initiative, July 2014. Part of the Cultivating Confidence Verification Series
2. Perrin A, Duggan M (2015) American's Internet Access: 2000–2015. Pew Research Center. http://www.pewinternet.org. Accessed 28 Feb 2019
3. Poushter J (2016) Smartphone Ownership and Internet Usage Continues to Climb in Emerging Economies. Pew Research Center. http://assets.pewresearch.org/wp-content/uploads/sites/2/2016/02/pew_research_center_global_technology_report_final_february_22__2016.pdf. Accessed 28 Feb 2019
4. Smith A, Anderson M (2018) Social Media Use in 2018. March 1, 2018 Pew Research Center. http://www.pewinternet.org/2018/03/01/social-media-use-in-2018/ Accessed 28 Feb 2019
5. Amazon Mechanical Turk: Working on HITs. https://www.mturk.com. Accessed 28 Feb 2019
6. The Science Behind Foldit. https://fold.it. Accessed 28 Feb 2019
7. Geo Tag-X: About. http://geotagx.org. Accessed 28 Feb 2019
8. Tomnod. http://www.tomnod.com. Accessed 28 Feb 2019
9. Stack Overflow. http://stackoverflow.com. Accessed 28 Feb 2019
10. Fabio M, (2018) In a Win for True Crime Podcasts Everywhere, Adnan Syed Granted New Trial. Forbes, March 30, 2018. https://www.forbes.com/sites/michellefabio/2018/03/30/in-win-for-true-crime-podcasts-everywhere-adnan-syed-granted-new-trial/. Accessed 28 Feb 2019
11. Locker M (2017) How 'Up and Vanished' Podcast Helped Solve Cold Murder Case. Rolling Stone, March 17, 2017. https://www.rollingstone.com/culture/how-up-and-vanished-podcast-helped-solve-cold-murder-case-w472620. Accessed 28 Feb 2019
12. Fiorella G, Leroy A (2018) Was Óscar Pérez Murdered? You Could Help Us Find Out. The New York Times, May 13, 2018. https://www.nytimes.com/2018/05/13/opinion/oscar-perez-venezuela-forensic-architecture.html. Accessed 28 Feb 2019
13. Besner A (2016) How Citizen Science Changed the Way Fukushima Radiation is Reported. http://voices.nationalgeographic.com. Accessed 14 April 2018
14. Introducing the Solarcast Nano (2017) Safecast, December 20, 2017. https://blog.safecast.org/2017/12/introducing-the-solarcast-nano/. Accessed 28 Feb 2019
15. Waag Society, Smart Citizen Kit. http://waag.org. Accessed February 19, 2016
16. Cherry S, Zein F (2015) Cellphones Track Voting in Southern Sudan: Citizens report from remote places where international monitoring doesn't reach. IEEE Spectrum. Available at http://spectrum.ieee.org. Accessed 28 Feb 2019
17. https://www.netflixprize.com/index.html; https://www.zillow.com/promo/zillow-prize/; https://uavchallenge.org. Accessed 28 Feb 2019
18. What we can learn from the epic failure of Google Flu Trends. http://www.wired.com. Accessed 28 Feb 2019
19. Lazer D, Kennedy R, King G, Vespignani A (2014) The Parable of Google Flu: Traps in Big Data Analysis. In: Science 343 (6176) (March 14): 1203–1205. Available at: https://dash.harvard.edu. Accessed 28 Feb 2019
20. Ditto Labs, Inc. http://blog.dittolabs.io. Accessed 28 Feb 2019
21. Google Cloud Vision. https://cloud.google.com/vision. Accessed 28 Feb 2019

22. Lavrinc D (2013) Why Almost Everyone in Russia has a Dash Cam. In: Wired. http://www.wired.com. Accessed 28 Feb 2019
23. Borovička J, Spurný P, Brown P, Wiegert P, Kalenda P, Clark D, Shrbený L (2013) The trajectory, structure and origin of the Chelyabinsk asteroidal impactor. http://www.nature.com. Accessed 28 Feb 2019
24. Popova O, Jenniskens P, Emel'yanenko V et al (2013) Chelyabinsk Airburst, Damage, Assessment, Meteorite Recovery, and Characterization. Science 342 (6162):1069–1073. doi: 10.1126/science.1242642
25. Red Balloon Challenge. http://archive.darpa.mil/networkchallenge/. Accessed 28 Feb 2019
26. Shachtman N (2012) U.S. Wants You to Hunt Fugitives with Twitter. Wired, March 1, 2012. https://www.wired.com/2012/03/u-s-wants-you-to-hunt-fugitives-with-twitter/. Accessed 28 Feb 2019
27. Firth N (2012) Social media web snares 'criminals.' New Scientist, April 4, 2012. https://www.newscientist.com/article/dn21666-social-media-web-snares-criminals/. Accessed 28 Feb 2019
28. Challenge: Faking Fissile Material. In: The Underhanded C Contest. The Register. http://underhanded-c.org. Accessed 28 Feb 2019
29. Williams C (2016) Winning Underhanded C Contest code silently tricks nuke inspectors. In: The Register. http://www.theregister.co.uk. Accessed 28 Feb 2019
30. Expression of Interest (EOI 27544-AI) to participate in IAEA Technology Challenge: Digital image processing for the Improved Cerenkov Viewing Device (ICVD). https://www.ungm.org/UNUser/Documents/DownloadPublicDocument?docId=451451. Accessed 28 Feb 2019
31. IAEA Robotics Challenge. https://challenge.iaea.org/challenges/2017-SG-Robotics. Accessed 28 Feb 2019
32. James Martin Center for Nonproliferation Studies (2016) Project on Crowdsourced Imagery Analysis. http://www.geo4nonpro.org. Accessed 28 Feb 2019
33. Dill C (2016) Development and Deployment of Geo4Nonpro. In: Social Media for Nuclear Nonproliferation Workshop. Santa Fe, NM, September 25, 2016
34. The James Martin Center for Nonproliferation Studies, Research. http://www.nonproliferation.org/research. Accessed 28 Feb 2019
35. Arms Control Wonk http://www.armscontrolwonk.com. Accessed 28 Feb 2019
36. Institute of Science and International Security. http://isis-online.org. Accessed 28 Feb 2019
37. Sullivan C. (2014) Nuclear Forensics Driven by Geographic Information Systems and Big Data Analytics. In: Information Analysis Technologies, Techniques and Methods for Safeguards, Nonproliferation and Arms Control Verification Workshop. Workshop Proceeding, Portland, OR, May 12–14, 2014
38. Lorenz T, Feldman Y (2014) The Efficacy of Social Media as a Research Tool and Information Source for Safeguards Verification. In: Information Analysis Technologies, Techniques and Methods for Safeguards, Nonproliferation and Arms Control Verification Workshop. Workshop Proceedings, Portland, OR, May 12–14, 2014
39. Fowler M, Khaled J, Sköld T (2016) Using Social Media Information in Analysis for IAEA Safeguards. In: Social Media for Nuclear Nonproliferation Workshop. Santa Fe, NM, September 25, 2016
40. Lee B, Zolotov M (2013) New Media Solutions in Nonproliferation and Arms Control: Opportunities and Challenges
41. Gastelum Z, Cramer N, Benz J, Kreyling S et al (2013) Identifying, Visualizing, and Fusing Social Media Data to Support Nonproliferation and Arms Control Treaty Verification: Preliminary Results.v Proceedings of the Institute of Nuclear Materials Management Annual Meeting, July 14–18, 2013, Palm Desert, CA
42. Henry M, Cramer N, Benz J, Gastelum Z, Kreyling S, West C (2014) Examining the Role and Research Challenges of Social Media as a Tool for Nonproliferation and Arms Control Treaty Verification. In: Information Analysis Technologies, Techniques and Methods for Safeguards, Nonproliferation and Arms Control Verification Workshop. Workshop Proceedings, Portland, OR, May 12–14, 2014

43. Barletta M, Fowler M and Khaled J (2017) Integrating Multimedia Information in IAEA Safeguards. In Proceedings of the American Nuclear Society Advanced in Nonproliferation Technology and Policy Conference, September 2016
44. Dupuy G, Feldman Y, Reed J and GastelumZ (2017) Enhancing the Use of Multimedia Information for IAEA Safeguards Analysis. In Proceedings of the Institute of Nuclear Materials Management Annual Meeting, July 2017
45. Wurmser D (2016) State Department Projects and Interests on Societal Mobilization to Support Nonproliferation Verification. In: Social Media for Nuclear Nonproliferation Workshop, Santa Fe, NM, September 25, 2016
46. Pabian F V, Renda G, Jungwirth R, Kim L K, Wolfhart E, Cojazzi G G M (2014) Open Source Analysis in Support of Nonproliferation Monitoring and Verification Activities: Using the New Media to Derive Unknown New Information. International Atomic Energy Agency 2014 Symposium on International Safeguards: Linking Strategy, Implementation, and People, Vienna, Austria October 20–24, 2014. https://www.iaea.org/safeguards/symposium/2014/home/eproceedings/sg2014-papers/000312.pdf. Accessed 14 April 2018
47. Lee B (2014) Assessing the potential of societal verification by means of new media. CCC PASCC Reports
48. Pogatchnik S (2009) Student hoaxes world's media on Wikipedia. http://www.nbcnews.com. Accessed 28 Feb 2019
49. Lipton E (2016) Man Motivated by 'Pizzagate' Conspiracy Theory Arrested in Washington Gunfire. The New York Times, December 5, 2016 https://www.nytimes.com/2016/12/05/us/pizzagate-comet-ping-pong-edgar-maddison-welch.html. Accessed 28 Feb 2019
50. Victor D (2017) Amateur Sleuths Aim to Identify Charlottesville Marchers, but Sometimes Misfire. New York Times, August 14, 2017. Available at: https://www.nytimes.com/2017/08/14/us/charlottesville-doxxing.html. Accessed 28 Feb 2019
51. Starbird K, Maddock J, Orand M, Achterman P, Mason R (2014) Rumors, False Flags, and Digital Vigilantes: Misinformation on Twitter after the 2013 Boston Marathon Bombing. http://faculty.washington.edu/kstarbi/Starbird_iConference2014-final.pdf. Accessed 28 Feb 2019
52. Eickoff C, De Vries AP (2011) How Crowdsourcable is Your Task? In Proceedings of Crowdsourcing for Search and Data Mining (CSDM 2011) February 2011
53. Hirth M, Hoßfeld T, Tran-Gia, P (2013) Analyzing costs and accuracy of validation mechanisms for crowdsourcing platforms. Mathematical and Computer Modelling 57:2918–2932
54. Lewis J (2009) Tourist Trip to Halabiye. http://www.armscontrolwonk.com. Accessed 28 Feb 2019
55. Gottemoeller R (2012) Arms Control in the Information Age. https://geneva.usmission.gov. Accessed 28 Feb 2019

Part IV
Technologies

Verification regimes will be driven by the need to adapt to a dynamic global international security and stability environment. Overcoming the competing requirements of greater transparency and protection of sensitive information is made possible by innovative advances in technology and capabilities. Forecasting and identifying new trends for potential technology applications and making the best use of existing (and improving) technologies is essential for choosing the most effective means of verification. For both treaty verification and state-level confidence, a few of the capabilities to detect and interpret nonproliferation and arms control indicators and signatures are reviewed here.

Futures Research as an Opportunity for Innovation in Verification Technologies

Joachim Schulze, Matthias Grüne, Marcus John, Ulrik Neupert, and Dirk Thorleuchter

Abstract Over the last 40 years, the verification of compliance with disarmament treaties for Weapons of Mass Destruction has steadily evolved using several technological solutions. Today's technological progress promises a variety of possible future and emerging technologies that may contribute to even better verification techniques. However, the latter might originate from technology areas far away from those usually monitored by the verification community. Scientifically-based futures research offers several validated approaches and methodologies that can provide orientation in this complex technology landscape and help to carve out the most probable technological future developments. We suggest to include a futures-research approach in the verification debate, complementing the expert community's findings and improving the awareness of technology-based opportunities and threats. Four different methodological approaches are described, which require specialized futurists to be exercised. The classical approach to technology foresight systematically studies appropriate scientific literature, looking for "known" and "unknown unknowns" and considering the interdependencies of the "technology complex". Bibliometrics can, although inherently retrospective, be used to characterise and forecast research topics and scientific networks. Modern text-mining tools can be used to extract unexpected information from the internet ("web mining"), potentially uncovering "unknown unknowns". Solution assessment by serious gaming can help to structure multi-perspective discussions.

J. Schulze (✉)
Consultant, Bad Neuenahr-Ahrweiler, Germany
e-mail: Joachim-Schulze@t-online.de

M. Grüne · M. John · U. Neupert · D. Thorleuchter
Fraunhofer Institute for Technological Trend Analysis INT, Euskirchen, Germany
e-mail: matthias.gruene@int.fraunhofer.de; marcus.john@int.fraunhofer.de; ulrik.neupert@int.fraunhofer.de; dirk.thorleuchter@int.fraunhofer.de

I. Niemeyer et al. (eds.), *Nuclear Non-proliferation and Arms Control Verification*, https://doi.org/10.1007/978-3-030-29537-0_13

1 Introduction: Technology Foresight of Verification Techniques

A wide array of nonproliferation and arms control verification systems exist. The Treaty on the Non-Proliferation of Nuclear Weapons entered into force in 1970. It is the first treaty on non-proliferation and disarmament of weapons of mass destruction. Verification of compliance has been implemented through the Safeguards Agreements to verify declarations of nuclear materials. The Comprehensive Nuclear-Test-Ban Treaty was opened for signature in 1996. Its International Monitoring System collects data from seismic, infrasound, hydroacoustic and radionuclide stations worldwide to verify compliance. Since 1993, the Chemical Weapons Convention has a Preparatory Commission and a Secretariat that execute on-site inspections. The verification systems were established years ago and, certainly, need upgrades. It is essential to consider new technologies in order to guarantee efficient verification of non-proliferation and disarmament treaty compliance for the future. In the case of the Biological and Toxin Weapons Convention (BWC), verification was discussed from 1992 until 1993. Results have been documented in the VEREX Report [1]. The inherent difficulties in implementing a verification regime for BWC finally resulted in rejection of any technical verification mechanisms, in 2001.

These arms control treaties and agreements rely on technological solutions for verification [2]. In the past decades, the already fast pace of technological progress has increasingly accelerated. Novel technology areas have arisen on the fringes of classical technology areas or by a convergence of technologies due to common paradigms and tools. A good example of this is nanotechnology. Cross-fertilization of technologies plays an important role [3]. A breakthrough in one technology can result in the feasibility of another technology, such as the advent of modern batteries enabling electric aircraft. From time to time, disruptive technologies can even act as game changers in one or many domains of technological application [4]. Certainly, the most pervasive example for this is the internet.

The first modern (i.e. scientifically-based) approaches to handle such rapid technological changes began shortly after the Second World War, building upon ideas from the first half of the century. In the United States, methods to systemize experts' opinions, influence factors and alternative options concerning the future were developed in the military sector [5]. This approach became known as *technology forecasting* [6], *technology foresight* [7, 8], *futures studies* or *futures research* [9]. The latter two terms emphasize the idea that the future is not determined but offers a wide area of plausible variants of development (it is described in French as "*les futuribles*" [10]). To cope with this uncertainty, alternative possible and plausible futures must be considered when preparing for future-oriented decisions. During the last quarter of the twentieth century, the United Kingdom, Germany, France, the Netherlands, and the Nordic countries established institutions and processes, in the military and civilian sectors, to prepare decision makers and planners in society, politics, and economy for coming technology-driven and technology-accompanied changes [6, 11].

A wide variety of methodological approaches and techniques has been developed in order to fulfill this task. There exists profound knowledge about innovation systems, repercussions, influence factors, etc., and information about the pace of development in diverse scientific fields has accumulated [12]. In the new millennium, the vast accessibility of a massive amount of data and information (especially scientific literature) via the internet has triggered the development of novel technology foresight techniques that complement the established ones (see e.g. Sects. 3 and 4). At the same time, the discussion within the futures-research community has deepened the understanding of the epistemological foundations of scientific futures research [13].

Improvements in the classical approaches to verification technologies are continuously debated, developed and monitored by an avid community of experts. However, scientific researchers are not familiar with nor necessarily equipped to forecasting their success on a timeline. We suggest that by including a futures-research approach (that is already state of the art in defence technology planning of the larger countries) the awareness of both opportunities and threats in the community can be improved.

To this end, we propose four different methodological approaches. The first one is the classical approach to technology foresight by studying the content of appropriate scientific literature (see Sect. 2). This can be used to obtain a professional forecast of probable futures of the technologies already applied and of the research directions already followed. Moreover, the interwovenness of invention and innovation processes can lead to novel aspects of technology development and usage. This can be accounted for by considering the enabling, complementary and cross-fertilising relationships between technological fields. Sometimes, even historically known approaches can become feasible only after future progress e.g. in enabling technologies like materials or information and communications technology (ICT) (see Sect. 2).

The second and third methods make use of the huge amount of information available via the internet and associated databases. Bibliometrics is the statistical analysis of publication-data subsets. This inherently retrospective method can, by pattern analysis, be used to characterise and forecast research topics and scientific networks. Thus, the classical approach can be complemented and augmented (see Sect. 3). In the third approach, modern text-mining tools can be used to extract unexpected information from the internet ("web mining"). This enables the identification of novel research topics, unknown to the community so far (even "unknown unknowns"; see Sect. 4).

Finally, decision-making can be supported not only in fact-finding but also in practice-oriented assessment of upcoming solutions. Serious-gaming approaches can help to structure multi-perspective discussion of the aspects "good guys vs. bad guys" and "scientists vs. practitioners" and the like (see Sect. 5).

In the following, we present these four approaches and some illustrative results.

2 The Classic Technology-Foresight Approach

The most straightforward approach is to forecast the future developments in research topics and in technological applications with the highest probability according to the observed trends in the respective research area(s). This is the basic task in the field of technology foresight as a whole. The process can be described in five steps: identification—structuring—forecast—assessment—communication of the results. The first step is to identify the relevant research topics and types of technological systems under development. This is accompanied by an analysis of its state of the art, research dynamics and internal structure to create orientational knowledge and a solid base for future-related projection. The forecast is then constructed by considering all relevant information and possible (foreseeable) obstacles. In the assessment step, the significance of the findings to the study purposes and to the client's interests is investigated. The final, important step is to communicate the results in an appropriate (and tailored) manner to ensure that the findings will be useful to the client's decision and planning processes. Even though these five stages are not mutually exclusive, to fulfil the demands of a scientific study, they should be addressed with different mindsets in a transparent process so the prerequisites that led to the results are understood.

Knowledge must be very thorough about the present state of the object under investigation in order to make assumptions about likely future developments. Additionally, to forecast what technological capabilities can be expected to exist in a certain time interval in the future, insight into the relevant technological trends (i.e. the conditions and the speed of progress in the respective technological areas) is needed. This is the most important ingredient for a technology-foresight study because real surprises are rare.

However, it is here where the highest probability of error occurs. Before the First World War, there were more electric cars on the streets than cars fuelled by gasoline or diesel. Today, the advent of massive electric road transport is a major issue of innovation policy. Who, in 1920, would have been able to forecast that this renaissance would take 100 years to begin? One reason for this difficulty is that research follows the money, or in other words, societal or economic demands. Besides, rapid progress in a competing technology can also hinder another technology's success very effectively. Additionally, progress of an enabling technology can strongly influence the development of a given technology, as was the case with railways and the quality of steel. These problems can be accounted for with an approach that takes the relevant framework conditions into consideration.

2.1 Scanning and Monitoring

The classical methodological approach to technology forecasting is systematic literature study. Traditionally, "scanning" and "monitoring" are considered different elements. "Scanning" denotes a broad, systematic search for all information about aspects so far unknown ("unknown unknowns"), whereas "monitoring" is the tracking of pre-defined aspects in a well-defined field over a long period of time to identify new aspects and significant changes ("known unknowns") [14]. The time horizon of the search is set according to the requirements and the character of the information available. Scanning and monitoring can be realised by continuous screening of so-called "key sources", meaning journals (sometimes even websites and conferences) which in their entirety cover all important subjects within a certain technological field, but of which none can be omitted. The validation of a set of key sources can be done e.g. by bibliometric methods but should also include the experts' experience accumulated during the process. Often, in order to structure, forecast and assess the technology under investigation, additional in-depth literature investigation follows, with a more focused search area.

Such studies can be conducted in two principal architectures [15, 16]. The foresighting team, who are the experts in methodology and process, can get the *technological* expertise from *external* experts or expert groups. This requires structured exercises (like roadmapping and other workshops, questionnaires, Delphi surveys, etc.) to retrieve this expertise in a form which can be processed further. However, if the foresight experts do not have an education in science and technology this process can lead to fundamental misunderstandings regarding concepts and models behind the technological issues under discussion. Moreover, external experts (researchers in their field) are often not used to thinking in terms of futures research and tend to think only in relatively narrow scientific domains.

This can be overcome if the required scientific and technological expertise is located *within* the foresighting team (or at least within its closer organizational surroundings). So the study would be conducted by scientists and engineers (preferably with a sound experience in scientific research), who have additionally acquired the competencies for futures research. In that case, the contents-related knowledge within the team (i.e. the fundamental understanding of models and concepts in the scientific domains under investigation, and the wide and profound knowledge about ongoing respective research) can assure the correctness of the results. This incorporates the *intra*-disciplinary expertise of the individual team member with the *inter*-disciplinary expertise of the team. The methodological expertise of the team ensures the scientific quality of the study, and the process-related competence can ensure that the results have adequate relevance for the client. This approach has been successfully followed by the Fraunhofer Institute of Technological Trend Analysis (INT, Euskirchen, Germany) for several decades in support of the German MoD's defence-research planning, but also on behalf of i.a. the German Parliament, the Federal Criminal Office, and the European Commission [15].

2.2 Cross-Fertilization of Future Technologies

As already mentioned, the future of technology does not develop linearly. It is influenced by the development of connected technologies, be it in a cross-fertilizing (enabling or complementary) or in a competing (sometimes superseding) manner. This has also been labelled "the technology complex" [17]. These interdependencies can even lead to a renaissance of a well-known research approach, which can suddenly become feasible due to the rising of an enabling technology. Additionally, societal demands (indicated by research funding), political or ethical considerations, and, not least, consumer or market demands (including life-style trends) can have an important, sometimes crucial, effect here. So to be able to forecast future technological developments properly, especially with a longer-term time horizon, a broader picture must be analysed.

This is carried out, in the classical scanning and monitoring approach, by establishing what sometimes is called a "360° technology radar"—meaning continuous monitoring in the whole width of the technological spectrum to obtain a situational overview. Metascanning, i.e. the systematic study of existing technology-foresight studies, can help to establish a top-down structuring of the search field, either in the form of a taxonomy or a more pragmatic description. Work-sharing monitoring done by experts who regularly connect their segment of the overview can reveal correlations of the technological domains. It is advisable to support this work by proper knowledge/information management tools.

The non-technological influence parameters can be regarded as boundary conditions. The client's interests determine the degree of the inclusion of these parameters. One approach is to systematically look at observable trends (possibly long-lasting "megatrends") in the "STEEP" fields: society, technology, economy, ecology, politics, sometimes amended by V for "values", M for "military", and maybe others. Such an analysis can be conducted in a much more elaborate way, using scenario techniques, to construct a set of scenarios, which can then be used as a standardised environment for the technology analyses proper.

3 "Trend Archaeology": Bibliometrics for Technology Foresight

One of the tacit assumptions of technology foresight is that tomorrow's technologies are based on today's scientific work. Thus the technology scanning and monitoring process established to identify and track emerging topics, must be based on a kind of *science observatory*, viz. the continuous surveillance of scientific research and technological development. Traditionally this is done by a systematic screening of key sources and a literature study performed by experts (see Sect. 2.1).

In recent years [18–20] there has been a growing interest in computationally-based support systems for foresight processes. One regularly discussed approach is the identification of emerging topics by bibliometric means, i.e. the statistical and quantitative analysis of publication data [for examples see [21, 22]]. Since most of these approaches are based on the analysis of an adequately chosen and constructed citation network, they admittedly face the major problem that bibliometrics is an inherently retrospective method based on an analysis of *previous* publications and citations. Additionally, it has been demonstrated [23] that the Fraunhofer INT's approach clearly outperforms a pure bibliometric approach in the identification of emerging technologies.

An ongoing research project at the Fraunhofer INT is investigating the feasibility of bibliometric methods for the process of technology foresight. Since the exclusive and automatic detection of emerging topics by bibliometric means appears highly questionable, the research focus has been expanded. In order to investigate, how eavesdropping into today's scientific communication by bibliometric means might support technology foresight, a procedure coined "*trend archaeology*" was introduced [24]. This approach examines historic scientific trends and seeks specific patterns and their temporal evolution. The proposed method is a multidimensional approach, to examine multiple aspects of a scientific theme using bibliometrics. "Trend archaeology" is also based on the synoptic inspection of different scientific themes, which might come from different fields, such as nanotechnology or materials science.The aim of this approach is to investigate whether the identification of specific patterns in the temporal development of a scientific theme allows one to establish a kind of classification scheme of such topics, solely based on the characteristics inherent in their dynamics and to analyse the predictive power of such a classification scheme.To this end a bibliometric workflow (described below) and a collection of software tools have been implemented by Fraunhofer INT. This allows for a comprehensive bibliometric analysis of any scientific topic.

3.1 The Bibliometric Workflow

At the Fraunhofer INT a bibliometric workflow has been established which consists of three different phases:

- **Phase 1** comprises the iterative formulation of a suitable search query, which is aimed at delineating the scientific research field, as accurately as possible.
- **Phase 2** addresses the preparation and cleansing of the bibliometric data.
- **Phase 3** conducts analysis and visualisation of the bibliometric data.

With the help of a software tool developed at the Fraunhofer INT it is possible to examine the process of scientific communication for any research theme by analysing a number of bibliometric observables with their time dependent characteristics. Some of these observable are described below.

One of the first observables considered is the **publication dynamics**—the number of published papers per year. Although this bibliometric observable is rather simple, it proves to be very useful since it is a direct measure of the relative scientific effort on a particular theme. Furthermore, it is possible to perform an *experience-based extrapolation* of the publication dynamics. Figure 1 presents an example of the number of publications over the course of about 40 years for nine scientific topics. The results clearly indicate the existence of at least two distinct growth patterns. The first is the (double) logistic growth as it is discernible in the case of the research on *fullerene, bulk metallic glasses, human enhancement, transcranial electric stimulation* and *cochlear implants*. For these areas, the initial publication activity phase show exponential growth, followed by inhibited growth, where the number of published papers grow only linearly. After this period of relative stagnation, another phase of exponential growth begins. Research on nanotubes and graphene can be interpreted as a special variation of this more general case, because they are still in the initial growth phase. A clearly distinct pattern is observed for *high-temperature superconductors* and *cold fusion*. After the ubiquitous initial growth phase a pronounced and prolonged decline in the publication numbers can be observed.

Another intriguing indicator is the time dependence of the **document types** used by scientists to publish their results. This observable allows one to gain insight into the preferred communication channels, which can also be used to characterise a scientific theme. Evidence has been presented that a significantly larger fraction of publications on cold fusion research, which is a prototypical example for pathological science,[1] were reported in fast publications (e.g. simple notes, letters etc.) [25].

Next, the time-dependent size of the **giant component of the co-author network** [26] will be introduced. This observable measures the degree of connectivity between the co-authors in the network. It can be interpreted as a measure of scientific maturity, since it directly maps the formation of a scientific community, which in turn can be linked to the process of establishing a scientific topic. In the example of the research on fullerene, nanotubes, or superconductors, a prototypical pattern is observed. Since the research on a scientific topic typically starts with only one or a few publications written by a handful of authors proposing a new scientific idea, the corresponding initial co-author network is small but highly interconnected. As more and more authors discuss the newly introduced idea, the number of authors increases but the interconnectedness of the network declines. If the scientific idea turns out to be worthy of continued working, authors start collaborating with each other and form larger and larger subnetworks. As the co-author network gets denser, the relative size of the giant component increases. Figure 2 illustrates how this process leads to networks that are highly interconnected, as is seen for fullerene, nanotubes, graphene, and superconductors, where approximately 90% of the authors are interconnected with each other. These topics can be considered to be established

[1]Pathological science is an area of research where "people are tricked into false results ... by subjective effects, wishful thinking or threshold interactions." The term was first used by Irving Langmuir, Nobel Prize-winning chemist, during a 1953 colloquium at the Knolls Research Laboratory. https://en.wikipedia.org/wiki/Pathological_science.

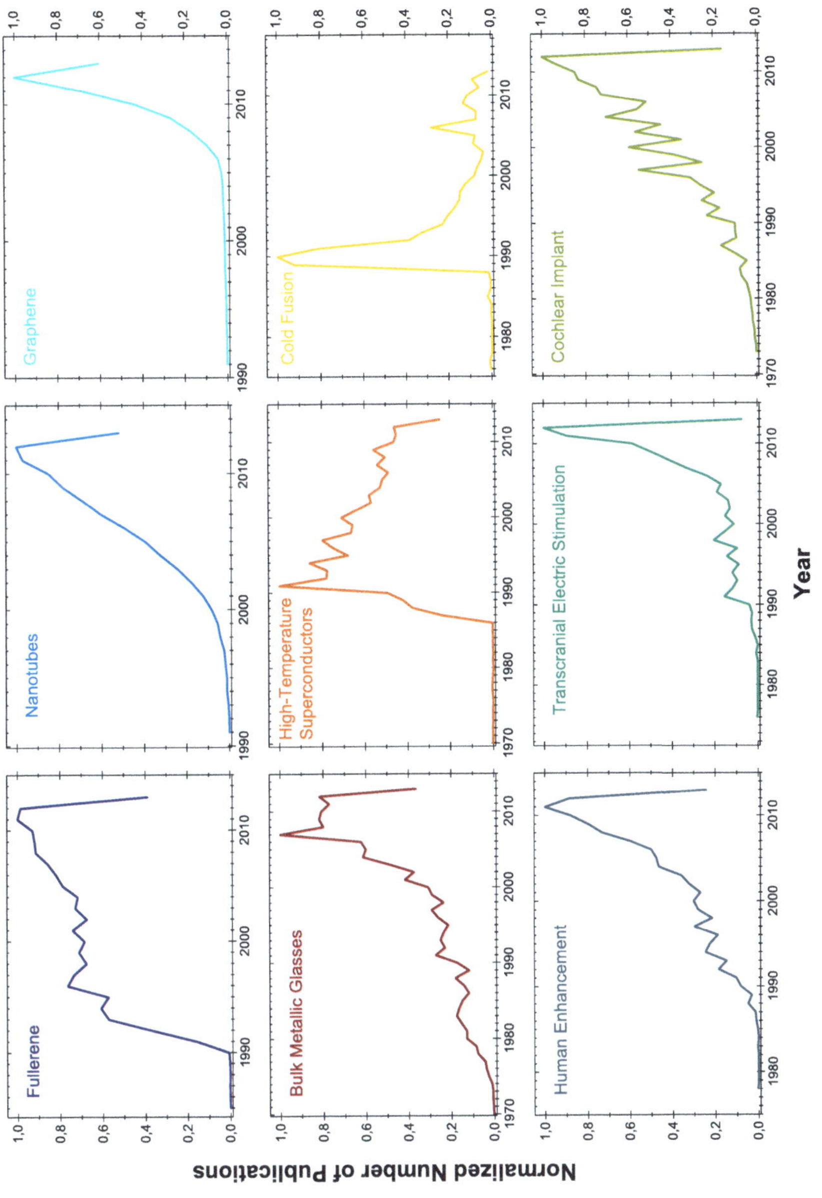

Fig. 1 Publication dynamics of nine scientific topics analysed using the software tool developed at the INT

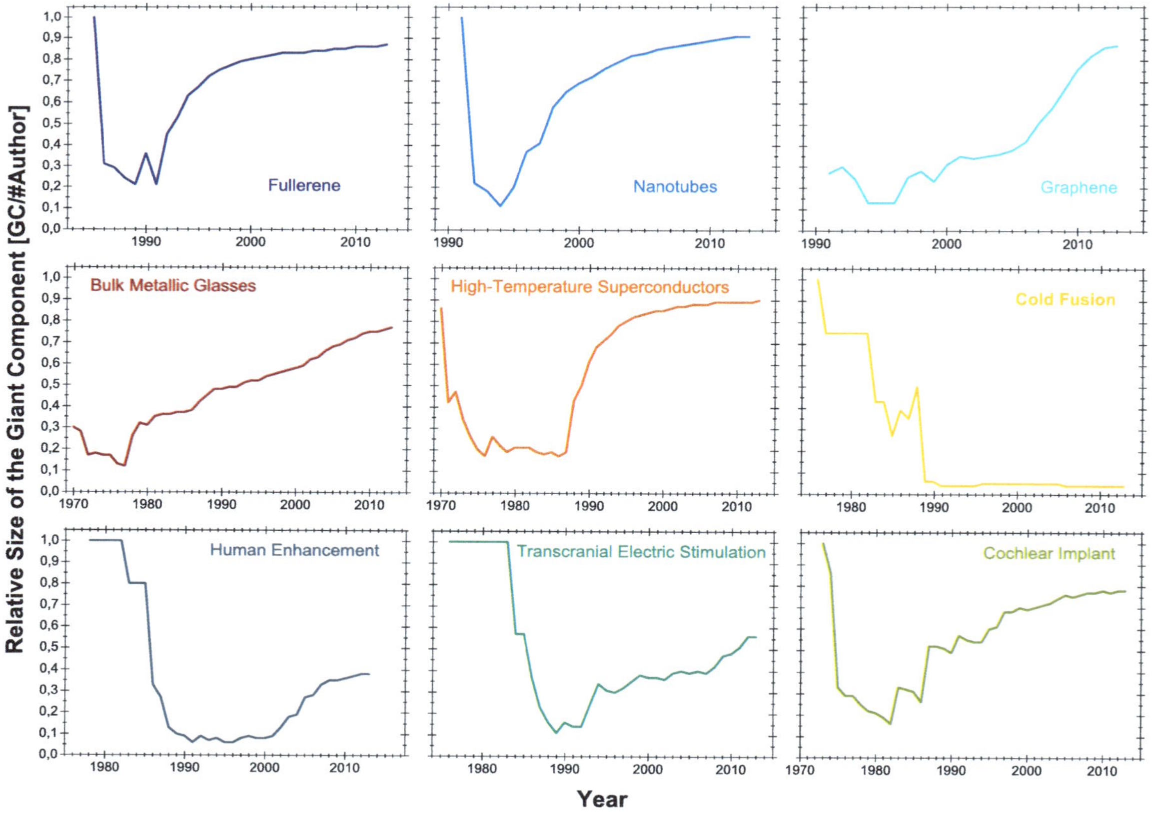

Fig. 2 Time-dependent size of the giant component of the co-author network of the nine scientific topics already shown in Fig. 1

or (scientifically) mature. For other topics, like the research on human enhancement or transcranial stimulation, the analysis shows that they are still in a consolidation period. The research on bulk metallic glasses and the cochlear implant appears to be in between these two phases. In contrast, in case of the research on cold fusion a scientific community has not been established (to-date), which confirms the pathological characteristic of this topic.

Finally, it is illustrative to take a look at the **activity pattern**, defined as the number of papers each country has published. With the help of the so called Herfindahl index it is possible to analyse how monopolistic the publication activities are. Is there a broad interest in this theme or is there only a small number of countries active? If so, which countries? Are there hot spots on a regional level, where research on a certain topic is concentrated? Additionally one might expand the analysis of the publication activities by country and take a closer look at their **patterns of cooperation**. A bibliometric analysis of this type, assisted by suitable visualisation tools addresses questions, such as: Do regional clusters of cooperation exist or does cooperation span the entire globe? Do a few actors dominate the cooperation networks? The questions addressed concerning the activity pattern and the patterns of cooperation are especially relevant in the context of research funding.

3.2 The Potential of Bibliometrics for Technology Foresight

Bibliometrics is a useful tool during all of the aforementioned steps of a typical foresight process. It might support an expert in conducting technology foresight, but it does not—and probably never will—replace the necessary expertise in science, technology and the methodology of futures research. Although using an automated (or at least semi-automated) bibliometric method to identify emerging themes might proof impossible, bibliometrics assists the underlying process of information retrieval. Although one of the first of such approaches dates back to 2005 [27], it is only recently that a systematic investigation of this topic has been initiated [28]. In addition to *experience-based extrapolation* of the publication dynamics, bibliometrics provides hints for *forecasting* the future development of scientific themes by possibly revealing typical patterns in historic trends, an approach coined trend archaeology [24]. Finally, since bibliometrics is a quantitative method it offers new opportunities for assessing and communicating through visualisation.

4 Novel Technological Solutions

4.1 *Prediction of New Technological Trends with Text/Web Mining and Clustering*

Predicting new technological trends is a demanding task. Literature shows that a possible solution is to analyse the content of existing information sources [29]. This is because new and upcoming technologies, as well as their potential application fields, are often discussed at an early stage. Currently, this represents weak signals [30] and can be found in various information sources. On the internet, a huge amount of information (e.g. books, patents, articles but also discussions and concepts), with a wide scope of different topics, can be found. The amount of information increases steadily and nearly 80% of this information is available as formatted or unformatted texts, and technologies are described within this information at an early stage. Thus, the internet is a valuable source of information to identify technological trends.

4.2 *Methodology*

Web mining approaches are very useful for the extraction of technological trends from the internet. The automated process scans all relevant internet websites [31], identifies relevant content from the websites, and crawls the content. Advanced programming interfaces (APIs) of internet search engines can be used to perform this task. A single search query normally does not totally cover a topic area, however, a set of search queries is suitable to cover all relevant aspects of the topic and can represent the topic area. The selection of the search query set should be done carefully because it directly influences the quality of the results. APIs can be used to restrict search results to a specific language. Alternatively, search queries can be translated to various languages and executed to obtain documents written in different languages. In the latter case, the documents will have to be translated to a target language for further processing. This also can be done automatically by APIs (e.g. Google translate API). The low quality of an automated translation has no significant effect on the results because for the subsequent text mining steps, a one-to-one translation of relevant terms is more important than a grammatically correct translation. The web mining steps as described above can be applied at several different points in time. This enables identification of time series within the data.

Analyzing web mining data is done with text mining and clustering. Text mining creates term vectors in a vector space model from each document. to prepare the text, the raw text is cleaned, the spelling is checked, tokeniziation (split text in a set of words) is applied, and the identified words are homogeneously converted (case conversion). Stop words and low-frequency words are discarded as well as words with a specific part of speech for text filtering. Words are summarized according to their stem. For term weighting, a term weighting scheme is built. Overall, text mining creates term vectors based on weighted term frequencies from web mining data.

Clustering is used to identify new and unexpected textual patterns within the term vectors. Latent semantic indexing (LSI) groups together related terms. It uses singular value decomposition (SVD) to form semantic generalizations. The co-occurrences of terms are used to indicate the relationships between them [32]. As a result, LSI creates an orthonormal, semantic, latent subspace where semantically related terms are grouped into classes. Classes are built based on their discriminatory power to other classes. LSI calculates the impact of each term on each class. Decision makers can use this information to interpret the content of the classes. LSI also calculates the impact of each document on each class. This information enables decision makers to distinguish between well-known/established classes and new/upcoming classes. Further, applying web mining, text mining, and clustering at several points in time enable to trace the development of classes.

4.3 Case Study

For a case study, the methodology was applied to identify new, upcoming technological trends in International Atomic Energy Aagency (IAEA) safeguards. Eleven search queries covering several aspects of this topic area were built: 'neutron detection', 'He 3 helium isotope', 'He 3 gas detector', 'laser induced breakdown spectroscopy LIBS', 'non-destructive assay measurement', 'non-destructive evaluation', 'non-destructive testing', 'fissionable material flow', 'fissile material', 'separation isotopes laser excitation', and 'atomic vapor laser isotope separation'. We did not use the word "safeguard" itself within the queries to prevent collection of established technologies directly used in safeguards environment. As the collected technologies are related to the search query terms, they are indirectly related to IAEA safeguards. Search queries results were restricted to the English language and executing search queries was done twice. In August 2014, the query resulted in 1191 documents, and 1160 documents were obtained in a second query run in January 2014. The increase of about 3% is in accordance with the increase of the overall number of internet documents and therefore not significant. Clustering was applied and parameters were selected [33]. The results consisted of 60 different clusters, the impacts of terms on the clusters, and the impacts of the retrieved documents on the clusters. An example for such a cluster was *tagging techniques*. Such techniques are frequently used worldwide however, they are also used for labeling of long-term stored waste (e.g. fissile materials). Even though they also can be applied to safeguards, they are seldom mentioned in the context of IAEA safeguards. Results show four classes related to tagging techniques. The four classes indicate the use of ultrasonic waves (class nr. 13), the Raman Effect (class nr. 34), light reflection (class nr. 14), and radio waves (class nr. 4) for tagging. Based on the data collection points in January 2014 and August 2014, the numbers of documents assigned to the three latter classes do not increase or decrease significantly. An exception is class nr. 13. Based on selected thresholds, the number of documents related to ultrasonic waves, related to tagging, and related to at least one of the search queries significantly increases from January to August. This indicates that *ultrasonic tagging* is possibly a new, upcoming technological trend in the context of IAEA safeguards and this should serve as a starting point for a further manual evaluation by human experts.

5 Including Experts and Practitioners: Participatory Approaches

To assure correctness of the results, the scanning and monitoring of technological developments in a certain field is ideally performed by a team of scientists and engineers who are competent in futures research. Moreover, the applicability of their findings depends on their insight into the operational requirements of the targeted field of application. To support this, participatory approaches that allow combining technological expertise and operational knowledge can be used.

A good example for this is the "Disruptive Technology Assessment Game" (DTAG) that was developed and validated within NATO's Research and Technology Organisation (RTO) in the subgroups 062 and 082 of the System Analysis and Studies Panel (SAS). The goal of these activities was to assess future technologies while they are still in a relatively immature state, in contrast to an assessment by war games that are usually assisted by computer simulation and deal with higher technology readiness levels [34–36].

These activities began by identifying new technological developments that were supposed to be of high relevance for the armed forces. In a second step, these technologies were used to create a number of "Ideas of System". The rationale was that it is impossible to directly assess the impact of technologies but rather of the systems and capabilities that they enable. The third step and core of this approach was the assessment game where the potential use of the system ideas was explored in a confrontational context by interaction of two teams of military players (Red and Blue), supported by technologists and analysts.

The game cycles started with a tactical situation briefing, upon which the players generated their course of action using current capabilities and doctrine. This was followed by a confrontation discussion, where all DTAG participants met and the Red and Blue teams briefed each other on their planned course of action. This was followed by a second planning phase where the same tactical situation was addressed, but now using the Ideas of System. Afterwards, all the DTAG participants met and the Red and Blue teams again briefed each other on their new course of action. In a mediated session especially the impact of the system ideas was discussed. It was assessed whether these systems had given benefits over current capabilities, if countermeasures could be envisaged and if there were any other implications. After this confrontation more detailed information was captured in a software tool through structured interviews of the Red and Blue teams. In addition the teams were asked to think of other future systems that might have helped them in the given tactical situations. These ideas were later analysed with respect to their technological feasibility and in some cases included in later games.

Each game cycle took approximately one day to conduct, the DTAGs were played over the course of one week so that different types of operations could be covered. The analysis of the gaming results started with the impact of the Ideas of System, but then moved back to the underlying enabling technologies with the goal of providing planning support for R&D. This approach allowed for an element of unconstrained "out-of-the-box"-thinking and created an opportunity for open interactive communication between military officers and scientists to discuss the usefulness of technological systems. By bringing technology and military experts together it could be avoided having to deal with unrealistic technological or tactical conclusions.

This methodology has subsequently been adopted and extended to address similar questions. For example, as part of the project ETCETERA ("evaluation of critical and emerging technologies for the elaboration of a security research agenda"), in the EU-FP7 programme, it was used to assess emerging technologies for the security domain [37]. The DTAG setting might also be used to assist the foresight of verification techniques. This red-and-blue-teaming approach is especially relevant for scenarios where inspectors act in a non-cooperative environment, such as during the aftermath of the 1991 Persian Gulf War. The inspectors in Iraq had to face systematic misrepresentation, deception and concealment. The task of verifying non-proliferation treaties might even become more difficult as technologies capable of supporting deception and denial efforts will become more widely available [38].

6 Conclusions

There are several ways technology-foresight methods and experience can be useful to anticipate future verification technologies. Forecasting the probable path a technological trend will develop can be supplemented by constructing a more comprehensive picture of the conditions, correlations and repercussions of such a trend. This can be used to consolidate or adjust existing expectations, as well as to provide new ideas "outside the box". Novel, computer-based methods can provide valuable insight into the character and structure of research topics and their correlations, but can also be used to identify novel approaches, "unknown unknowns", in the respective research area. Finally, best-practice assessment of future technological options needs the involvement of the prospective end-users with sufficient operational experience.

By this means, valuable recommendations can be developed to offer a scientific basis to decision-makers and planners. Of course, it must not be forgotten that predictions, in the proper sense, are not possible. Rather, technology-oriented futures research cannot unveil more than today's expectations with the highest probability, taking into account all facts and factors that are presently available. Nevertheless, interesting results can be expected from applying this scientific planning-support approach to the area of technologies for verification of treaties on weapons of mass destruction.

References

1. Conference on Disarmament, Geneva. www.opbw.org/verex/docs/CONFIII-VEREX-9.pdf. Accessed 28 Feb 2019
2. Avenhaus R, Kyriakopoulos N, Richard M, Stein G (eds.) (2006) Verifying Treaty Compliance. Limiting Weapons of Mass Destruction and Monitoring Kyoto Protocol Provisions. Springer, Berlin
3. Agar J (2009) On the origin of technology. Nature 461(7262):349. https://doi.org/10.1038/461349a
4. Bower J L, Christensen C M (1995) Disruptive Technologies: Catching the Wave. Harvard Business Review 73(1):43–53. https://hbr.org. Accessed 28 Feb 2019
5. Campbell V (2004) How RAND Invented the Postwar World. Satellites, Systems Analysis, Computing, the Internet – Almost All the Defining Features of the Information Age Were Shaped in Part at the RAND Corporation. Invention and Technology, Summer:50–59.
6. Zweck A, Holtmannspötter D (2002) Monitoring of technology forecasting activities in Europe. ESTO project report; working document. Dusseldorf (Zukünftige Technologien / Future Technologies, 37).
7. Martin B R (2010) The origins of the concept of 'foresight' in science and technology: An insider's perspective. Technol. Forecast. Soc. Change 77(9):1438–1447. https://doi.org/10.1016/j.techfore.2010.06.009
8. Miles I (2010) The development of technology foresight: A review. Technol. Forecast. Soc. Change 77(9):1448–1456. https://doi.org/10.1016/j.techfore.2010.07.016
9. Masini E (2006) Rethinking futures studies. Futures 38(10):1158–1168. https://doi.org/10.1016/j.futures.2006.02.00
10. Jouvenel B (1965) Futuribles. RAND Corp., RAND Paper P-3045. http://www.rand.org/content/dam/rand/pubs/papers/2008/P3045.pdf. Accessed 28 Feb 2019
11. Steinmüller K (2012) Zukunftsforschung in Deutschland. Versuch eines historischen Abrisses (Teil 1). Zeitschrift für Zukunftsforschung 1(1):6–19. urn:nbn:de:0009-32-34116
12. Technology Futures Analysis Methods Working Group (2004): Technology futures analysis: Toward integration of the field and new methods. Technol. Forecast. Soc. Change 71(3):287–303. https://doi.org/10.1016/j.techfore.2003.11.004
13. Grunwald A (2013) Wissenschaftliche Validität als Qualitätsmerkmal der Zukunftsforschung. Zeitschrift für Zukunftsforschung 2(1):22–33. urn.nbn:de:0009-32-36941
14. Kerwin A (1993) None Too Solid: Medical Ignorance. Knowledge: Creation, Diffusion. Utilization 15(2):166–185. https://doi.org/10.1177/107554709301500204
15. Grüne M (2013) Technologiefrühaufklärung im Verteidigungsbereich. in: Popp R, Zweck A (eds): Zukunftsforschung im Praxistest. Springer Fachmedien Wiesbaden (Zukunft und Forschung 3):195–230. https://doi.org/10.1007/978-3-531-19837-8_9

16. Burbiel J, Schietke R (2014) ETCETERA - Evaluation of Critical and Emerging Security Technologies for the Elaboration of a Strategic Research Agenda. Final Report. Fraunhofer INT, Euskirchen. http://www.etcetera-project.eu. Accessed 13 Apr 2016 and http://www.burbiel.de/Publikationen/ETCETERA_Final_Report.pdf. Accessed 12 Nov 2019
17. Geschka H, Hahnenwald H (2013) Scenario-Based Exploratory Technology Roadmaps - A Method for the Exploration of Technical Trends. In: Moehrle MG., Isenmann R, Phaal R (eds): Technology Roadmapping for Strategy and Innovation. Springer Berlin Heidelberg, pp 123–136. https://doi.org/10.1007/978-3-642-33923-3_8
18. Von der Gracht H, Bañuls VA, Turoff M, Skulimowski AMJ, Gordon TJ (2015) Foresight support systems: The future role of ICT for foresight. Technological Forecasting and Social Change 97:1–6. https://doi.org/10.1016/j.techfore.2014.08.010
19. Keller J, Von der Gracht H (2014) The influence of information and communication technology (ICT) on future foresight processes – Results from a Delphi survey. Technol. Forecast. Soc. Change 85:81–92. https://doi.org/10.1016/j.techfore.2013.07.010
20. Zhang Y, Porter AL, Hu Z, Guo Y, Newman NC (2014) Term clumping for technical intelligence: A case study on dye-sensitized solar cells. Technol. Forecast. Soc. Change 85:26–39. https://doi.org/10.1016/j.techfore.2013.12.019
21. Schiebel E, Hoerlesberger M, Roche I, François C, Besagni D (2010) An advanced diffusion model to identify emergent research issues: the case of optoelectronic devices. Scientometrics 83(3):765–781. https://doi.org/10.1007/s11192-009-0137-4
22. Chen C (2006) CiteSpace II: Detecting and visualizing emerging trends and transient patterns in scientific literature. Journal of the American Society for Information Science and Technology 57(3): 359–377. https://doi.org/10.1002/asi.20317
23. Wepner B, Huppertz G (2014) Report on the Comparative Analysis of Three Methods to Assess Emerging Technologies. EU FP7 SEC project ETCETERA, deliverable 4.2. http://www.etcetera-project.eu/deliverables/documents/ETCETERA_Deliverable4.2_ComparativeAnalysis.pdf. Accessed 13 Apr 2016
24. John M, Fritsche F (2013) Bibliometric classification of emerging topics. in: Hinze S, Lottmann A (eds.) Translational twists and turns: Science as a socio-economic endeavor, iFQ-Working Paper, iFQ, Berlin, pp 181–184. http://www.forschungsinfo.de/STI2013/download/STI_2013_Proceedings.pdf. Accessed 12 Nov 2019
25. John M, Fritsche F (2013) Fullerene and cold fusion: Bibliometric discrimination between normal and pathological science. In: Gorraiz J, Schiebel E, Gumpenberger C, Hörlesberger M, Moed HF(eds.) Proceedings of ISSI 2013 Vienna, AIT - Austrian Inst. of Technology, Vienna, pp 1989–1991. http://www.issi2013.org/proceedings.html. Accessed 12 Nov 2019
26. Bettencourt LMA, Kaiser DI, Kaur J (2009) Scientific discovery and topological transitions in collaboration networks. Journal of Informetrics 3(3):210–221. https://doi.org/10.1016/j.joi.2009.03.001
27. Kostoff RN, Shlesinger MF (2005) Cab: Citation-assisted background. Scientometrics 62(2):199–212. https://doi.org/10.1007/s11192-005-0014-8
28. Mayr P, Schaer P, Scharnhorst A, Larsen B, Mutschke P (eds.) (2014) Proceedings of the First Workshop on Bibliometric-enhanced Information Retrieval, CEURWorkshop Proceedings, vol. 1143. CEUR-WS.org. urn:nbn:de:0074-1143-7
29. Abebe M, Angriawan A, Tran H (2010). Chief executive external network ties and environmental scanning activities: An empirical examination. Strategic Management Review 4(1):30–43
30. Ansoff IH (1975) Managing strategic surprise by response to weak signals. California Management Review 18(2):21–33
31. Kosala R, Blockeel H (2000) Web research: A survey. ACM SIGKDD Explorations Newsletter 2(1)
32. Jiang J, Berry MW, Donato JM, Ostrouchov G, Grady NW (1999) Mining consumer product data via latent semantic indexing. Intelligent Data Analysis 3(5):377–398
33. Thorleuchter D, Van den Poel D (2013) Weak signal identification with semantic web mining. Expert Systems with Applications, 40(12):4978–4985

34. NATO (2010) Assessment of Possible Disruptive Technologies for Defence and Security. Report RTO-TR-SAS-062 (NATO unclassified)
35. NATO (2012) Disruptive Technology Assessment Game - Evolution and Validation. Report RTO-TR-SAS-082 (NATO unclassified)
36. Neupert U, Römer S, Wiemken U, Rademaker JGM (2009) Assessment of potentially disruptive technologies for defence and security, Fraunhofer Symposium Future Security, 4th Security Research Conference, Karlsruhe, Germany
37. EU FP7 SEC ETCETERA project homepage. http://www.etcetera-project.eu. Accessed 13 Apr 2016
38. Tucker JB (1996) Monitoring and verification in a noncooperative environment: Lessons from the U.N. experience in Iraq, The Nonproliferation Review 3(3). https://doi.org/10.1080/10736709608436633

Attribute Information Barriers

Malte Göttsche and Gerald Kirchner

Abstract While arms control verification nowadays refers to the verification of delivery vehicles, there seems to be fairly broad agreement that verification should include nuclear warheads and military fissile material. Due to the sensitivity of information that could be gained from measuring these items, information barriers may likely be required. One type of system utilises the attribute approach: The inspected party would declare attributes that characterise the treaty accountable item to be authenticated. Criteria for proposing appropriate attributes are preventing leakage of sensitive information, minimising the possibility of cheating, robustness and a low complexity of measurements/analysis. Measurement issues that arise because inspectors may not know the configuration of the items and possible shielding must be understood as they introduce systematic uncertainties. Measurements of dismantled warhead components may be more reliable than measurements of fully assembled warheads where less information on shielding will likely be available. In any case, false positives and negatives should be as low as technically achievable. Procedures to cope with uncertainties are repeated measurements to deal with statistical uncertainties, measurements based on complementary measurement techniques to deal with systematic uncertainties and consultations between inspector and host. Despite remaining challenges, attribute information barriers may significantly add to the confidence in disarmament when embedded in a web of additional verification measures.

M. Göttsche (✉)
Aachen Institute for Advanced Study in Computational Engineering Science and III. Institute of Physics B, RWTH Aachen University, Aachen, Germany
e-mail: goettsche@aices.rwth-aachen.de

G. Kirchner
Carl Friedrich von Weizscker-Centre for Science and Peace Research, University of Hamburg, Hamburg, Germany
e-mail: gerald.kirchner@uni-hamburg.de

I. Niemeyer et al. (eds.), *Nuclear Non-proliferation and Arms Control Verification*,
https://doi.org/10.1007/978-3-030-29537-0_14

1 Disarmament Verification

Nowadays, verification of nuclear arms control refers to the verification of delivery vehicles. This is the case for instance in the New START Treaty between Russia and the United States [1]. Warheads are counted indirectly via the delivery vehicles they are associated with. Regarding the longer-term future of nuclear arms control and disarmament, there seems to be fairly broad agreement that verification should become more intrusive and that direct verification of warheads—deployed as well as non-deployed—could play a vital role in this to measure progress in dismantlement [2].[1]

To achieve confidence in the irreversibility of disarmament, the verification of weapons-usable materials could inform about the potential of building new warheads. Weapons-usable materials take a variety of forms, for example warhead components, naval fuel, fissile material resulting from warhead dismantlement or materials in civilian use.[2] Civilian materials could principally be covered by Safeguards inspections in nuclear weapon states, but all other weapons-usable materials, which are referred to as military materials in this chapter, require new approaches. In the following, the term 'treaty accountable item' refers to all of the above items, both warheads and materials, that could be covered by a verification regime.

Such verification could concern inventories of nuclear warheads and materials and their changes throughout the disarmament process. In addition to informing about available stocks, verification could build confidence that existing material stocks are currently not transferred to warhead production. The dismantlement process itself could be verified, including the warhead entering the process and the components that leave the process, possibly including their disposition or conversion to materials used for peaceful purposes.

Such verification regime were most robust if it entailed three components, authentication, unique identification and continuity of knowledge [3, p. 27]. Authentication is the process during an on-site inspection through which it is assessed by measurements whether a specific item is the nuclear item it is declared to be. Uniquely identifying treaty accountable items allows tracking them. Continuity of knowledge is the process of monitoring treaty accountable items throughout time and processes.

---

[1]One example of a current effort in this regard is the International Partnership for Nuclear Disarmament Verification, a multinational body discussing future verification challenges including warhead verification.

[2]In [3, p. 52], the different forms of weapons-usable materials are discussed in more detail.

2 Why an Information Barrier?

States declaring their nuclear arsenals will very likely not allow inspections that reveal information of their warheads or military materials which they consider sensitive. The Nonproliferation Treaty's Articles I and II set boundaries when verification involves non-nuclear weapon state participants. However, due to reasons of national security, nuclear weapon state participants may not have access to significantly more information [3, p. 38]. National security issues also explain, why military materials such as naval fuel are sensitive. Assuming a likely case where verification measures are decided upon in a cooperative manner, mutual agreement on verification activities can only be reached in the case where nuclear-armed states are confident that their sensitive information is not at stake. The inspecting party—whose interest is to gain maximum confidence—would likely prefer rather intrusive and comprehensive verification measures. The goal is to create a regime which builds confidence while preventing unacceptable levels of intrusion that could leak information the inspected state is unwilling to share. This appears to be somewhat contradictory.

For the authentication of treaty accountable items, a solution is to take potentially intrusive measurements containing sensitive information, but to automatically process the measurement information via an algorithm so that the only output visible to the inspector would be of non-sensitive nature (e.g. a green, yellow or red light indicating "specified item", "inconclusive measurement" or "not specified item"). Preventing the leakage of sensitive information would be the task of a so-called information barrier.

Main requirements of such an authentication system are (a) that automatic measurement and analysis function properly so that the best possible indication of authenticity is given and (b) that no sensitive information is released. (a) is in the interest of both an honest host and inspector, as, for example, the honest host would like to get credit for having dismantled a true warhead and prevent false alarms, while the inspector would like to ensure that he is not cheated upon. Therefore, ideally, the system should accept all specified treaty accountable items and reject all other items, showing low probabilities of false positives and false negatives. (b) is certainly in the host's, but also in the inspector's interest since all parties must adhere to the Nonproliferation Treaty's Articles I and II. Authentication systems must be designed in a way that both host and inspector can ensure that the host has not built in a capacity to manipulate measurement results (equipment authentication), and that the inspector has not built in a capacity which could enable assessing sensitive information (equipment certification) [4].

3 Attribute Approach

One type of authentication system utilizes the attribute approach: The inspected party would declare attributes that characterize the treaty accountable item to be authenticated. It is essential that attributes do not include sensitive information.

For obtaining maximum confidence, the attributes should be chosen in a way to minimize the possibility of cheating, meaning that all treaty-accountable items will pass the attribute test, while no other will. This idealistic concept may not hold in reality. Instead, the challenge is to define attributes and analysis algorithms in such a way that the analysis results in reasonable confidence. An attribute information barrier only analyses whether the item under investigation meets the defined attributes; the value of an information barrier then depends on the attribute definitions and algorithms as well as on the quality of the measurement techniques. These are the main factors that may limit confidence, as discussed below.

Attributes might be of qualitative or quantitative nature. Attributes based on quantitative thresholds provide the benefit that it is not necessary to declare the actual values which might be sensitive. The following list contains a variety of published attributes developed by different initiatives, including one proposed measurement technique per attribute:

- Presence of plutonium (gamma spectroscopy [5])
- Plutonium ratio Pu-240/Pu-239 (gamma spectroscopy [5])
- Plutonium mass threshold (combination of gamma spectroscopy and passive neutron multiplicity counting [5])
- Plutonium age (gamma spectroscopy [6])
- Absence of oxide (combination of gamma spectroscopy and neutron multiplicity counting [6])
- Item symmetry (neutron measurements [6])
- Presence of U-235 (active neutron multiplicity counting [5])
- Uranium enrichment (active neutron multiplicity counting [5])
- U-235 mass (active neutron multiplicity counting [5])
- mass of chemical explosive (neutron-induced gamma spectroscopy [5])

In [7], a feasibility analysis of a range of various nuclear measurement techniques per attribute is provided. In particular for uranium, but also for plutonium-based warheads, the measurement techniques require much more research; for various attributes there is a need of the measurement techniques to show higher reliability. With the exception of the chemical explosive's mass, these attributes should also be relevant for warhead components. For other military materials, if an information barrier is required due to sensitivity issues, some attributes would differ.

3.1 Attribute Definition Issues

State actors attempting to deceive the inspectors have very large amounts of resources available they may choose to invest emphasising the need to minimize loopholes. The dilemma is that inspectors may not know much about the item's properties, in particular in the case of non-nuclear weapon state inspectors. The challenge is then to choose attributes that can create this confidence. Inspectors should have a sufficient level of confidence that the host did not propose attributes allowing significant cheating. Attributes must be carefully chosen and their reliability/robustness extensively tested by both the host and inspector. Still, there is no easy solution to this dilemma, some questioning may remain making it difficult for inspectors to obtain full confidence from this single verification measure. Clear communication on both sides can relieve the issue somewhat, for instance explanations of the host why they consider the chosen attributes well-suited. Discussions on the attributes should be as open as possible within the given constraints. In the end, the host would like to receive credit from the inspector, so it is in the host's interest to be as transparent as possible.

While the choice of some of the attributes is intuitive, one might come up with different attributes nevertheless. Attributes must take into consideration the availability of measurement techniques and the item's shielding properties. For instance, shielding of an item with a certain plutonium mass may result in a lower measured mass, if that shielding is unknown to the inspector. The mass threshold would then need to be lowered, accordingly.

The output of the information barrier is usually considered to be a "yes" or "no". On the one hand, a binary output is the best solution from the point of view of not revealing sensitive information, while on the other hand, more detailed outputs might be preferred for giving the inspector sufficient information to enable him to reach a conclusion regarding the authenticity of the presented item.

A systematic procedure for selecting appropriate attributes should take into account the following criteria:

- Sensitive information: Does the output prevent the leakage of sensitive information that cannot be shared?
- Meaningfulness: Do proposed attributes minimize the possibility of cheating and can inspectors base confidence in them?
- Robustness: Does the outcome of the measurement by a specific technique critically depend on the unknown properties of the item configuration (see below)?
- Complexity: Are measurements/analysis of such a high complexity that they produce an additional risk of failure?

3.2 Measurement Issues

Besides the attribute definitions, the challenge of information barriers in general is the impossibility of inspectors to closer analyze the measurement results, if the inspector is in doubt for whatever reasons. The problem is even worse as an information barrier may not even inform the inspectors of an uncertain result. In contrast to other situations where radioactive samples are characterized, only inadequate knowledge exists prior to the measurements. The geometry and other properties will be unknown, it will also be impossible to calibrate the detector with representative standards. This raises the problem that the designed system must authenticate an item whose configuration such as geometry remains largely unknown. Therefore the authors propose that in case of availability of various techniques the measurement system should be chosen which is least dependent on calibration with materials representative of the warhead or component and which requires a minimum amount of assumptions regarding the nature of the item [7]. Reducing the dependency of measurements on geometry and shielding would increase the measurement accuracy and could allow more meaningful attribute thresholds as close as possible to the real values, as the host can rest assured that the likelihood of false negatives remains very low. As a result, the level of confidence gained from such a system will increase.

Part of this issue is the potential undeclared presence of shielding. In the case of fully assembled warheads, shielding might arise from material surrounding the fissile component or self-absorption of radiation within the fissile material itself. Furthermore, most nuclear warheads are stored in containers for safety reasons that could act as further shielding. This may similarly apply to containers for the storage of warhead components and other military materials. Such issues could have an impact on the measurement and therefore on the output of the attribute analysis. It shall be recognized that it is in the interest of an honest inspected country to ensure a positive outcome of the authentication activity. Thus, the host could provide information on shielding causing a strong impact on the analysis where possible[3]—to incorporate this information in the attribute assessment—or offer an alternative measurement method. This could prevent false information barrier outputs and allow better attribute thresholds. Still, it is advisable (and one might find it crucial) to use measurement methods where the inspecting party has confidence in its functionality apart from assurances, e.g. regarding shielding, by the inspected party. Specific information on the influence of sample configuration and shielding for plutonium measurements is provided in the text box.

This issue can be conceptualised as a systematic uncertainty. For inspectors and hosts to develop an understanding of the reliability of specific measurement methods and attribute analyses, it is important to test (for example using simulation tools) how large deviations between real and measured values become as item and shielding properties vary in plausible manners. This will allow estimating the magnitude of systematic uncertainty the inspector faces when it is impossible to obtain further information on the item. When the systematic uncertainty is assessed to be very high, the inspector may discuss such an issue with the host—it may be the host's interest then to relieve it by further information. With a low systematic uncertainty, the inspector may assert pressure on the host to declare more plausible attribute thresholds closer to expected actual ranges.

---

[3]When considering the possibility of declaring shielding, the Nonproliferation Treaty Art. I and II must be considered. Even if specific shielding information is not considered proliferative, it may currently be classified. For allowing effective verification, it may, however, be in the interest of the host to review whether certain information can be declassified. In [3, p. 81], further information is provided.

Gamma Spectroscopy and Neutron Multiplicity Counting for Plutonium
Plutonium-239 has a wealth of gamma lines, plutonium-240 has significantly less with significant lines at 104.2, 160.3 and 642.4 keV. In all of these cases, lines of plutonium-239 are in proximity (for example 103.0 keV; 160.2, 161.5 keV; 640.0, 645.9 keV). Gamma shielding can be an issue, especially at energies below 200 keV as can be seen from the measured spectra (shown below) of a 12.5 g plutonium sample ("PM1", 95.4% Pu-239, 4.5% Pu-240) in an unshielded configuration, with lead and with a container that is suited to contain warheads ("AT-400"). The shielded spectra reduce or suppress relevant plutonium peaks. This is due to the very small lead thickness required to strongly attenuate photons in this energy region, as displayed in the table below. At higher energies (600 keV), attenuation is weaker, so analysing the plutonium isotopic composition in this region may be more feasible.

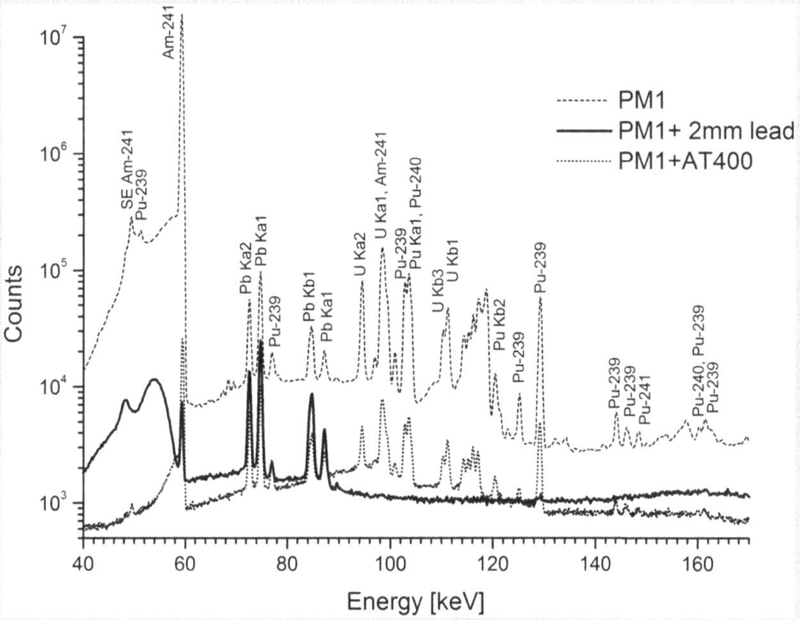

Gamma spectra of an unshielded and shielded plutonium sample [7]

A further issue is that gamma radiation from the inside of the fissile material is attenuated to a large extent before reaching the surface so that an isotopic composition assessment does not give the average isotopic

(continued)

composition of the volume if it is heterogeneous, but rather at the volume's surface. The importance of this effect is illustrated by the following table:

Gamma travel length until 50% are absorbed[8]

	Lead	Plutonium
150 keV	0.3 mm	0.1 mm
600 keV	5.2 mm	2.4 mm

Self-absorption of gamma radiation by plutonium will result in only measuring a portion of the total plutonium mass. Therefore, this attribute could rather be assessed by neutron counting. To achieve this, the spontaneous fission rate would need to be measured from which the plutonium-240 mass could be deduced. Then, the total plutonium mass can be calculated when the isotopic composition is known, e.g. from gamma spectroscopy. Neutron multiplicity counting is a technique that can distinguish neutrons from spontaneous fission, induced fission and other sources. Multiplicity counting is however dependent on the sample configuration and shielding between fissile material and detector. For example, neutrons may be absorbed in shielding and therefore reduce the detection efficiency. They may also scatter elastically, which has an effect on the neutron energy spectrum, which also influences the detection efficiency. Induced fission events produce additional neutrons, the number of which depends on both fissile mass and geometry. Without further knowledge on the configuration, the measured fissile mass may be biased. This bias can be reduced, however, if certain information is available [9].

4 How to Authenticate Nuclear Warheads?

In terms of unknown properties, nuclear warheads pose a significant challenge among the various types of items discussed. How to establish confidence if high uncertainties remain? One option is to shift the issue of warhead authentication to the authentication of dismantled warhead components. In the case of warhead component authentication, a reasonable approach may be to share the container design or elements of it with the inspecting party [3, p.33] and/or include the storage container in the calibration of the detector. For fully assembled warheads, the issues will be much more difficult to resolve due to the sensitivity of shielding that is part of the warhead. Without sufficient shielding information, none of the quantitative attribute assessments appear to be robust.

Therefore, a concept of taking authentication measurements after dismantlement may be favored. The challenge then is to guarantee that the continuity of knowledge is sufficient for giving confidence that (1) the authenticated component belongs to the original warhead and that (2) no fissile material was introduced or diverted during the dismantlement process when inspectors will most likely not be present. Theoretically, this should be sufficient since—through continuity of knowledge—the authentication result post dismantlement would also be valid pre-dismantlement. This would put a heavy burden on procedures and techniques supporting continuity of knowledge which will not be addressed here in detail.[4] Given the restrictive access to high security facilities where nuclear weapons are handled, the challenge to undermine possibilities for cheating may be significant since the host remains the custodian and controls inspector movement and actions.[5] Therefore, it may be argued that authenticating warhead components shifts the challenge from authentication to continuity of knowledge. Future research is needed for elucidating the consequences of this shift on the overall process.

In addition, there are political considerations. The time between an initial warhead declaration and its dismantlement are unclear. A baseline inventory declaration does not necessarily need to be given at the time of warhead dismantlement. Postponing authentication, however, might not be acceptable, but immediate verification might be requested. Taking these considerations into account, a procedure could be:

1. Performing authentication measures on fully assembled warheads upon their declaration to provide an initial level of confidence
2. Performing authentication measures on the warhead components post dismantlement to provide a higher level of confidence than what was possible prior to dismantlement, for example due to more realistic attribute thresholds.

If both tests produce green lights, the item passes. If the first assessment is positive (green), but the second negative, the inspector has reason to question the first assessment. This procedure could not only provide both timely and reliable verification, but could also discourage cheating: While cheating is more likely prior to dismantlement due to the technical measurement challenge, it would be much harder post dismantlement since the measurement would be more reliable. A nuclear weapon state would be deterred from cheating in the beginning because of the substantive risk of detecting cheating later.

[4]A detailed discussion of the advantages and disadvantages of performing authentication measurements at various stages can be found in 1[10].

[5]The UK-Norway Initiative has conducted exercises to study the confidence inspectors can gain while restrictive access procedures are in place [11].

5 When the Light Is Red ...

All imperfections of information barriers discussed above may result in weak attributes not suited to create sufficient confidence. But also if powerful attributes have been agreed, measurement issues remain very relevant as they may lead to false positives and false negatives. Developing a robust authentication system should be prioritized due to the importance of correctly authenticating warheads which becomes clear by realising the devastating destruction potential. The potential implications would be immense, therefore both types of uncertainties should be reduced as far as possible. For inventory declarations, false positives would deceive the inspector of the military capabilities which may result in (re-)armament. False positives during warhead dismantlement would allow the host to "dismantle" fake warheads, while keeping the real ones and presenting the warheads at a later date when other countries have reduced their arsenals. False negatives, on the other hand, may result in immense political implications, although the host did not cheat. Several options may be possible to cope with this issue:

- Repeated measurements: Both false positives and false negatives may be the result of statistical uncertainty. Kütt et al. have developed strategies for repeated measurements with minimum probabilities of false positives and false negatives [12]. If the indication of the first measurement is negative, but the indications of further measurements are positive, confidence can be established. In fact, Kütt et al. propose repetitions for all authentication cases. However, this procedure is likely to fail if the mis-classification is not caused by purely statistical uncertainty but includes systematic bias, e.g. due to undeclared shielding.
- An additional measurement using information barriers based on complementary assessments: Repeating measurements with the same information barrier does not solve systematic uncertainties. However, the same attributes may be assessed using measurement techniques that are based on different physical processes. This is a way of identifying false positives and negatives of the first measurement. An example is a calorimetric measurement of the fissile mass in addition to neutron measurements. Another example is using different information barrier concepts, for example testing items with an information barrier based on the template approach which is described in the subsequent chapter.
- Consultation with the host: In the case of false negatives, the host should be interested in resolving the issue. In consultations with the inspecting party, the host may try to offer explanations why the information barrier yields a negative response despite measuring the declared item. The amount of credibility given to such statements depends on the wider political context and evaluating the host's incentives to cheat, but also on the willingness to accept additional measurements with a different technique.

Generally, a binary information barrier output is being conceptualised. However, we have come to the conclusion that, due to the physical reality of systematic and statistical uncertainties, a red or green light indication that an item is as declared may be too simplistic and may cause misinterpretations. Concepts requiring further research include the introduction of a yellow light to inform the inspector that a measurement has been inconclusive or even the output of a confidence index, which yields an assessment of how likely it is that an item is as declared. Bayesian statistics may offer a convenient framework for such an approach [13].

Overall, this chapter has introduced the concept of the attribute information barrier. While this concept is not new, many challenges still need to be resolved. This is, however, true for most verification techniques. In our understanding, it is the combination of different approaches and elements that yield confidence, and an attribute information barrier would be embedded in a web of additional measures, in particular related to unique identification and continuity of knowledge. Even with remaining limitations it may turn out to be an invaluable instrument in adding to the overall confidence that needs to be established for nuclear disarmament.

The current status of research is well characterized by the key judgment of the International Partnership for Nuclear Disarmament Verification in its Phase I Summary Report: "While tough challenges remain, potentially applicable technologies, information barriers, and inspection procedures provide a path forward that should make possible multilaterally monitored nuclear warhead dismantling while successfully managing safety, security, non-proliferation, and classification concern in a future nuclear disarmament agreement" [14].

Acknowledgement We thank the German Foundation for Peace Research for funding our research activities.

References

1. Russian Federation and United States of America (2010) Treaty between the United States and the Russian Federation on measures for the further reduction and limitation of strategic offensive arms
2. Shea TE (2014) The Trilateral Initiative: IAEA verification of weapon-origin plutonium in the Russian Federation and the United States. In: Proceedings of the 2014 International Atomic Energy Agency Symposium on International Safeguards, Vienna, 20–24 October
3. Nuclear Threat Initiative (2014) Innovating verification: New tools & new actors to reduce nuclear risks: Verifying baseline declarations of nuclear warheads and materials, Washington DC
4. MacArthur D, Hauck D, Thron J (2012) Simultaneous authentication and certification of arms-control measurement systems. In: Proceedings of the Institute of Nuclear Materials Management 53rd Annual Meeting, Orlando, 15–19 July 2012
5. Warren G, Archer DE, Cunningham M, McConchie S, Thron J (2012) Concepts for the measurements subsystems of the third generation attributes measurement system. In: Proceedings of the Institute of Nuclear Materials Management 53rd Annual Meeting, Orlando, 15–19 July 2012
6. Fissile Material Transparency Technology Demonstration, Technical Overview of Fissile Material Transparency Technology Demonstration (2016) Executive Summary. http://www.lanl.gov/orgs/n/n1/FMTTD/presentations/pdf_docs/exec_sum.pdf. Accessed 28 Feb 2019

7. Göttsche M, Kirchner G (2014) Measurement techniques for warhead authentication with attributes: Advantages and limitations. Science and Global Security 22(2):83–110
8. Storm E, Israel HI (1967) Photon cross sections from 0.001 to 100 MeV for elements 1 through 100. LA-3753, Los Alamos Scientific Laboratory
9. Göttsche M, Kirchner G (2015) Improving neutron multiplicity counting for the spatial dependence of multiplication: Results for spherical plutonium samples. Nucl. Inst. Meth. Phys. Res. A 798:99–106
10. MacArthur D, Hauck D, Smith M (2013) Confirmation of nuclear treaty limited items: Pre-dismantlement vs. post-dismantlement, ESARDA Bulletin 50:116–123
11. Backe S, Enger E, Hustveit S, Hoibraten S, Kippe H, Mykkeltveit S, Reistad O, Sekse T, Sidhu RS, Waters C, Chambers D, White H, Russell I, Allen K, Collinson A (2012) The United Kingdom - Norway initiative: Further research into managed access of inspectors during warhead dismantlement verification. In: Proceedings of the Institute of Nuclear Materials Management 53rd Annual Meeting, Orlando, 15–19 July 2012
12. Kütt M, Philippe S, Barak B, Glaser A, Goldston RJ (2014) Authenticating nuclear warheads with high confidence. In: Proceedings of the Institute of Nuclear Materials Management 55th Annual Meeting, Atlanta, 20–24 July 2014
13. Zähringer M, Kirchner G (2008) Nuclide ratios and source identification from high-resolution gamma-ray spectra with Bayesian decision methods. Nucl. Instr. Meth. Phys. Res. A 594:400–406
14. The International Partnership for Nuclear Disarmament Verification (2017) Creating the Verification Building Blocks for Future Nuclear Disarmament. Phase I Summary Report. https://www.ipndv.org/reports-analysis/phase-1-summary/. Accessed 28 Feb 2019

Minimally Intrusive Approaches to Nuclear Warhead Verification

Alexander Glaser and Yan Jie

Abstract Future arms control treaties may place limits on the total number of warheads in the arsenals of weapon states, which would introduce qualitatively new verification objectives, including confirming numerical limits on declared nuclear warheads and confirming the authenticity of nuclear warheads prior to dismantlement. Meeting these objectives would require on-site inspections at new types of facilities, including warhead storage sites, which could put at risk highly sensitive information both related to military operations and warhead design. Weapon states may be reluctant to consider some of the anticipated procedures. As a way to address this challenge, in this paper, we examine the potential of verification approaches that emphasize non-intrusiveness from the outset. Relevant examples include innovative tagging approaches and hashed declarations to confirm the correctness of warhead declarations and novel types of inspection systems to confirm the authenticity of nuclear warheads, while satisfying the different and sometimes conflicting requirements by the host and the inspector. New international R&D efforts could usefully focus on non-intrusive technologies and approaches, which may show more promise for early demonstration and adoption. If demonstrated, such non-intrusive verification approaches could be particularly important for moving forward discussions about expanding the scope of current agreements and facilitating discussions with weapon states that have so far not been part of formal nuclear arms control agreements.

A. Glaser (✉)
Program on Science and Global Security, Princeton University, Princeton, NJ, USA
e-mail: alx@princeton.edu

Y. Jie
China Academy of Engineering Physics, Mianyang, China
e-mail: yanjie@caep.cn

I. Niemeyer et al. (eds.), *Nuclear Non-proliferation and Arms Control Verification*,
https://doi.org/10.1007/978-3-030-29537-0_15

1 Background

Verification of future arms control agreements is likely to face some fundamentally new and complex verification challenges, which may require trusted, secure, and potentially non-intrusive technologies and protocols designed to support baseline declarations, information exchanges, and inspection activities. This chapter explores and illustrates with examples two verification tasks that are likely to be central: confirming numerical limits on declared nuclear warheads and confirming the authenticity of nuclear warheads before they are dismantled.

Given the complexities of protecting sensitive information that may be exposed to an inspecting party during these activities, here, we emphasize non-intrusive verification approaches and explore, in particular, the concept of *not* acquiring sensitive information in the first place. This approach may alleviate reservations and concerns that some weapon states may have with regard to on-site inspections and encourage them to actively engage in R&D and discussions at an early time. Ideally, these initial efforts could over time evolve and expand as parties become more familiar and confident with the process.

Here and in the following, for simplicity, we refer to nuclear warheads when discussing verification options and challenges. More generally, the concepts could also apply to warhead components or any other treaty-accountable item whose design or location is considered sensitive for nonproliferation or security reasons.

2 Confirming Numerical Limits on Declared Nuclear Warheads

Procedures and techniques to confirm upper limits on the number of nuclear warheads will become a key verification objective should future arms control agreements place limits on the *total* number of nuclear weapons in the arsenals. Verifying such agreements would then require the ability for inspectors to "count" individual warheads (rather than launchers). In principle, this can be accomplished by tagging treaty-accountable items with unique identifiers (UIDs), which transforms a numerical limit into a ban on untagged items [1, 2]; but, as further discussed below, this may be difficult to implement in practice because the host may have safety, performance, or other concerns, and inspections would necessarily be highly intrusive.

Depending on the type and deployment status of a nuclear weapon, confirming numerical limits on warheads faces rather different verification challenges. In the following, we distinguish three categories and then focus the discussion on non-deployed warheads, which is the most difficult category to address.

Retired Warheads (and Warhead Components) In the case of nuclear warheads that are in long-term storage without being moved and are awaiting dismantlement, monitoring could be relatively straightforward and based on standard containment and surveillance (C/S) methods already used by the International Atomic Energy Agency [3]. Treaty-accountable items in this category could also include warhead components in classified or unclassified form. In principle, these items could be placed in containers, which could then be tagged and sealed. Containers may not have to be accessed over long periods of time, and their storage locations may not be considered sensitive.

Deployed Strategic Warheads At the other end of the spectrum are deployed strategic weapons. Given their military significance, verification procedures are complex, but detailed procedures have been successfully developed and used as part of the INF, START, and the New-START treaties. The types and amount of information exchanged between the state parties has significantly increased over time, even if relatively little is made public. Verification of the START treaties fundamentally relies on counting launchers (e.g. missile silos), delivery vehicles, and the warheads associated with them [4]. If necessary, neutron detectors can be used to identify non-nuclear objects that may be present on missiles or bombers. Over the years, the United States and Russia have accumulated a vast and diverse experience with verifying limits on deployed strategic weapons; this experience could be shared with other parties when needed [5].

Non-deployed (Inactive) Warheads and Non-strategic Weapons Arguably the most difficult category of nuclear warheads to monitor include non-deployed, inactive warheads, and non-strategic weapons. In U.S. terminology, inactive warheads are part of the nuclear arsenal, but they are maintained at depots in non-operational status [6]. As such, they are, of course, not associated ("mated") with delivery vehicles. In the case of the United States, inactive warheads are part of a reserve (or hedge), but other weapon states may primarily have warheads that fall into this category as warheads and delivery vehicles are often stored separately. Being able to monitor non-deployed warheads is therefore particularly important if progress toward multilateral arms control is a priority. Some weapon states also have non-strategic (tactical) weapons, and very much the same verification challenges apply to them. The locations and movements of warheads and weapons in this category would generally be considered sensitive information, and many of the techniques and approaches available for the first two categories do not apply here.

Our discussion below is based on some important premises that acknowledge the interests of a host state not to disclose or risk disclosure of sensitive information; these premises also emphasize simplicity of potential verification approaches, which could help facilitate early implementation. Over time, additional and perhaps more intrusive provisions could be added.

Premise 1: No Need for (Early) Authentication Confirming the authenticity of nuclear warheads is perhaps the most challenging verification task in this area, and it is discussed separately below. It is often assumed that warheads would have to be authenticated *before* they can be counted, but this may be an unnecessary constraint. Viable inspection systems for warhead authentication are years away from demonstration or deployment, and it may be unhelpful to postpone the "counting task" until these systems become available. Instead, one could simply count declared items. In practice, this could mean that "bogus items" are being tracked, i.e., that a party is overstating its warhead inventory. Allowing for this ambiguity, however, is potentially an advantage for countries that do not want to reveal the true initial inventory or want to retain the option of increasing the inventory within the limits of a treaty later on.[1]

Premise 2: No Need for Direct Tagging Tagging is a powerful tool to track accountable items of almost any type, and it is routinely used for IAEA safeguards. It is also used under the New START Treaty to identify delivery vehicles [4]. Directly tagging nuclear warheads, however, may be considered problematic or impractical for several reasons. Confirming the integrity and authenticity of the tags would require direct access to the warhead, which may raise a variety of security concerns. Operationally, it would be extremely difficult to implement. If tags are applied to containers associated with non-deployed warheads, then they are of little use unless the containers are securely sealed. This may not be practical for warheads in storage as they may be undergoing maintenance or other procedures that require direct access; it is hard to imagine that states would accept procedures that could interfere with military operations. Overall, a verification approach that does not rely on direct tagging of nuclear warheads would be much easier to implement.

Premise 3: No (Early) Need for Continuous Chain of Custody Chain of custody seeks to ensure that "a declared warhead, once identified, is continually accounted for and tracked from deployment to dismantlement" [7]. Research efforts in the United States and the United Kingdom have placed strong emphasis on chain-of-custody approaches for nuclear arms control [8]. A recently completed 3-year project evaluated, for example, the use of a "real-time system for counting items of inspection using radio-frequency identification (RFID) tags" [9]. In practice, chain of custody combined with real-time monitoring could provide, for example, continuous information about the presence of an item or the state-of-health of a seal, which could in turn reduce the number of on-site inspections that would otherwise be deemed necessary. Implementing continuous chain of custody in a treaty context, however, has to be considered a difficult (and perhaps also controversial) undertaking as warheads move through a complex and varied operational environment within the military complex.[2] Even if these obstacles could be overcome, and while recognizing that chain-of-custody concepts would be very valuable in making verification approaches more robust, they are neither necessary nor sufficient to confirm numerical limits on nuclear warheads. Overall, it appears more useful to

[1]Allowing for bogus items to be counted may require a procedure at the authentication stage to confirm that they have been invalid all along; presenting an invalid item, for example, at the time of verified dismantlement, could raise concerns about non-compliance.

[2]Moreover, existing RFID technology has not been designed for security applications and carries the risk of being spoofed: *"RFIDs are inventory devices, not security devices, and should not be used to make determinations about nuclear theft, diversion, tampering, or espionage"* [10].

focus on the critical elements of possible verification approaches first; chain-of-custody could then be used or added where considered particularly useful, for example, to follow a warhead through a dismantlement facility.

Similar to chain of custody, researchers at Los Alamos have developed the "provenance" concept, which seeks to verifiably answer the question: *Has the item undergone movements or come from a location consistent with being a warhead?* [11]. This approach has potentially great merit, but it could raise concerns that are similar to those of chain-of-custody concepts.[3]

Verification approaches that can be implemented based on these three premises may be considered particularly non-intrusive. They can achieve a high level of non-intrusiveness because potentially sensitive information and operational details are never acquired, and they should therefore be easier to implement at an early stage. Below we discuss two examples that meet these criteria. Of course, many other implementations may be possible.

Proposal 1. Hashed Declarations The concept of hashed declarations as a basis for verifying limits on the number of nuclear warheads in the arsenals has been previously recognized. In general, the hash (or "message digest") is much shorter than the message itself, but the underlying cryptographic hash function is designed such that it is extremely difficult to find a valid message for a given hash or to construct two different messages that produce the same hash (collision resistance) [12]. A 2005 National Academies study discussed the applicability of the concept in some detail [13]. Here, for illustrative purposes, we provide a slightly modified and simplified example.

Under such a framework, treaty parties would regularly exchange hashed declarations with entries for each declared nuclear warhead. These declarations could even be made public as they do not contain any information besides the total number of declared items. In preparation for an on-site inspection and based on the most recent available declaration, the inspecting party could randomly pick one entry and request the cleartext for that entry revealing *inter alia* the location of one warhead (Fig. 1). At that point, a stand-down for the relevant site would take effect, which could be verified by national technical or other means. The host party would then reveal the entries for all other warheads that are stored at the same site. The inspecting party could confirm that the cleartext produces the correct message digests and proceed to inspect the site itself. Once the inspecting party arrives at the site, the host would present the declared number of warheads. Procedures would have to be available to confirm that additional objects that could be mistaken for nuclear warheads are in fact not; this is not necessarily a trivial task, but potentially much easier than confirming the authenticity of a warhead, which is discussed in the

[3]It is also worth noting that a capability to demonstrate to an inspector the provenance of a nuclear warhead could ultimately provide sufficient confidence in its authenticity so that a dedicated authentication step (e.g. using radiation measurements combined with attribute or template methods) may be considered unnecessary.

	Basic Information (Warhead type and location)	Additional Information 1 (for potential later release)	Additional Information 2 (for potential later release)
#0001	330E3EFAF5CDC3C1 49D7948186BDECA7 992EC42AF67D7526 0ECD8ED245F6EC10	FCF8203ACEDE4E10 B3666E3F483EEFE6 649528EB1D63BBB2 B15B226869177E2A	0D41CD73CFF738C2 618D32F9EB756418 17FF0FF4F30204EE 54329F29B0CFFD3B
#0002	AB9DCD2AAB42F17D 0DA6987EBDAC0D15 BAE8378876627C28 0B5360504A695287	B8B50D3B69ADB200 17B04F5D30702682 9AF20AB9D572AD3A 145D21F7AC1417A6	6BC70957F2AF770C 11B41BEFFA3D5FF6 005F7A557EDACEBB 10C21FE091E82A4E
#0003	13D1EBA7C8666343 3888FCE752AAD556 7DE9DF29F863085D D31A188D37D4DBEF	1CACE90DCB1FD968 C08C4FF9F02119C1 35BC8BBEAFEE2041 6B6CD1405CCB6706	6056F5458255AA7E 80E9D372A92C67B6 D9C174DF074B4EAC D1BF6E1081E3642F

<table>
<tr><th></th><th>Basic Information
(Warhead type and location)</th><th>Additional Information 1
(for potential later release)</th><th>Additional Information 2
(for potential later release)</th></tr>
<tr><td>#0001</td><td>8C+oNYGZ%|x)>Ly1Tfw5
Zel > W-99 < j]5
@-t > +40.3492 < 58]
EIa > -74.6595 < w_w
NQJ=k9kNd*`(>mgu@SG[</td><td>FCF8203ACEDE4E10
B3666E3F483EEFE6
649528EB1D63BBB2
B15B226869177E2A</td><td>0D41CD73CFF738C2
618D32F9EB756418
17FF0FF4F30204EE
54329F29B0CFFD3B</td></tr>
<tr><td>#0002</td><td>AB9DCD2AAB42F17D
0DA6987EBDAC0D15
BAE8378876627C28
0B5360504A695287</td><td>B8B50D3B69ADB200
17B04F5D30702682
9AF20AB9D572AD3A
145D21F7AC1417A6</td><td>6BC70957F2AF770C
11B41BEFFA3D5FF6
005F7A557EDACEBB
10C21FE091E82A4E</td></tr>
<tr><td>#0003</td><td>O]z$'^(SD,`iT<]q!TCZ
Fsu > W-99 < ^17
BB7 > +40.3492 <)uP
%ym > -74.6595 < Ph2
H{iW+ORHNM)-']|KKgI<</td><td>1CACE90DCB1FD968
C08C4FF9F02119C1
35BC8BBEAFEE2041
6B6CD1405CCB6706</td><td>6056F5458255AA7E
80E9D372A92C67B6
D9C174DF074B4EAC
D1BF6E1081E3642F</td></tr>
</table>

Fig. 1 Hashed declarations to confirm numerical limits on nuclear warheads. The inspector is allowed to see basic-information cleartext for one of the listed entries. The host then reveals the selected entry and all other entries for the same site. The inspector confirms the digests and can then inspect the site. Alternatively, the inspector could request cleartext for all entries of one particular location (for example, a known warhead storage site). Extra columns could contain additional information (e.g. serial numbers) that may or may not be revealed at a later date. Shown message digests are for illustration purposes only; actual digests may have to look differently to meet security requirements. *Graphics: Alexander Glaser*

next section. Inspectors may also be allowed to access other areas to gain confidence in the absence of undeclared accountable items at the site.

Note that, in this most basic scheme using hashed declarations, no tags whatsoever are required; and yet, hashed declarations could over time provide high confidence in the correctness of warhead declarations made by the parties of a treaty. Hashed declarations could also include extra columns for additional information that is not necessarily revealed in the initial stages of an agreement (as illustrated in the figure; such as serial numbers or perhaps even the amounts of special nuclear material in particular warhead types). Disclosure of these entries in the future could further increase the confidence in the correctness and completeness of the declarations, potentially reaching back for years or decades.

Proposal 2. Buddy-Tag Concept Buddy tags are tokens that are used to prove that one owns an object without having to present it [14]. In an arms-control context, for warhead monitoring, each treaty partner would receive a number of buddy tags, nominally one for each treaty-accountable item. While each accountable item has exactly one "buddy" that is associated with it, the tag itself is not directly attached to or even necessarily nearby the item of interest. During a short-notice inspection, the treaty partner must be able to present exactly one buddy tag for each declared item at the site (Fig. 2). Inspectors could then count and authenticate the buddy tags, which would be particularly difficult to counterfeit. Sensors in the tag, which can itself serve as a tamper-indicating enclosure (TIE), would show that it has not been moved to the site after the inspection was called. National Technical Means or other surveillance measures might be used to ensure that treaty-accountable items are not moved in a stand-down situation in the meantime.

Fig. 2 Buddy tags for minimally intrusive on-site inspections at a hypothetical storage site. On the left: Inspectors have access to agreed areas where buddy tags are stored and confirm the authenticity of the tags based on unique identifiers affixed to them. Inspectors can then visually confirm that an identical number of treaty-accountable items are also present at the site; they cannot, however, directly access or inspect them. Confirmation of items (e.g. warheads) may only occur at a later stage, e.g., when selected warheads are offered for dismantlement. On the right: close-up of a buddy tag, about 15 cm on each side, with a unique identifier (reflective particle tag) and LED indicators in a tamper-indicating enclosure [5]. *Graphics: Tamara Patton, Princeton University*

Once the treaty-accountable item and its tag are physically separate, much more flexible and non-intrusive verification approaches for on-site inspections become possible. In particular, as in the case of the hashed declarations, immediate access to sensitive treaty-accountable items (e.g. nuclear warheads) may no longer be required and agreed inspection activities can be much less intrusive than they otherwise would have to be. Inspection of the tag does not put any sensitive information at risk. In particular, the tag may not necessarily reveal the precise location of the item it is associated with, and it may help protect some other operational details that the host might consider sensitive. Since buddy tags cannot be quickly transferred between sites to match the number of warheads at the selected site without detection, the concept can provide confidence that the party is in compliance even if only one site is inspected at a time.

Note that the buddy-tag concept does not involve any declarations at all; in many ways it is complementary to the concept of hashed declarations discussed above, and both concepts could be combined to provide a more robust verification approach.

3 Confirming the Authenticity of Nuclear Warheads

At some point prior to dismantlement, and even if verification arrangements (seeking to confirm numerical limits on nuclear warheads) have been in place for extended periods of time, the inspecting party will require reassurance that declared warheads are authentic and reductions in the arsenals credible.

There are several plausible cheating scenarios that an inspecting party might worry about. First, the inspected party might present objects that are similar to genuine warheads, but prior to inspection some fissile material is removed or replaced by other nuclear materials (e.g. natural uranium); the main objective of this strategy would be to withhold fissile material, presumably for future use in weapons. A second scenario would involve presenting objects that might or might not resemble real warheads, but they could be designed primarily to deceive the inspection system. These objects would presumably contain some fissile material in order to pass an inspection based on radiation measurements. The main objective of this strategy would be to withhold real warheads or extra amounts of fissile material.

Nuclear warheads contain kilogram quantities of plutonium or highly enriched uranium and often both. Even when placed inside a missile or a storage container, these materials emit detectable neutron and gamma radiation signatures that are unique for special nuclear materials, the respective amounts of these materials, and their configuration.

Two fundamental concepts have been proposed to confirm the authenticity of nuclear warheads: the attribute approach and the template (or template-matching) approach. In a nutshell, the attribute approach examines a set of properties that are considered characteristic for nuclear weapons; this can include qualitative criteria, such as the mere presence of a special nuclear material (e.g. the presence of plutonium), and quantitative criteria, such as meeting agreed threshold values for mass or isotopics (e.g. a maximum ^{240}Pu content). In contrast, the template approach does not seek to determine absolute or relative attributes of the inspected item; instead, it compares a unique radiation signature or "fingerprint" against a previously recorded template generated with a reference item that is known or believed to be authentic.[4]

The radiation signatures acquired for both the attribute and the template method are considered highly sensitive and cannot be revealed to an inspecting party seeking to confirm the authenticity of a nuclear warhead.[5] Primarily for this reason, warhead inspections would necessarily involve complex measurement techniques and procedures. To enable such measurements, the concept of the information barrier has been developed since the 1990s. An information barrier processes the acquired radiation signatures, but displays the outcome of the analysis in a simple pass/fail manner. The critical functional requirements for the barrier are threefold: First, the inspected party must be assured that classified information is protected so that under any circumstances, i.e., even when the equipment is malfunctioning or operated incorrectly, only non-sensitive information is presented to the inspecting party ("certification"); second, the inspecting party must be confident that the inspection system measures, processes, and presents the conclusion drawn from the data in an accurate and reproducible manner ("authentication"); and third, the system must reliably distinguish valid from invalid items ("confirmation") [11]. Simultaneously certifying and authenticating information barriers has been the most serious obstacle to demonstrating the concept as a viable verification technology [15].

Joint development of the information barrier, including of both its hardware and software, has generally been a key strategy to facilitate mutual confidence and trust in the inspection equipment [16]. Even then, however, the question remains who supplies or has last custody of the equipment that will be used in an actual inspection. Project participants have usually concluded that the information barrier would *defacto* be host-supplied. This makes certification relatively straightforward, but authentication for the inspector extremely difficult. From the perspective of information security, host-supplied equipment appears preferable but, from the outset, the inspector is at a disadvantage given the difficulty of authentication equipment in a host-controlled environment. This is particularly difficult, and perhaps impossible, given recent advances in stealthy insertion of Hardware Trojans, which could trigger hidden switches to produce non-genuine results during an

[4]In principle, and in addition to nuclear radiation signatures, a fingerprint used for the template method can also include mechanical, thermal, electrical, or acoustical properties. These non-nuclear signatures cannot be sensitive to isotopics, however.

[5]Some efforts have been underway to determine which signatures could be declassified. Namely, *"the United States is conducting a nuclear warhead modeling and measurement campaign to establish a comprehensive nuclear warhead and component signature set. The resulting data will support assessment of sensitive information that could be revealed as a result of future treaty verification activities, and will further guide future R&D in the areas of radiation detection and information protection"* [8].

inspection.[6] There has been some interest in challenging the premise of host-supplied equipment in favor of equipment that would be provided by the inspector instead [17, 18].

Finally, both attribute and template systems face some additional challenges that are characteristic for each approach. In the case of attribute measurements, the question arises what types of attributes should be selected and what the exact threshold values for those attributes should be. A party may even refrain from proposing certain attributes worrying that the mere suggestion might reveal weapon features that the other party is unaware of. In general, the more representative the attributes (and, when applicable, their threshold values) are, the more robust the verification approach would be; but more information about the inspected warhead would necessarily be revealed also. In the case of template measurements, qualitatively different challenges exist; these include: how to establish the authenticity of the template in the first place, how to protect sensitive design information that the template contains, how to account for differences between valid items (e.g. manufacturing tolerances, age of material) and how to store the template between measurements so that the inspector remains confident in its authenticity.

In all traditional approaches to nuclear warhead verification, the origin of the problem is due to the acquisition and storage of sensitive information, which has then to be protected, removed, or destroyed. This requires an information barrier that both the host and the inspector simultaneously trust. In contrast, if sensitive information is never measured at all, many of the challenges outlined above disappear.

Confirmation of Warhead Authenticity Using a Zero-Knowledge Approach One way to resolve the difficulty of obtaining a trusted information barrier is to remove complex equipment at critical points in the measurement process altogether, i.e., to avoid the use of an engineered information barrier that both parties have to trust. This may offer a fundamentally different approach to warhead verification that avoids the measurement of sensitive information at the outset as a way to address concerns about its potential leakage (Fig. 3).

The core of the idea is the concept of a "zero-knowledge proof," a class of interactive proof systems developed by mathematicians working in cryptography and intended to prove an assertion while yielding nothing beyond the validity of the assertion being proved. Although widely used today in the digital domain, this concept has so far not found practical real-world physical applications [21].

Among the many possible strategies to implement a zero-knowledge proof for confirming the authenticity of an item offered for inspection, here we use a particularly simple concept to demonstrate the proof-of-principle. The marble example in the figure illustrates the fundamental idea. Similar to this example,

[6]Insertion can occur at *"any stage during the design and manufacturing process, including the specification, design, verification and manufacturing stages"* [19]. Recent research has examined (and demonstrated) the insertion of Trojans at the sub-transistor level, which is resistant to most detection techniques that are currently in use [20].

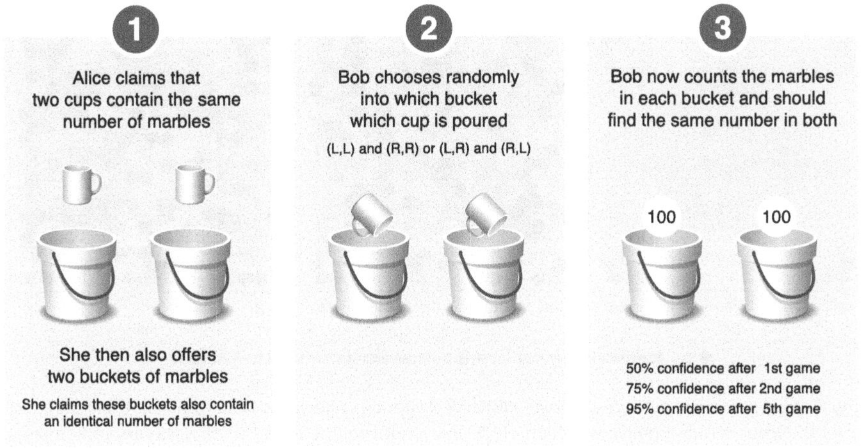

Fig. 3 A simple example of a zero-knowledge proof to demonstrate that two cups contain the same number of marbles. Alice (the host) has two small cups both containing X marbles, where X is some number between 1 and 100. She wants to prove to Bob (the inspector) that both cups contain the same number of marbles, without revealing to him what this number X is. To do so, Alice prepares two buckets, which she claims each contain (100-X) marbles. Bob now randomly chooses into which bucket which cup is poured. Once this is done, Bob verifies that both buckets contain 100 marbles [22]. *Graphics: Alexander Glaser*

pairs of neutron detectors (or pairs of detector arrays) will be preloaded by the host prior to inspection. During the inspection, the inspector randomly chooses which detector board is used with which item, i.e., the reference item or the item offered for inspection. The host has chosen the preloads such that, in the actual inspection and using the agreed experimental setup and procedures, the neutron count in all detectors reaches a previously agreed value if a valid item (identical to the reference item) is presented. Just as in the case illustrated in the figure, the inspector's confidence the honesty of the host increase as more rounds of inspection are carried out. Technically, this concept is based on the following two key elements:

Differential Neutron Radiography Measurements 14-MeV neutrons from a deuterium-tritium (DT) neutron generator illuminate the inspected item making what in effect are differential measurements of both neutron transmission and neutron emission. This method is particularly sensitive to changes in geometry and elemental composition (see Fig. 4).

Preloadable Non-electronic Detectors To overcome the difficulties of certifying and authenticating electronic equipment such as that connected to a radiation detector for signal processing or used in information barriers, non-electronic detectors offer additional advantages when combined with the proposed setup. So-called superheated droplet (or "bubble") detectors, which consist of small (approx. 0.05 mm) fluid droplets that are suspended in a liquid matrix. Neutrons interacting with the superheated fluid in these droplets produce ions through recoil interactions,

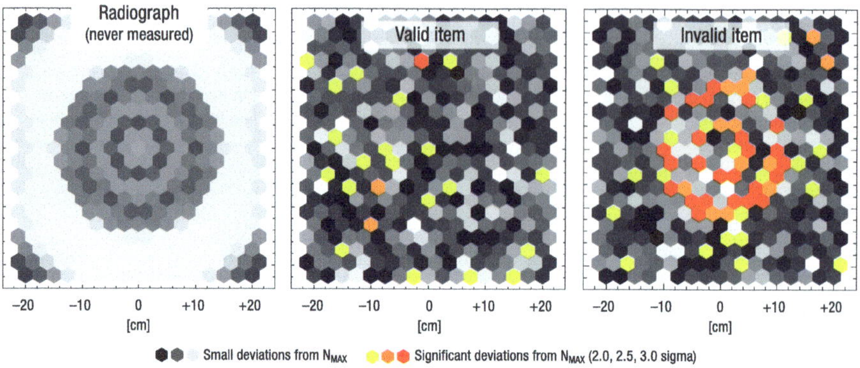

Fig. 4 How can one be convinced that a nuclear weapon is real without learning anything about its design? Neutron radiography with non-electronic preloaded detectors offer a way to implement a zero-knowledge protocol to accomplish this task. If a valid item is presented, then no information will be registered in the detector bank; in contrast, an invalid item will both fail the inspection and leak information—an additional incentive for the host not to cheat [22]. *Graphics: Alexander Glaser*

which deposit their energy in the droplet, evaporating the liquid, and creating detectable macroscopic bubbles with a diameter of about 0.5 mm. The number of bubbles produced in the detector directly scales with the neutron fluence, and their detection threshold can be tuned to any desired minimum neutron energy. For radiography with 14-MeV neutrons, one can set this threshold, for example, to 10 MeV, which minimizes the impact of scattered neutrons that would otherwise "blur" the radiograph. Bubble detectors also are completely insensitive to gamma radiation, which is another important advantage for this application.

If implemented properly, this approach achieves a proof that is complete, sound, and zero-knowledge. It is complete because, if the items are identical and both host and the inspector follow the protocol, then the inspector will accept with probability $p \geq 1 - (\frac{1}{2})^n$; it is sound because, if the items are different and the inspector follows the protocol, then, no matter *what the host does*, the inspector will reject with probability $p \geq 1 - (\frac{1}{2})^n$; and it is zero knowledge because, as long as the host follows the protocol and presents matching items, the inspector gains no knowledge during their interaction except for the fact that the items match. The system can also be tuned to desired false-positive rates (valid item flagged as invalid) and false-negative rates (invalid item passing the inspection). We have developed the mathematical formalism to understand how both the host and inspector can use repeated measurements to achieve desired inspection targets in terms of false-positive and false-negative rates [23].

4 Conclusions

Future arms control treaties may place limits on the total number of warheads in the arsenals of weapon states, which would introduce qualitatively new verification challenges, including confirming numerical limits on declared nuclear warheads and confirming the authenticity of nuclear warheads prior to dismantlement. Meeting these objectives would require on-site inspections at new types of facilities, including warhead storage sites, which could put at risk highly sensitive information both related to military operations and warhead design. Weapon states may be reluctant to consider—or at least cautious to embrace—some of the anticipated procedures. As a way to address this challenge, in this paper, we have examined the potential of verification approaches that emphasize non-intrusiveness from the outset. Relevant examples include innovative tagging approaches and hashed declarations to confirm the correctness of warhead declarations and novel types of inspection systems to confirm the authenticity of nuclear warheads, while satisfying the different and sometimes conflicting requirements by the host and the inspector.

New international R&D efforts could usefully focus on non-intrusive technologies and approaches, which may show more promise for early demonstration and adoption. Ideally, these concepts could be demonstrated at selected sites or "test beds" and be designed in ways that allow future upgrades if desired. Efforts could and should involve non-nuclear-weapon states and non-governmental organizations such as the International Atomic Energy Agency. Agreeing on "universal test objects" that contain special nuclear materials as a stand-in for nuclear warheads would facilitate open discussions and enable comparison of different technologies and inspection systems. If demonstrated, such non-intrusive verification approaches could be particularly important for moving forward discussions about expanding the scope of current agreements (e.g., to include tactical or non-deployed weapons) and facilitating discussions with weapon states that have so far not been part of formal nuclear arms control agreements and may currently have concerns about inspections that the United States and Russia are already used to.

References

1. Garwin T (1980) Tagging Systems for Arms Control Verification. AAC-TR-10401/80, Analytical Assessment Corporation, Marina del Rey, California
2. Fetter S, Garwin T (1989) Using Tags to Monitor Numerical Limits in Arms Control Agreements. In: Blechman BM (ed) Technology and the Limitation of International Conflict, Washington, DC, pp 33–54
3. Fuller J (2010) Verification on the Road to Zero: Issues for Nuclear Warhead Dismantlement. Arms Control Today, December 2010
4. Ifft E (2014) Verification Lessons Learned from the INF, START I and New START Treaties. Paper presented at the 55th Annual INMM Meeting, Atlanta, Georgia, 20–24 July 2014
5. Patton T, Podvig P, Schell P (2013) A New START Model for Transparency in Nuclear Disarmament. United National Institute for Disarmament Research, Geneva

6. U.S. Department of State (2014) Transparency in the U.S. Nuclear Weapons Stockpile, Factsheet, Washington, DC
7. Rhoades D, DeLand S (2011) Warhead Monitoring: Technical Challenges to Maintaining a Chain of Custody. Presented at the INMM Workshop on Preparing for Nuclear Arms Reductions Technical Transparency and Monitoring Challenges, Monterey, California, 4–5 May 2011
8. U.S. Department of Energy (2015) Joint U.S.-U.K. Report on Technical Cooperation for Arms Control, Washington, DC
9. United Nations (2015) Actions 5, 20 and 21 of the Action Plan of the 2010 Review Conference of the Parties to the Treaty on the Non-Proliferation of Nuclear Weapons, NPT/CONF.2015/38, 2015 NPT Review Conference, United Nations, New York, 1 May 2015. https://undocs.org/NPT/CONF.2015/38
10. Warner JS, Johnston RG (2010) Why RFID Tags Offer Poor Security. Paper presented at the 51st Annual INMM Meeting, Baltimore, Maryland, 11–15 July 2010
11. MacArthur DW (2014) Warhead Confirmation: A Follow-on to New START. LA-UR-14-21945, Los Alamos, New Mexico
12. Easttom C (2016) Modern Cryptography: Applied Mathematics for Encryption and Information Security. McGraw Hill, New York
13. National Academy of Sciences (2005) Monitoring Nuclear Weapons and Nuclear-Explosive Materials: An Assessment of Methods and Capabilities, Washington, DC
14. Jordan SE (1991) Buddy Tag's Motion Sensing and Analysis Subsystem. Sandia National Laboratory, Albuquerque, New Mexico
15. Yan J, Glaser A (2014) Nuclear Warhead Verification: A Review of Attribute and Template Systems. Science & Global Security, 23(3):157–170
16. United Kingdom – Norway Initiative (2015) Trust in Verification Technology: A Case Study Using the UK-Norway Information Barrier. Non-Paper, United Nations, New York
17. Philippe S, Barak B, Glaser A (2015) Designing Protocols for Nuclear Warhead Verification. Paper presented at the 56th Annual INMM Meeting, Indian Wells, California, 12–16 July 2015
18. Tolk K (2015) Authentication in Arms Control. Paper presented at the 56th Annual INMM Meeting, Indian Wells, California, 12–16 July 2015
19. Beaumont M, Bradley Hopkins B, Tristan Newby T (2011) Hardware Trojans: Prevention, Detection, and Countermeasures (A Literature Review). DSTO-TN-1012, Department of Defence, Australian Government
20. Becker GT, Regazzoni F, Paar C, Burleson WP (2013) Stealthy Dopant-Level Hardware Trojans. In: Bertoni G, Coron J-S (eds), Cryptographic Hardware and Embedded Systems (CHES 2013), Lecture Notes in Computer Science 8086, Springer, Heidelberg, pp 197–214
21. Fisch B, Freund D, Naor M (2014) Physical Zero-Knowledge Proofs of Physical Properties. In: Garay JA and Gennaro R (eds), Advances in Cryptology (CRYPTO 2014), Lecture Notes in Computer Science 8617, Springer, Heidelberg, pp 313–336
22. Glaser A, Barak B, Goldston RJ (2014) A Zero-knowledge Protocol for Nuclear Warhead Verification, Nature 510:497–502
23. Philippe S, Barak B, Kütt M, Goldston RJ, Glaser A (2020) Optimum Testing Strategies for Confirming the Authenticity of Nuclear Weapons (in preparation)

Advances in Seismic and Acoustic Monitoring

Jürgen Altmann

Abstract Seismic and acoustic (infrasound) monitoring form important parts of the International Monitoring System (IMS) for verification of the Comprehensive Nuclear Test Ban Treaty (CTBT). Of the globally planned 170 seismic plus 60 infrasound stations, 87% have been certified. Understanding of the propagation in the Earth and the atmosphere has improved markedly, instruments and evaluation algorithms have become more sophisticated. Thus the detection threshold has decreased to values below 0.1 kt TNT equivalent for teleseismic signals, and below 0.5 kt for infrasound, much better than the IMS design goal of 1 kt. A seismic aftershock monitoring system (SAMS) can be set up during on-site inspections by the CTBT Organization. Semi-automatic evaluation of the SAMS signals is used to localise the hypocentre of an underground explosion precisely. Other research focuses on the reduction of periodic disturbances e.g. from aircraft. Acoustic and seismic sensors can detect land and air vehicles in the monitoring of peace-keeping agreements, and ballistic-missile launches for improved early warning. Local systems of seismic sensors have promise for safeguards of the International Atomic Energy Agency (IAEA) for final repositories of spent nuclear fuel.

1 Introduction

Seismic monitoring has a long tradition in the detection of underground nuclear explosions, first as a side result of earthquake monitoring, but as nuclear tests went underground, they were researched as a focus on its own (e.g. [1]). New impetus in particular in academic research of geophysics came with the negotiations in the Conference on Disarmament (CD) in Geneva, where a Group of Scientific Experts (GSE) was formed. The motive was to counter the earlier argument that a Comprehensive Nuclear Test Ban Treaty (CTBT) could not be concluded because

J. Altmann (✉)
Experimentelle Physik III, Technische Universität Dortmund, Dortmund, Germany
e-mail: juergen.altmann@tu-dortmund.de

I. Niemeyer et al. (eds.), *Nuclear Non-proliferation and Arms Control Verification*,
https://doi.org/10.1007/978-3-030-29537-0_16

seismic signals from underground explosions could not be distinguished from those of earthquakes. The success of the geophysicists in solving this problem by a variety of criteria, such as source depth, relative strengths of surface versus body waves and the direction of the first P-wave motion, as well as the agreement in the GSE about this, were important preconditions for the conclusion of the CTBT in 1996. Questions about the verifiability were raised afterwards in some quarters, but were answered satisfactorily by new or repeated scientific studies [2, 3].

Infrasound monitoring expanded at the time of atmospheric nuclear explosions (1950s and early 1960s), but interest decreased as testing mostly went underground following the Partial Test Ban Treaty of 1963. Infrasound research expanded again with the signing of the CTBT, applying the methods also to general studies of the atmosphere.

The CTBT Organization (CTBTO) has built up most of its International Monitoring System (IMS) of which the seismic stations represent the biggest part (50 primary and 120 auxiliary stations); in addition the IMS comprises 60 infrasound, 11 hydroacoustic and 80 radionuclide stations. Of the total 321 stations plus 16 laboratories, 294 (87%) have been installed and certified by May 2018 (Table 1). The IMS was designed with the goal of providing 90% detection probability for any explosion releasing energy of 1 kt TNT equivalent or more, at three or more stations so that localisation is possible. In order to assess its global detection capability and to improve it, the CTBTO holds biannual Science and Technology Conferences where research results are being presented [5]. To support such research the CTBTO can make data available, often by special agreements.

After some basics of acoustics and seismology (Sect. 2), advances in teleseismic and infrasound monitoring are presented in Sects. 3 and 4, respectively. Seismic methods are also to be used over short ranges during on-site inspections (OSIs) of the CTBTO; research for making the interactive detection of such events easier is presented in Sect. 5. Acoustic and seismic monitoring can also be used for other tasks in the context of war prevention, disarmament and non-proliferation, for example for monitoring of vehicle movements and for early warning of ballistic-

Table 1 Status of the IMS as of May 2018 [4]

	Certified	Installed	Under construction	Planned	Total
Primary seismic	44	1	1	4	50
Auxiliary seismic	107	8	2	3	120
Infrasound	50	2	1	7	60
Hydroacoustic	11	0	0	0	11
Radionuclide	69	1	2	8	80
of which noble gas	25	5	10	0	40
Radionuclide labs	13	0	0	3	16
All facilities	294	12	6	25	337

The planned total consists of 321 stations and 16 radionuclide laboratories. Of the planned 230 seismic and infrasound stations, 201 (87%) have been certified missile launches, both are sketched in Sect. 6. A newer area is monitoring of underground final repositories, this is treated in Sect. 7. The final Sect. 8 gives concluding remarks.

2 Basics of Acoustics and Seismology

Acoustic and seismic monitoring are methods to gain information about a distant source. This information is transported by waves that propagate in an elastic medium. Inside a fluid, there is only a compressional, that is longitudinal, wave type whereas within a solid, due to its resistance to shear deformation, there is a second type, a shear or transversal wave. Because their wave speed is higher (often by a factor $\sqrt{3}$), the compressional waves arrive earlier and in seismology are called primary (P) waves, the transversal ones secondary (S) waves. At layer boundaries additional wave types can occur, for example Rayleigh or Love seismic waves at a surface, propagating still slowlier.

The amplitude of a wave decreases with the distance r from the source due to several effects. Geometric spreading causes a decrease proportionally to $1/r$ for spherical expansion into a volume and to $1/\sqrt{r}$ for circular waves propagating in a surface. In addition there is attenuation (due to absorption or scattering out of the original direction), usually following an exponential decrease with distance. The absorption coefficient in the exponent often increases linearly with the frequency for seismic waves, and generally with the square of the frequency in the case of sound waves in air. These are the reasons why in both cases detection at large distances has to rely on very low frequencies, from millihertz to a few hertz.

In air the wave speed is around 0.34 km/s, slowly increasing with temperature; the seismic-P-wave speed can be as low as 0.1 km/s in loose sand at the surface, but varies between 5 and 14 km/s in the Earth interior from the crust to the core. Where the wave speed changes, be it a sharp jump or a continuous variation, reflection and refraction occur, in a solid also (partial) conversion from one wave type to the other.

Whereas in the Earth there are often sharp layer boundaries, in the atmosphere continuous change of the wave speed plays a big role because the temperature and wind vary smoothly with altitude. With the normal, negative temperature gradient an originally horizontal sound wave is refracted upward, the same holds for propagation opposite to the wind direction. Conversely, downward refraction occurs with a temperature inversion, as sometimes in the low atmosphere and constantly in the stratosphere (20–50 km height) and the thermosphere (above 100 km), and for propagation with the wind. The latter effects can turn up-going sound waves eventually back down, hitting the ground beyond a shadow zone. Figure 1 shows how rays are refracted back down in one or the other layer depending on the profile of the effective sound velocity that results from the temperature and the wind vector.

Because the stratospheric wave guide depends on the wind it shows seasonal variation. With additional reflection(s) at the ground very long propagation distances can be observed, strong events travel around the globe several times. Diurnal as well

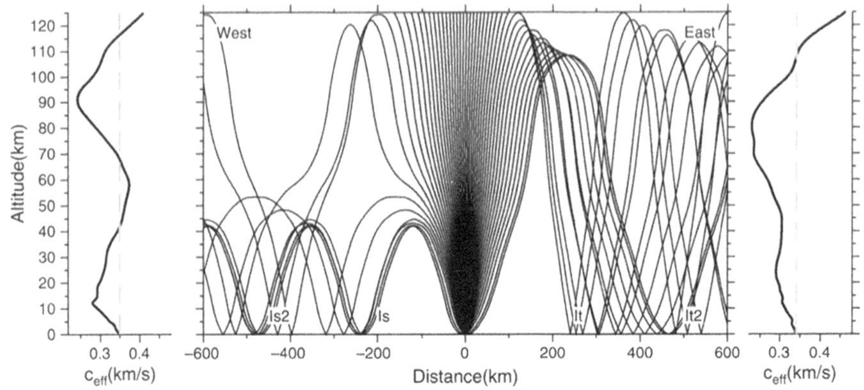

Fig. 1 Simulation of infrasound propagation to the west and the east under summer conditions, with an easterly jet stream, using the effective-sound-velocity profiles for west (left margin) and for east (right margin). From the source at the surface rays at each 4° are traced. In the west mainly stratospheric returns occur, in the east only thermospheric ones (Source: [6], reprinted by permission)

as irregular variation is introduced by the weather. Absorption and, at high altitudes, dispersion play a role, too. The Earth surface is another layer boundary. Body waves pass from one medium to the other with refraction, but due to the large difference in density and wave speed the energy of the transmitted wave is about four orders of magnitude lower, with Rayleigh surface waves the portion is higher. Earthquakes produce infrasound, and atmospheric events excite Earth vibration. For very long waves the whole atmosphere is involved, up to the ionosphere. The atmosphere is a much more complex medium than the Earth, thus understanding infrasound propagation and deriving source properties from recorded signals are more difficult than in seismology.

3 Advances in Teleseismic Monitoring

When the IMS was designed, a detection limit of an explosion with yield 1 kt TNT equivalent served as the goal. In hard rock this yield translates to a seismic body-wave magnitude of $m_b = 4$.[1] Several studies have investigated what the actual detection threshold is around the globe. It has improved over the years as the IMS was built up and seismology advanced; Table 1 contains the IMS evolution over time. Already in 2002 the threshold was estimated at $m_b = 3.5$, corresponding to 0.1 kt TNT equivalent, in some former testing areas even at 0.01 kt [2, 7]. Ten years later, significant advances were noted with new seismic methods for regional

[1]Depending on the host rock, variations by up to 0.5 m_b units are possible.

Table 2 North-Korean nuclear tests and their detection by the CTBT IMS

Date	Magnitude	No. of IMS stations established at the time	No. of IMS stations that detected the event	Radionuclides detected
9 Oct 2006	4.1	180 (53%)	22	Yes, 2 weeks later by the IMS radionuclide station at Yellowknife, Canada
25 May 2009	4.52	252 (75%)	61	No, neither by the CTBTO nor any other organization
12 Feb 2013	4.9	286 (85%)	96	Yes, 55 days later by two IMS radionuclide stations in Takasaki, Japan, and Ussuriysk, Russia
6 Jan 2016	4.85	277 (82%)	77	No clear evidence
9 Sept 2016	5.1	277 (82%)	≥100	No
3 Sept 2017	6.1	281 (83%)	132	No

Source: [8–13], used by permission

distances (below 1600 km), more high-quality seismic stations and increases in computing power and data-storage possibilities [3]. With regional evaluation the m_b threshold is 2.8 corresponding to around 0.02 kt TNT equivalent. Transparency measures at test sites would allow 0.005 kt TNT equivalent. Higher-frequency data allow a better discrimination of explosions from earthquakes and the detection of evasion attempts by decoupling, that is, exploding the device in a cavity.

Important test cases were the underground test explosions done by North Korea 2006, 2009, 2013, January and September 2016 and 2017 [8–13]. Table 2 shows how the number of active stations evolved over time; correspondingly the number that picked up the signals increased from 22 to 132. All six tests were detected immediately and evaluated automatically. Within 2 h the CTBTO sent the first automatic analyses to the Member States, with preliminary evaluation of the time, location and seismic magnitude of the event. Atmospheric radionuclides were detected at IMS stations in 2006 and 2013. Manual evaluation by experts followed immediately. The source locations were very close to each other; with the increased number of seismic stations the location uncertainty was reduced from about 30 km in 2006 to about 15 km in 2009 to 2017 (Fig. 2). The tests condemned by the international community were announced by North Korea, but detection occurred before and independently. Beside providing test cases for the IMS these events serve to improve scientific understanding; analysis is on-going (e.g. [14–16]).

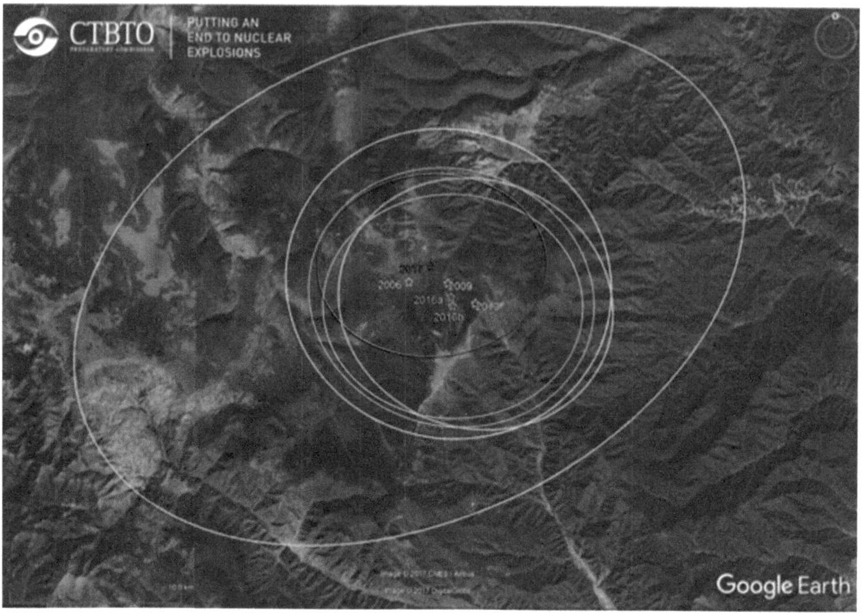

Fig. 2 Event locations and respective error ellipses from evaluation of teleseismic detections for the North-Korean nuclear tests of 2006, 2009, 2013, January and September 2016, and 2017. REB: Reviewed Event Bulletin (Source: [13], reprinted by permission)

4 Advances in Infrasound Monitoring[2]

Following the cutoff of human hearing, infrasound has frequencies below 20 Hz; relevant signals have frequencies down to a few millihertz. Atmospheric explosions produce strong infrasound, but underground explosions (as well as earthquakes) can act as sources, too. Of the 60 infrasound stations stipulated in the Treaty 50 have been updated by May 2018 (see Table 1). For nuclear-test detection, the primary monitoring passband is 0.4 to about 1.2 Hz, mostly for stratospheric signals. The secondary passband is 0.04–0.1 Hz, immediately below the band of background-noise from microbaroms (created by ocean swells, 0.1–0.5 Hz), best for thermospheric returns. The sampling frequency is 20 Hz. Work has focused on several areas [18]. An optimal sensor-array configuration of eight elements reduces spatial aliasing at high frequencies, thus the design is being changed from the earlier four-element one. For best signal correlation between the elements the array aperture should be 1 km with a 0.25- or 0.30-km triangular subarray. Wind contributes the strongest background noise. The traditional method of spatial averaging using pipe arrays can be improved markedly by adding a porous screened enclosure that reduces turbulence, but further research is needed [19]. A sophisticated method for automatic and interactive data processing has been

[2] A comprehensive treatment is found in the chapters of [17].

developed [20]. This Progressive Multichannel Correlation (PMCC) searches by cross correlation for planar waves passing the array, using an increasing number of elements. Each elementary detection is then processed over time in 11 frequency bands between 0.07 and 4.0 Hz, and identified either as "phase" or as "noise". Then fitting arrivals from several stations are associated, and potential source locations found. Atmospheric modelling is included but needs to be improved. As in the seismic case software-supported interactive analysis is needed to determine the site and characterise the type of the source. Figure 3 shows the evaluation for the signals at one IMS station from an explosion accident. Concerning the network detection capability, a simulation study using improved specifications of stratospheric winds and of wind noise concluded that explosions with 0.5 kt TNT equivalent would be detected at two stations at any time in the year [21].

5 Research for Seismic Monitoring During CTBT On-Site Inspections

When IMS signals indicate an underground explosion of potential nuclear origin, the CTBTO can send an inspection team to the area of interest for up to 130 days. Actual inspections can only take place after entry into force of the Treaty, but they are actively prepared. The localization uncertainty of an underground explosion from teleseismic signals is 10 to 30 km (see Fig. 2). In order to prove the nuclear character, radionuclides need to be detected. If a large-scale release to the atmosphere can be prevented and no detections at radionuclide stations of the IMS occur, gas samples need to be taken from the soil very close to the hypocentre. This requires knowledge of the latter with an uncertainty of 100 m or less. Among the methods to achieve such accuracy mentioned in Part II of the Protocol to the CTBT is passive seismological monitoring. A local seismic array can be set up in the inspection area of up to 1000 km^2 to sense the very faint signals from aftershocks, created by rocks falling down in the explosion-created cavity or by rock fractures developing around it [23]. This seismic aftershock measurement system (SAMS) can comprise several dozen sensor positions with e.g. mini-arrays of four geophones each. The location, including depth, of the respective source can then be gained from the times of arrival of the different waves at the various positions. A SAMS was set up at both Integrated Field Exercises (IFEs) of the OSI Division of the CTBTO, 2008 in Kazakhstan and 2014 in Jordan.[3]

Because many events can be detected most of which originate outside of the inspection area, some pre-screening and semi-automatic analysis is helpful. In parallel to the preparations for OSI by the respective Division of the CTBTO, including various directed exercises and the large-scale Integrated Field Exercises,

[3]For IFE 2014 in general see posters T2.1-P1 to T2.1-P19 presented at Comprehensive Nuclear-Test-Ban Treaty: Science and Technology Conference 2015, Vienna, Austria, 22–26 June 2015, available via https://www.ctbto.org/specials/snt2015/poster-presentations/. The SAMS experience is treated in [24].

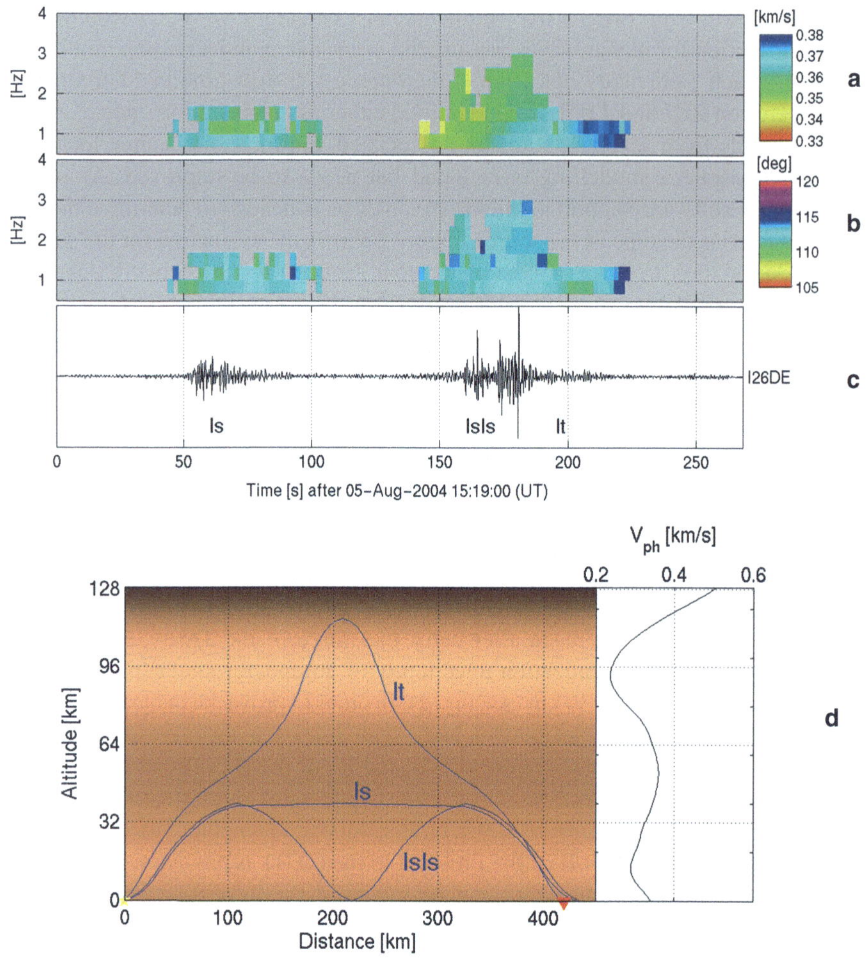

Fig. 3 Infrasound detection at the IMS station I26DE (southeast Germany) of the explosion in a fireworks factory near Budapest on 5 August 2004 around 17:55 (CEST). PMCC analysis shows that the apparent velocity across the array was between 0.34 and 0.38 km/s (**a**) and that the back azimuth was around 110° (**b**), fitting to the direction to the source. The sum trace of the five sensors (**c**) shows three arrivals that correspond to the wave paths indicated in (**d**): Is (stratospheric return), IsIs (double stratospheric return) and It (thermospheric return), derived from the effective sound-speed profile indicated on the right (Source: [22] reprinted by permission © BGR Hannover)

since many years research has been done to allow fast, interactive screening of the signals to allow daily processing [25, 26]. Detecting events at low signal-to-noise ratio, localising the sources in three dimensions and estimating the magnitude are made easier by several tools, including variable filtering, sonograms (time-frequency presentations with adaptive muting of noise in the spectral domain),

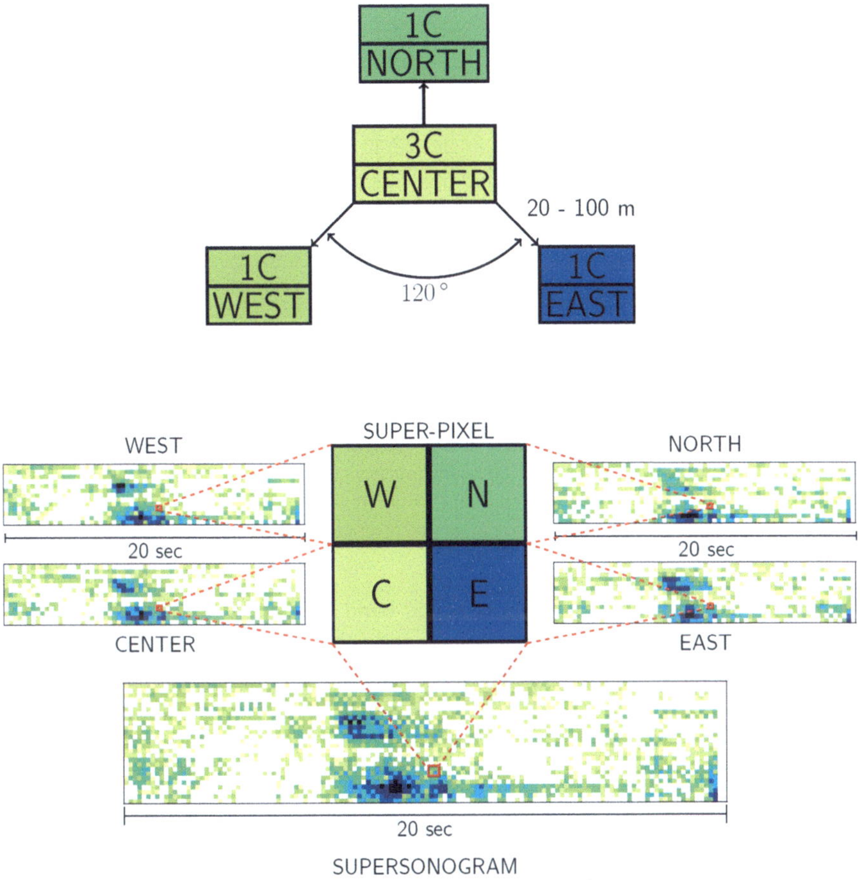

Fig. 4 In a supersonogram the pixels of the sonograms from the four positions of a mini-array are combined to form so-called superpixels for each time and frequency band (Source: [25], reprinted by permission)

model signals computed with ground-layer properties. So-called supersonograms, combining the data of a four-sensor mini-array, improve the visualisation and thus the detection of events versus noise (Fig. 4). Multi-sensor signals from a fairly long time period (hours) are presented in a special condensed visual display that emphasizes times and frequency bands of relevant events. If an event of interest has been selected, the wave onsets can be picked and the source location computed interactively.

In the detection of the weak aftershock signals a special problem is posed by disturbances that can mask them. Such disturbances can be caused by other vibration sources, including airborne sound that couples into the ground. Many artificial interfering sources produce periodic noise, for example car engines or aircraft propellers. Their spectra consist of peaks at discrete frequencies whereas

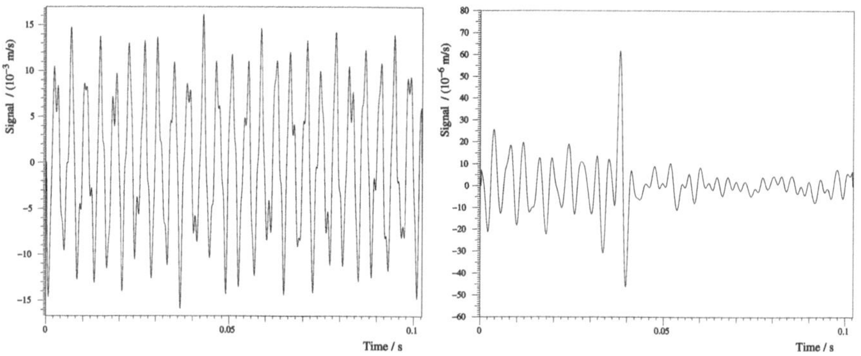

Fig. 5 Time signal of a seismic impulse (vertical velocity from a firecracker at about 100 m distance, peak-to-peak value 0.11 mm/s). Left: Six synthetic sines of various frequencies and amplitudes added, total peak-to-peak value 34 mm/s. Right: Remaining signal after fitting the sine parameters and subtraction of the sines. Some disturbances remain, but the pulse is clearly visible, even though the disturbances were more than two orders of magnitude stronger (Source author, from [27])

the impulsive aftershock signals have broad-band spectra. Research has found out how the peaks can be characterized so exactly that they can be subtracted from a disturbed spectrum, leaving the weak broad-band portion more or less intact [28]. Figure 5 shows an example where periodic disturbances stronger by more than two orders of magnitude than an impulsive signal, completely burying it, were successfully removed. Because in real disturbance sources the frequencies can change over time—the engine rotation rate is varied, or the motion-caused Doppler shift changes—recently the work has been extended to frequencies that change with time linearly [29]. Because several disturbances are acoustic in the first place, research is also done on how acoustic waves produce seismic excitation [30, 31]. This work may lead to recommendations for the placement of SAMS sensors. To expand the signal analysis to acoustic-seismic coupling, it may be advisable to deploy acoustic sensors in addition to the seismic ones.

6 Applications for Peace Keeping and Early Warning

Acoustic and seismic sensors can be used to detect land and air vehicles over short distances (dozens of metres to several kilometres) [32, 33]. The vehicle engines produce sound. For land vehicles usually the exhaust is the strongest source if the vehicle is bigger than a car. This part of the vehicle sound shows the periodicity of the engine, it is visible in acoustic spectra as a harmonic series of peaks at a fundamental frequency and its integer multiples. The fundamental frequency is tied to the engine rotation rate. Tracked vehicles have another important periodic acoustic source: the sprocket wheel driving the track elements. Here the fundamental frequency is proportional to the vehicle speed.

Seismic waves are mainly excited by the interaction between the wheels and the road/ground. Varying forces to the ground are exerted because of the roughness of the road and the tire profile, additionally influenced by the vehicle suspension. For tracked vehicles, the track acts as a sort of road with periodic roughness, so that for them the track frequency and its harmonics dominate the seismic spectrum. But the acoustic waves propagating along the ground exert forces as well, producing additional seismic excitation; depending on the decrease of the seismic amplitude with distance, the acoustically produced seismic reaction can become stronger than the direct one. Because there are several types of seismic waves, some with dispersion, and the mentioned superposition of direct and acoustically coupled parts, seismic signals and spectra are not well suited for the recognition of vehicle type. Acoustic signals, on the other hand, mostly maintain their form during propagation; recognition of engine and thus vehicle type worked well using the relative strengths of the engine harmonics. Also finding the direction to a source by beam forming with a sensor array is easier with acoustic signals. Detection ranges using acoustic and seismic sensors are about 50 m for medium trucks, 100 m for heavy road tractors and markedly above 100 m for tracked vehicles. Aircraft produce harmonic series by propellers or rotors. Jet aircraft, on the other hand, show mostly broad-band spectra. Direction finding and triangulation with three-dimensional microphone arrays worked well. Due to the extreme sound pressures that excite ground vibration, too, accelerated movement for take-off can be detected and characterised by a sensor chain along the runway, but landing and taxiing can be registered by both modalities, too.

Another possible application of acoustic and seismic sensors is detection of launches of intercontinental ballistic missiles [34]. For improving early warning by providing an additional confirmation that a missile has not been launched, seismic sensors could be deployed co-operatively near each missile silo. The broad-band noise from the rocket-engine exhaust gases is extremely loud—sound pressures are hundreds of pascals, that is levels around 140 dB, at a few hundred metres. Where this hits the ground it produces corresponding ground motion, with a proportionality constant that depends on the soil, but usually is between 10^{-6} and 10^{-4} m/(s Pa). Thus ground velocities on the order of mm/s arise, to be compared with sensitivities of geophones or seismometers that are four to five orders of magnitude lower, and seismic background that—depending on weather and other close sources such as traffic—is two to three orders of magnitude lower. Figure 6 shows the expected sound pressures from a ballistic-missile launch, and other sources that could principally produce similar signals, versus the respective source-sensor distance. Amplitudes from the other sources could only become comparable if the latter are sufficiently close to the sensor. Reliable discrimination from a missile launch would be possible by looking at the amplitude time course and by evaluating spectral characteristics. One to three seismic early-warning sensors should be deployed at 100–500 m from each silo, with a secure connection that provides continuous real-time communication to the monitoring side. At a cost of around $50,000 per

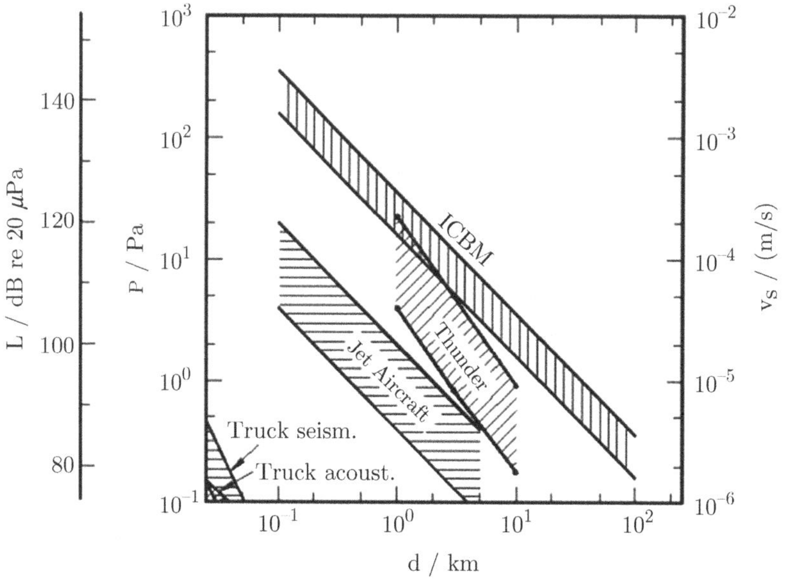

Fig. 6 Estimated maximum root-mean-square sound pressure p from various sources versus distance (horizontal from the launch point for missiles, slant for the other sources). Also given are the sound-pressure level L (left) and the resulting soil-vibration velocity v_S (right ordinate) assuming a typical acoustic-seismic coupling coefficient of 10^{-5} m/(s Pa) (Source author, from [34])

missile (that itself costs tens of millions of $) the co-operation partners would get an additional confirmation that no attack had started, potentially helpful in a situation when other early-warning channels had failed or showed erroneous indications of attack.

7 Research for Safeguards at Underground Final Repositories

If spent nuclear fuel is put into an underground final repository directly, without reprocessing, the plutonium bred from uranium-238 is still contained in it and could principally be recovered and diverted for use in nuclear weapons at some later time. Thus the material should remain under safeguards of the International Atomic Energy Agency (IAEA). Whereas safeguards for existing nuclear installations are well established, underground final disposal of nuclear waste presents a new challenge. During operation of the repository with mining, emplacing and backfilling, traditional methods such as video cameras have limited utility in the parts of the mine that are open. But detecting the creation of undeclared cavities and the undeclared re-opening of areas already backfilled needs other methods. The same holds for the very long period after the emplacement phase, when drifts and shafts will have been closed. Undeclared access to the spent fuel requires some form of mining. Mining and other underground operations produce seismic vibrations, directly as well as via acoustic noise. Seismic excitation propagates through the ambient medium and can thus be used to detect activities remotely. The main question with seismic monitoring is whether signals from undeclared activities can be detected, that is separated from signals from different sources and from other background noise. In order to find answers, at first measurements were carried out at a potential repository site [35]. Afterwards, numerical modelling of the seismic propagation was done [36].[4] Both studies focused on the Gorleben salt dome in Germany that had been explored—as the sole site—since the mid-1980s.[5]

In the measurements geophones and microphones were put at many positions underground (at 840 m depth) and on the surface. Signals from typical mining activities as well as background noise were recorded. The most important results were the following. The seismic-velocity signal strengths from the same source at comparable distance showed considerable variability, up to a factor 2–5. The general amplitude decrease with distance varied among the sources; if fitted by a power law the exponent was between -2.2 and -0.8 (that is, the decrease was mostly stronger than in a homogeneous medium where the theoretical exponent is -1). The source strengths differed markedly; nominal seismic peak-to-peak velocities at 100 m distance were 0.4 μm/s for the weakest source, a hand-held chainsaw, 5–70 μm/s for most tools and vehicles, and 10^5 μm/s for blast shots (Fig. 7). Seismic spectra from geophones showed broad-band noise to several kilohertz, plus harmonic components for periodic sources, such as vehicle engines or a pick hammer. The detection ranges resulting from a simple amplitude criterion, with background values of 0.5–10 μm/s, are many kilometres for blasts, many hundreds of metres for heavy vehicles, and 100 m to a few times this value for weaker sources. After closure of the mine the background—for sensors deployed in boreholes of several 100 m depth—will be lower, plausibly at 0.1–0.2 μm/s. Thus most medium-strength activities probably could be detected at 1 km. Since the measurements could only be done inside the exploratory mine, that is, close to the potential repository, and at the surface, no information could be gained about the strength of the signals and the background outside of the salt dome as well as larger distances inside of it, at positions 0.5–1 km from the potential repository that could be used for monitoring.

To roughly assess the signal strength at such positions before an expensive drilling and testing program, numerical calculations of the wave propagation were done on a computer cluster using a spectral-finite-element code (SpecFEM3D [37, 38]). This three-dimensional modelling took into account the geological strata of the salt dome and its surroundings, using realistic values of the respective seismic

[4]Both projects were tasked by the German Support Programme to the IAEA.

[5]Since 2013 exploration has been stopped there and a new site search is to be done without prejudice, Gorleben remains an option.

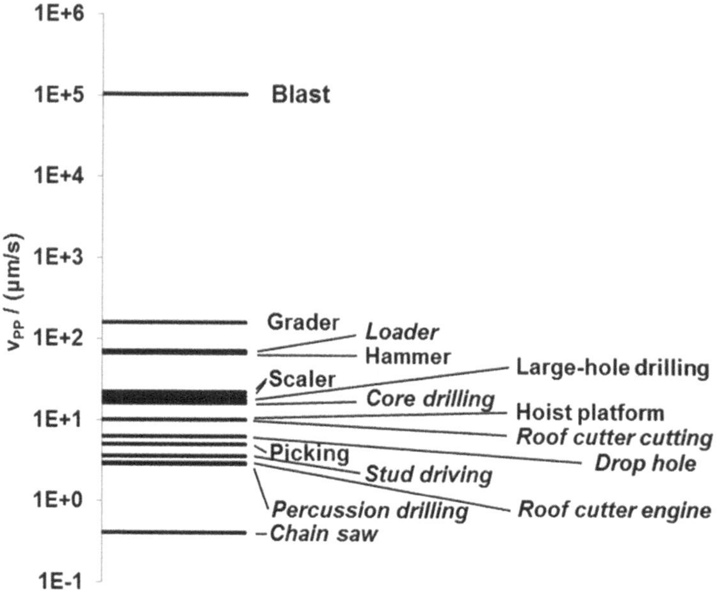

Fig. 7 Nominal peak-to-peak value of vertical wall velocity in 100 m distance from the source for various mining activities, logarithmic scale. These values are uncertain by a factor 2, and more if the distance interval used for the trend ends markedly below 100 m (italics) (Source author, from [35])

parameters. Due to insufficient knowledge of the three-dimensional structure in the dome and because of memory and computing-time limitations some simplifications were needed. Because of the limited space and time resolution the model signals contained frequencies up to a few times 100 Hz, about one tenth of the spectral range of the geophone measurements. Two types of pulsed excitation were used as input: a unidirectional blow, described by a force with a Gaussian time course, and a spherical explosion, described by a seismic moment with an error-function time course. Model signals were gained at several hundred positions. For an electrohydraulic pick hammer, Fig. 8 compares the peak-to-peak values of measured seismic velocity with the model values; for a good fit the latter, gained with pulses of 184 N peak force and 5 ms half duration, had to be multiplied by 700. Probably the measured pulses were stronger and shorter.

These are the most relevant results: The seismic-velocity signals were generally dominated by the wave with the shortest path, components reflected at boundaries within the salt dome, at its margin or at the surface could be recognized, but had markedly lower amplitude. Different ways of extruding a two-dimensional cross section into the third dimension did not produce strong differences in the signals. Superposition of single-pulse signals to simulate repetitive sources led to varying signal envelopes depending on the phasing; amplitude increases up to a factor 3 were seen. The maximum peak-to-peak values versus distance showed considerable

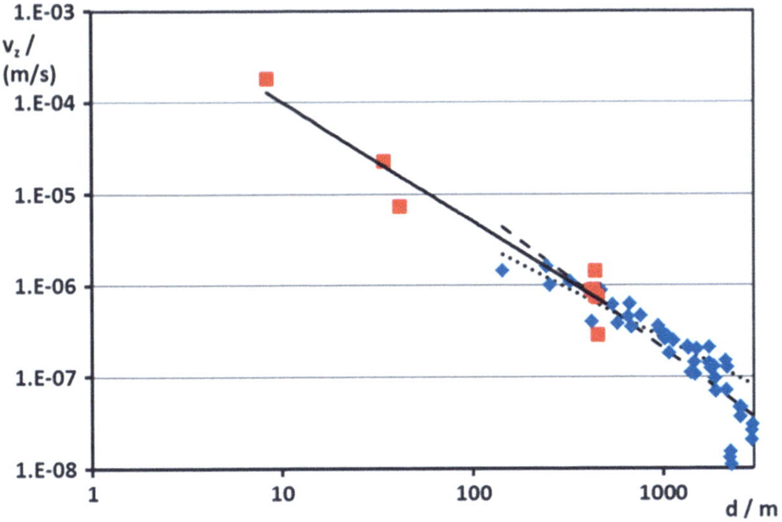

Fig. 8 Measured peak-to-peak values of vertical seismic velocity for picking (red squares), and model results from 100 blows with 23 ms spacing, multiplied by 700 (blue diamonds) to fit to the measurements, versus distance from the source. Each model blow was a Gaussian force pulse with maximum 184 N and half duration 5 ms. The lines are power-law fits to the respective subset (Source [36])

scatter; power-law fits gave exponents of −2 to −1.1, similarly to the measurements. Amplitude-transmission coefficients from salt to the surrounding media fitted to simple theoretical values, with numbers from 1.1 to 1.5 no significant influence of the boundaries on detectability is expected. Signal strengths of an explosion with a seismic moment of 1 Nm were lower than the ones from a force pulse of 1 N of the same duration by a factor 4400, the value of the used P-wave speed in m/s, as predicted theoretically for a homogeneous medium. This and the transmission coefficients close to 1 suggest that concerning detection of activities the boundaries do not play a big role. Consequently it seems plausible that homogeneous-media theory will give meaningful results for other media than salt such as clay or granite, too, if the appropriate seismic parameters will be used.

Two open questions remain: The source time functions for force pulses or explosions are not known, in particular the amplitudes; they could not be measured during the measurements. Thus the absolute seismic amplitudes from the model cannot be compared directly with the measured ones. Additional investigations are needed here. They are also needed for the second issue, namely the seismic background at the monitoring positions. Thus some more research will be needed before a well-founded decision can be made to pursue the seismic-monitoring route further. But already now the outlook for the utility of seismic monitoring for safeguarding an underground final repository seems generally good.

8 Conclusions

Seismology and infrasound science have made great progress since the signing of the CTBT in 1996. Seismic and acoustic monitoring continue to be fields of active research. For the verification of compliance with the CTBT improvement of the global detection capabilities of the international monitoring system remains relevant. The second strand of CTBT verification is on-site inspections. Here local seismic detection of very weak signals makes high demands, and research for improvements gains in importance. Acoustic and seismic monitoring can be used in other contexts of war prevention, disarmament and non-proliferation. Examples are sensors for peace keeping, early warning of ballistic-missile launches, and safeguards for underground nuclear repositories. Whether their potential actually will be used depends on political developments, similarly to the situation with on-site inspections in the context of the CTBT.

References

1. Dahlman O, Israelson H (1977) Monitoring Underground Nuclear Explosions. Elsevier, Amsterdam
2. Committee on Technical Issues Related to Ratification of the Comprehensive Nuclear Test Ban Treaty (2002) Technical Issues Related to the Comprehensive Nuclear Test Ban Treaty. National Academies Press, Washington DC
3. Committee on Reviewing and Updating Technical Issues Related to the Comprehensive Nuclear Test Ban Treaty (2012) The Comprehensive Nuclear Test Ban Treaty: Technical Issues for the United States. National Academies Press, Washington DC
4. Comprehensive Nuclear Test Ban Treaty Organization (2018) Interactive Map, IMS categories. https://www.ctbto.org/map/. Accessed 14 April 2018
5. Comprehensive Nuclear Test Ban Treaty Organization (2015) Science and Technology – The Conference Series. https://www.ctbto.org/the-organization/science-and-technology-the-conference-series/. Accessed 30 April 2019
6. Evers LG, Haak EW (2010) The Characteristics of Infrasound, its Propagation and Some Early History. In: Le Pichon A, Blanc E, Hauchecorne A (eds) (2010) Infrasound Monitoring for Atmospheric Studies. Springer, Dordrecht, pp 3–27
7. Hafemeister D (2007) Progress in CTBT Monitoring Since its 1999 Senate Defeat. Sci Glob Secur 15(3):151–183
8. Comprehensive Nuclear Test Ban Treaty Organization (2015) The CTBT Verification Regime: Monitoring the Earth for nuclear explosions. https://www.ctbto.org/fileadmin/user_upload/public_information/2015/Verification_Regime_final_2015_final.pdf. Accessed 30 April 2019
9. Comprehensive Nuclear Test Ban Treaty Organization (2007) The CTBT verification regime put to the test – the event in the DPRK on 9 October 2006. https://www.ctbto.org/press-centre/highlights/2007/the-ctbt-verification-regime-put-to-the-test-the-event-in-the-dprk-on-9-october-2006/. Accessed 30 April 2019
10. Comprehensive Nuclear Test Ban Treaty Organization (2013) On the CTBTO's detection in North Korea. https://www.ctbto.org/press-centre/press-releases/2013/on-the-ctbtos-detection-in-north-korea/. Accessed 30 April 2019
11. Comprehensive Nuclear Test Ban Treaty Organization (2016) Technical Findings. https://www.ctbto.org/the-treaty/developments-after-1996/2016-dprk-announced-nuclear-test/technical-findings/. Accessed 30 April 2019

12. Comprehensive Nuclear Test Ban Treaty Organization (2016) 9 September 2016 – North Korea – Announced Nuclear Test – Technical Findings. https://www.ctbto.org/the-treaty/developments-after-1996/2016-sept-dprk-announced-nuclear-test/technical-findings/. Accessed 30 April 2019
13. Comprehensive Nuclear Test Ban Treaty Organization (2017) 3 September 2017 – North Korea – Announced Nuclear Test – Technical Findings. https://www.ctbto.org/the-treaty/developments-after-1996/2017-sept-dprk/technical-findings/. Accessed 30 April 2019
14. Murphy JR, Stevens JL, Kohl BC et al (2013) Advanced Seismic Analyses of the Source Characteristics of the 2006 and 2009 North Korean Nuclear Tests. Bull Seismol Soc Amer 103(3):1640–1661
15. Carluccio R, Giuntini A, Materni V et al (2014) A Multidisciplinary Study of the DPRK Nuclear Tests. Pure Appl Geophys 171(3–5):341–359
16. Kim SG, Gitterman Y, Lee S et al (2017) Accurate Depth Determination and Source Characterization of the DPRK Nuclear Tests (2006, 2009, 2013, 2016J (01/06/2016) and 2016S (09/09/2016)) Using Regional and Teleseismic Arrays. In: Comprehensive Nuclear-Test-Ban Treaty: Science and Technology Conference 2017, Vienna, 26–30 June 2017. https://ctnw.ctbto.org/DMZ/event/3239/programme/2017-06-27. Accessed 30 April 2019
17. Le Pichon A, Blanc E, Hauchecorne A (eds) (2010) Infrasound Monitoring for Atmospheric Studies. Springer, Dordrecht
18. Christie DR, Campus P (2010) The IMS Infrasound Network: Design and Establishment of Infrasound Stations. In: Le Pichon A, Blanc E, Hauchecorne A (eds) (2010) Infrasound Monitoring for Atmospheric Studies. Springer, Dordrecht, pp 29–75
19. Walker KT, Hedlin MAH (2010) A Review of Wind-Noise Reduction Methodologies. In: Le Pichon A, Blanc E, Hauchecorne A (eds) (2010) Infrasound Monitoring for Atmospheric Studies. Springer, Dordrecht, pp 141–182
20. Brachet N, Brown D, Le Bras R et al. (2010) Monitoring the Earths's Atmosphere with the Global IMS Infrasound Network. In: Le Pichon A, Blanc E, Hauchecorne A (eds) (2010) Infrasound Monitoring for Atmospheric Studies. Springer, Dordrecht, pp 77–118
21. Le Pichon A, Vergoz J, Blanc E et al (2009) Assessing the performance of the International Monitoring System infrasound network: geographical coverage and temporal variabilities. J Geophys Res 114 (D8):D08112
22. Bundesanstalt für Geowissenschaften und Rohstoffe (2016) Explosion in Feuerwerksfabrik bei Budapest, Ungarn. https://www.bgr.bund.de/DE/Themen/Erdbeben-Gefaehrdungsanalysen/Seismologie/Kernwaffenteststopp/Verifikation/Infraschall/Quellen_Phaenomene/Besondere-Ereignisse/Explosion-Budapest/budapest_inhalt.html?nn=1558738. Accessed 30 April 2019
23. Ford SR, Labak P (2016) An Explosion Aftershock Model with Application to On-Site Inspection. Pure Appl Geophys 173(1):173–181.
24. Gestermann N, Sick B, Häge M et al (2015) The Seismic Aftershock Monitoring System (SAMS) for On-Site Inspection: Experience from IFE14. In: Comprehensive Nuclear-Test-Ban Treaty: Science and Technology Conference 2015, Vienna, Austria, 22–26 June 2015. https://www.ctbto.org/fileadmin/user_upload/SnT2015/SnT2015_Posters/T2.1-P18.pdf. Accessed 30 April 2019
25. Sick B, Walter M, Joswig M (2014) Visual Event Screening of Continuous Seismic Data by Supersonograms. Pure Appl Geophys 171(3–5):549–559
26. Joswig M (2008) Nanoseismic monitoring fills the gap between microseismic networks and passive seismic. First Break 26(6):117–124
27. Altmann J, Gorschlüter F, Liebsch M (2012) Investigations of Periodic Disturbances on Seismic Aftershock Recording. In: European Geosciences Union General Assembly, Vienna, 23–27 April 2012. http://presentations.copernicus.org/EGU2012-11993_presentation.pdf. Accessed 30 April 2019
28. Gorschlüter F, Altmann J (2014) Suppression of Periodic Disturbances in Seismic Aftershock Recordings for Better Localisation of Underground Explosions. Pure Appl Geophys 171(3–5):561–573

29. Gorschlüter F (2014) Sinusoids with linear frequency shift in time series – precise characterisation and removal. PhD Dissertation, Technische Universität Dortmund
30. Liebsch M, Altmann J (2016) Acoustic-seismic coupling for a wide range of angles of incidence and frequencies using signals of jet-aircraft overflights. J Sound Vib 385:202–218
31. Liebsch M (2017) Acoustic-Seismic Coupling of Broadband Signals – Support for On-Site Inspections under the Comprehensive Nuclear Test-Ban-Treaty. PhD Dissertation, Technische Universität Dortmund
32. Altmann J (2004) Acoustic and Seismic Signals of Heavy Military Vehicles for Co-operative Verification. J Sound Vib 273(4–5):713–740
33. Blumrich R, Altmann J (2000) Medium-range localisation of aircraft via triangulation. Appl Acoust 61(1):65–82
34. Altmann J (2005) Acoustic-Seismic Detection of Ballistic-Missile Launches for Cooperative Early Warning of Nuclear Attack. Sci Glob Secur 13(3):129–168
35. Altmann J (2013) Seismic Monitoring of an Underground Repository in Salt – Results of the Measurements at the Gorleben Exploratory Mine. ESARDA Bull 50:61–78
36. Altmann J (2015) Modelling Seismic-Signal Propagation at a Salt Dome for Safeguards Monitoring. ESARDA Bull 52:60–79
37. Komatitsch D, Vilotte JP (1998) The spectral-element method: an efficient tool to simulate the seismic response of two-dimensional and three-dimensional geological structures. Bull Seismol Soc Amer 88(2):368–392
38. Komatitsch D, Tromp J (1999) Introduction to the spectral element method for three-dimensional seismic wave propagation. Geophys J Int 139(3):806–822

Radiation Detectors and Instrumentation

Martin Williamson and Jeffrey Preston

Abstract Direct measurements of radiation signatures can be utilized as a key functional component to verify a country's commitments in a transparency agreement. Special nuclear material comprises isotopes of uranium and plutonium, where each has unique signatures that differentiate it from the other isotopes when measured with radiation detection equipment. Depending on the type of confirmatory information desired, a wide range of radiation detection methods are available including several nondestructive methods of gamma spectroscopy, neutron and gamma counting, multiplicity counting, and imaging, and some destructive methods of alpha spectroscopy and mass spectrometry. Each method provides information indicative of material composition, isotopic composition, or processing method where radiation detection equipment is often utilized in the laboratory and field to perform these analyses. This chapter aims to provide a brief introduction to the radiation signatures of interest for radiation detection, an overview of how these signatures are utilized for standard radiation detection-based verification techniques utilized in non-proliferation and arms control monitoring and verification, as well as a description of some of the challenges associated with implementation of these techniques into international agreements.

1 Introduction

One of the key functional components in a transparency agreement is the capability to perform confirmatory analysis to verify a country's commitments through direct measurements of radiation signatures at nuclear sites [1]. Radiation detection-based verification techniques measure the radiation signature of an item, thus utilizing the intrinsically unique nature of nuclear reactions and decays from each type of radioactive material. Verification and monitoring based upon radiation detection

M. Williamson (✉) · J. Preston
Consolidated Nuclear Security, Y-12 National Security Complex, Oak Ridge, TN, USA
e-mail: martin.williamson@cns.doe.gov; jeffrey.preston@cns.doe.gov

I. Niemeyer et al. (eds.), *Nuclear Non-proliferation and Arms Control Verification*,
https://doi.org/10.1007/978-3-030-29537-0_17

techniques are especially useful when access and/or range limitations must be overcome, such as the case where inspectors are tasked to identify or confirm the contents of a sealed container. This chapter aims to provide a brief introduction to the radiation signatures of interest for radiation detection, an overview of how these signatures are utilized for standard radiation detection-based verification techniques utilized in non-proliferation and arms control monitoring and verification, as well as a description of some of the challenges associated with implementation of these techniques into international agreements.

2 Special Nuclear Material Signatures

Both destructive and nondestructive analysis (NDA) methods of measurements can utilize radiation signatures and features associated with either a single isotope or a mixture of isotopes to determine material properties of isotopic composition, chemical purity, chemical form, processing methods, separation efficiencies, and other aspects. The radiation signatures of different types of Special Nuclear Material (SNM), defined in Title 1 of the Atomic Energy Act of 1954 as plutonium, uranium-233, or uranium enriched in isotopes uranium-233 or uranium-235, vary significantly and are summarized in the following subsections. Uranium properties are discussed first given its widespread usage in both reactor fuels and as a weapons material, whereas plutonium is less commonly utilized for reactor fuel and is primarily considered a weapons material.

2.1 Uranium Signatures

Uranium is a complex element that naturally consists of three isotopes: uranium-234, uranium-235, and uranium-238. Uranium-233 is produced via neutron irradiation of thorium-232 in a nuclear reactor. Table 1 provides a brief summary of uranium signatures of various types that can be utilized to gain information about a sample being inspected. Uranium enriched in the isotope of uranium-233 has been studied for use for both nuclear power reactors and weapons, but uranium enriched in the isotope uranium-235 is commonly utilized for these applications [2]. During a uranium-235 enrichment process by centrifuge, gaseous diffusion, or atomic vapor laser isotope separation (AVLIS), the lighter isotopes are preferentially separated with a characteristic separation factor that indicates the process method. Isotopic ratios between the three naturally occurring atoms can be led to forensically indicate which separations process was utilized. For example, highly enriched uranium of 93% enriched uranium contains 93% uranium-235 by atom percent with the balance comprised of uranium-234, in greater than natural abundance, and uranium-238.

Due to the low spontaneous fission rates from uranium isotopes, gamma and alpha radiation are the primary radiation signatures of interest. The challenge

Table 1 Selected uranium properties of interest

	Uranium-233	Uranium-234	Uranium-235	Uranium-238
Natural abundance [3]	–	0.0054%	0.7204%	99.2742%
Atomic weight (amu)	233.0396355	234.0409456	235.0439231	238.0507826
Model of decay	Alpha (159.2E3 years)	Alpha (245.5E3 years)	Alpha (703.8E6 years)	Alpha (4.468E9 years)
Decay energy (MeV)	4.909 MeV	4.859 MeV	4.679 MeV	4.270 MeV
Thermal fission cross section (0.0253 eV)	531.2 b	6.218 mb	584.4 b	11.77 μb
Average number of neutrons per fission [4, 5]		–	2.432+0.066E (0<E<1)	–
			2.349+0.15E (E>1)	
Significant gamma lines (keV) [6]	42.44 (0.0862%)	53.20 (0.123%)	109.16 (1.54%)	1001.7 (0.838%)
	54.699 (0.0182%)	120.9 (0.0342%)	143.76 (10.96%)	1737.73 (0.0211%)
	97.134 (0.0203%)		163.330 (5.08%)	1831.3 (0.0172%)
	146.345 (0.00657%)		185.715 (57.2%)	1867.68 (0.0092%)
	164.522 (0.00623%)		202.11 (1.08%)	1874.85 (0.0082%)
	291.354 (0.00535%)		205.31 (5.01%)	1911.17 (0.0063%)
	317.16 (0.00776%)			1937.01 (0.0029%)

associated with these particles is that they are easily shielded or masked, such that detection in sealed containers can present a significant challenge. In these cases, multiple methods of analysis are utilized to either extrapolate or directly measure the material composition depending on the extent of shielding (including self-shielding) and accessibility of the material. For shielded objects, active interrogation techniques using neutrons to induce fission in uranium-235 can also be utilized to generate highly penetrating fission neutrons that can be detected.

Monitoring and verification techniques leveraging these uranium signatures are discussed in more detail in the following sections.

2.2 Plutonium Signatures

Plutonium signatures provide another layer of complexity beyond uranium as the man-made isotopes nearly always exist in varying isotopic combinations of plutonium, curium, americium, and decay products. Chemical processes on spent fuel are designed to separate the fuel from fission products, then separate the uranium from the higher actinides, and finally separate the neptunium, curium, and americium from the plutonium isotopes. However, these methods are not perfectly efficient due to chemical similarities between these actinides. As such, the features shown in Table 2 aid in forensic examination of the chemical separations process, reactor configuration utilized to produce the material, and length of time after a chemical separations process in addition to the expected isotopic compositions.

Gamma spectroscopy is not readily performed with plutonium due to the dearth of gammas and contamination with americium-241 at 59.5 keV that limits the effective measurement range above the X-ray region to the next nearest, abundant gamma. A common difficulty with bulk material plutonium gamma measurements is the presence of neutrons stemming from both (alpha,n) reactions with plutonium and americium-241 and the high spontaneous fission probability, where smaller samples reduce the neutron background and corresponding gamma background considerably. A preferred method of plutonium analysis is with alpha spectroscopy, whereas with uranium, each isotope has a unique alpha energy that is easily discerned in an alpha energy spectrum. Neutron coincidence counting is also possible with plutonium isotopes and provides some information for the isotopic split between plutonium-239 and plutonium-240. These isotopes have similar fission and fast neutron cross sections, but very different thermal cross sections. Plutonium has a high spontaneous fission probability that is often utilized as a heuristic when detecting americium-241 gammas. Utilizing these differences, the isotopic ratio can be obtained. Plutonium-241 is primarily a beta emitter as it transitions to americium-241 quickly, releasing the characteristic americium-241 alpha and 59.5 keV gamma.

Table 2 Selected plutonium properties of interest

	Plutonium-239	Plutonium-240	Plutonium-241
Natural abundance [3]	–	–	–
Atomic weight (amu)	239.0521565	240.053807	241.0568453
Model of decay	Alpha (24,110 years)	Alpha (6561 years)	14.325 years β: 100.00% α: 2.5E-3%
Decay energy (MeV)	5.245 MeV	5.256 MeV	5.140 MeV
Thermal fission cross section (0.0253 eV)	747.4 b	58.77 mb	361.5 b
14 MeV fission cross section	2.338 b	2.253 b	0.1050 μb
Average number of neutrons per fission [4, 5]	2.874+0.138E		

3 Radiation Detection Monitoring and Verification Techniques

A wide range of radiation detection techniques are available for a given application, the choice of which is dictated by several factors including the operational scenario and the type of information desired (e.g., verification of SNM presence, enrichment, mass, etc.). For example, one operational scenario may have restrictions on the detector system such that it must be portable and easily operated in limited-access environments, while another may be suited for a laboratory. Cost, sensitivity, performance, and directional capability are also common characteristics of concern.

This section summarizes a selection of radiation detection monitoring and verification techniques typically utilized for nonproliferation and arms control purposes. Types of radiation detection equipment (RDE) available for each technique are presented, as well as the type of information they can provide to add confidence for monitoring and verification purposes.

3.1 Radiation Pulse Counting

In pulse counting systems, radiation is converted into an electronic signal that, when above a set threshold or within a given amplitude range, will trigger a switch to cause a counter to increment by 1. Combined with a timer, the counter can be utilized to determine a count rate that may be related back to a source's decay rate. For a given source to detector geometry, the number of counts per second can determine the presence or absence of a given gamma or neutron emitting isotope. Pulse counting is common for detector materials where the pulse output is not necessarily indicative of the incident particle energy (e.g., helium-3 gas-based neutron detectors and Geiger-Mueller detectors). These detectors provide information as to the dose rate, decay rate, or isotopic quantity of a given source.

One application for pulse counting is for spent fuel identification, where the high activity of the fission products typically preclude spectroscopic instruments and dose rate measurements are made as a function of time spent outside of a reactor. The dose rate present is modeled as a function of decay time, and measurements with various absorber thicknesses offer multiple data points for verification of a declaration of origin, where spent fuel composition is a function of reactor operating history, and quantity of spent fuel to ensure against material diversion. Count rates are also utilized for measuring active length of fresh reactor fuels as a means to validate the mass of SNM present in fuel bundles or assemblies. For isotopes with a high probability of spontaneous fission or induced fission, neutron emission offers another means to validate the mass present, where neutron counting provides information to modeling codes that attempt to reconstruct the geometry and isotopic composition. While pulse counting is useful in many operational scenarios, the method does not give any indication as to the energy of the incident particles, which offers unique capabilities for material identification when measurement criteria of lower counting rates and detector proportionality are met.

3.2 Spectroscopy

A widely used method of analysis for radiation detection is spectroscopy, which may be defined here as the analysis of characteristic energies from a system quantized in a histogram by frequency of occurrence. This definition is intended to cover a range of common spectroscopic methods that involve traditional light spectroscopy utilized in analysis of chemicals, conversion of gamma ray energies into electronic signals, to mass spectrometry that does not involve light emission but rather kinetic energy of a given element or atom. In modern spectroscopic radiation detection systems, a detector apparatus converts the particle energy into an electronic signal that is digitized and added to the histogram, commonly referred to as the energy spectrum. Each bin or channel in the energy spectrum may be thought of as a pulse counter for that specific energy. An energy spectrum for a given isotope is compiled by a multichannel analyzer (MCA) that show peaks characteristic of a specific isotope type that aid in identification of an unknown material, stoichiometric determination, and quantization of isotopic ratios. It is important to note the difference in the underlying assumption that the detector of interest in spectroscopy provides an output that is indicative of the particle energy, contrasted with pulse counting in which this is not necessarily true.

Gamma Spectroscopy

One of the primary non-destructive tests for material quantity and enrichment verification is gamma spectroscopy as it passively detects the radiation emitted naturally from materials of interest. The fundamental principle is that radioisotopes have a characteristic set of gammas with energies, relative intensities, and emission angles that combine with the characteristic half-life that are akin to a fingerprint in terms of uniqueness, which is a primary means for isotopic identification. Gamma spectroscopy is useful as an NDA method when using an analysis code that compares peak ratios for a given detector material, detector efficiency, and collimation. Verification of declared isotopic compositions is a primary function of inspectors as it indicates whether material is being diverted or if undeclared materials are being utilized. Nondestructive analysis with gamma spectroscopy is a common tool that is performed with both medium resolution scintillation instruments and high resolution semiconductors. Depending on the measurement method, different analysis software packages are utilized to provide the isotopic composition. Enrichment verification and plutonium isotopic measurements with semiconductor detectors (e.g., high purity germanium) are typically performed with software packages such as FRAM, MGA, MGAU, or NaIGEM. FRAM performs a decomposition of the X-ray region of a spectrum and identifies composition of decay chain daughter products of various actinides in order to determine enrichment. The MGA and MGAU codes deconvolute the low energy regions with a multi-group analysis (MGA) approach that determines isotope compositions of various plutonium isotopes, uranium-235, neptunium-237, and americium-241. The MGAU code specializes on the uranium isotopes for enrichment verification. The NaIGEM code utilizes gamma lines from uranium-235 and uranium-238 to determine enrichment relative to a calibration standard. Each of these codes are utilized with samples both in an out of containers as they include compensation for material attenuation and shielding.

Gamma spectrometers are generally divided between scintillator and semiconductor detectors that have different detection efficiency, Z-value, energy and time resolution, and cost, which may be a determining factor in selection for a given detection application [7]. The semiconductor class of instruments include high purity germanium (HPGe), which offers the highest resolution (0.5% at 662 keV) and is generally a confirmatory analysis tool, but has several limitations for applications. For confirming uranium isotopic composition with 185–1001 keV ratios, plutonium/uranium ratios, holdup monitoring, and general gamma spectroscopy, HPGe offers the highest confidence results but requires constant cryogenic cooling that requires significant power to operate with mechanical coolers or dewars for liquid nitrogen. Alternative semiconductors, including the room-temperature operating cadmium zinc telluride (CZT), are often utilized in small, handheld spectroscopic instruments that can provide resolutions between 1 and 3.5% at 662 keV, which enable high confidence for isotopes that emit radiation at lower energies (e.g., uranium-235 emits at 185 keV). For room temperature semiconductor detectors, physics often limits the ability to fabricate or operate detectors over certain volumes or thicknesses due to crystal growth constraints or unrealistically high bias voltages, resulting in limited sensitivity at higher energies (e.g., uranium-238 daughter emits at 1001 keV). Therefore, CZT is often utilized for low energy detection where scintillation detectors typically have overlapping peaks due to lower resolution (between 3%-8% at 662 keV), as is the case with SNM in cases of heavy shielding or masking. As an example with uranium isotopic composition, a thick-walled drum containing UF6 causes a significant amount of scattering that can inhibit gamma response as the 185 keV peak is located in a competing region. High resolution detectors allow for peaks to be more readily visible with fewer counts as the energy is not spread over so many channels, which effectively raises the signal to noise ratio. For our example, this translates to shorter counting times for inspectors, lesser amounts of collimation, reduced need for pin hole apertures, and more stable results; however, semiconductors have difficulties with high temperature operation (e.g., $>40\,^{\circ}C$) as noise dominates the lower energies.

Scintillation detectors offer the most cost effective performance and are the most widely deployed. Compared to semiconductors, scintillators may range from small to very large volume detectors with high sensitivities. Each scintillating material has an effective Z-value, density, light color or wavelength of emission, decay time, and temperature dependence that aid in selecting where the application is most useful

[8]. Spectroscopic scintillation detectors, including sodium iodide (NaI(Tl)) and lanthanum bromide (LaBr3:Ce), are available from many vendors and are utilized in many overlapping areas as semiconductors, and extend into wide area monitoring, dosimetry, and other applications. Because scintillation detectors are relatively inexpensive compared to the semiconductor detectors, operations typically utilize the scintillation detectors first and confirm with semiconductors if necessary. Holdup measurement systems often utilize scintillation detectors as vendors have devised extension poles to reach inaccessible locations where holdup may accumulate. Uranium composition through software comparing the relative magnitudes of the 185–1001 keV gammas are often utilized on scintillation instruments. Large volume inorganic scintillators are often utilized for unattended area monitoring in storage areas to ensure material is not removed during standard operations in absence of inspections. Very large volume, high sensitivity, low resolution plastic scintillators are often utilized for portal monitors to ensure material is not removed from the premises. Collimated detectors are often utilized for active length measurements in fuel fabrication facilities and power plants to measure and verify uranium-235 content in fuel assemblies.

Figure 1 presents a comparison of gamma spectra collected from a 19.4% uranium-235 metallic standard using a suite of gamma spectrometers and highlights the differences in resolution capabilities at low energies [9].

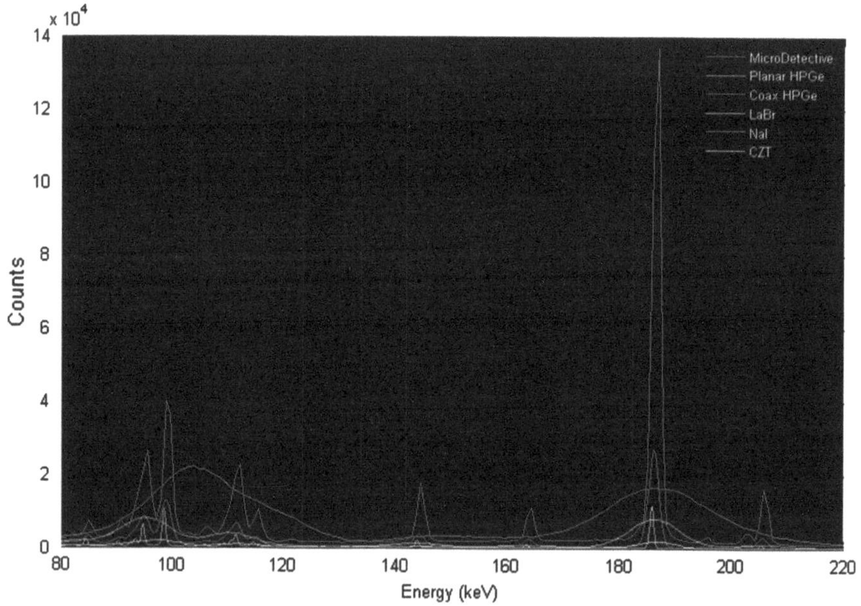

Fig. 1 Comparison of gamma spectrometer results for a 19.4% uranium-235 enriched metal standard. Source: [9]

Alpha Spectroscopy

As with gamma spectroscopy, alpha particles emitted during radioactive decay of actinides have characteristic energies that aid in identification of the isotope. Alpha spectroscopy is typically a destructive analysis method as a small powder, pressed sample, metallic electroplating, or other sample form is required to be placed inside of a small vacuum chamber to remove effects of energy attenuation in air. Detectors are often energy calibrated with a set of isotopically purified alpha standards comprising electroplated sources of uranium-235, uranium-238, neptunium-237, plutonium-239, and americium-241. The chamber is loaded with a tray containing the sample at a fixed distance from a semiconductor detector, and the chamber is brought to low vacuum. A sample acquired in the field emits alphas of an unknown energy and distribution depending on the isotopic composition, which generates an energy spectrum complete with peaks corresponding to alpha energies. An energy calibrated detector may show alpha particles overlapping with those present in the standards, or a decay catalog may be utilized to identify unknowns [3]. In the case of measuring uranium-234 content, the gamma spectroscopy comparison of uranium-234 to other isotopes is difficult due to the low emission rate, lack of highly intense gammas, and low energy gammas that are readily shielded by containers; however, the alpha energy is quite unique in the alpha spectroscopy field. As a destructive analysis, alpha spectroscopy provides information not observable with passive gamma detection outside of a storage container. As with uranium enrichment methods, the ratios of peaks in the spectra are indicative of isotopic composition, but do not suffer from effects of non-constant energy absorption as is with gamma detection; thus, alpha spectroscopy is a higher confidence measurement when possible. Relative composition is able to be obtained from a complete analysis, which is then utilized to confirm a declared material or verify presence of an undeclared isotopic composition or isotope.

Mass Spectrometry

A second destructive method is mass spectrometry, which also provides information from small samples and swipes. Because the atomic masses are different for each uranium isotope, the mass spectrometer provides proportionally different kinetic energy during the acceleration process that creates the detection event. The resulting mass spectrum illustrates the differences in masses in a quantifiable method similar to alpha spectrometer, but without the difficulty of maintaining ideal sample geometry for consistent efficiency between measurements.

3.3 Neutron Detection

For SNM isotopes exhibiting spontaneous or induced fission, neutron detection provides a non-destructive testing method. Due to neutrons being neutral particles and the fact that, unlike gammas, they don't interact directly with surrounding electrons, neutrons are much more penetrating than alphas, betas, and gammas. Unlike gammas, neutrons do not aid in isotopic identification through their energy, and for many detector designs, neutrons must typically be fully absorbed in the detector material in order to be counted. Neutron proportional counters are metal tubes typically filled with pressurized helium-3 gas that absorbs neutrons and emits electronic signals proportional to the energetic response of the absorption reaction. Spectroscopy of the helium-3 detector is not useful for determining the energy of the incident neutron as the energy spectra only provide information to the two charged particles created from the absorption; thus, these detectors are typically operated as counting devices as discussed earlier in the chapter. These detectors are commonly utilized in handheld radiation detectors, portal monitors, area monitors, and multiplicity counters. Figure 2 shows an image of a typical helium-3 filled neutron proportional counter from GE.

Scintillation-based neutron counters are also common as zinc sulfide (ZnS) scintillating sheets combined with a lithium-6 containing compound. Lithium-6 is the neutron absorbing medium that emits an alpha and triton into the scintillator, initiating the counting event. These detectors can scale to large sizes for portal monitoring and can also be utilized in compact, handheld detectors.

A common application of neutron detection is in active well counters that determine the average number of neutrons released per fission in a sample, which is indicative of the specific isotope. This method is a primary method for determining isotopic composition of bulk samples of fissile and fissionable materials. A neutron multiplicity measurement in a well counter utilizes neutron detectors arranged in concentric rings around a given sample, which are analyzed in groups for coincident neutron detection events. A fission event will yield multiple neutrons simultaneously in different directions, and with a short coincidence window, the number of excited

Fig. 2 Typical helium-3 neutron detector. Source: [10]

detectors per event is corrected by the detection efficiency of the well and is utilized to determine the number neutrons per fission of a sample. A mixed isotope sample will provide multiplicity numbers between the two values for a pure sample, which is interpolated as the isotopic ratio. Well counters may be passive or active, where passive monitors spontaneous fission and active monitors induced fission from a neutron source. Active Well Coincidence Counters or AWCCs contain an Americium-Beryllium or californium-252 source in a container above and/or below the sample. Background neutron multiplicity is measured prior to addition of the sample material. Once the sample is introduced, a subcritical assembly is created that contains both the background and sample multiplicity. A background corrected number of neutrons per event is collected over a duration that reduces uncertainties of the collection. This method is particularly useful for uranium-235 and plutonium-239 composition measurements. Figure 3 shows picture of an active well neutron coincidence counter from Canberra.

3.4 *Imaging Techniques*

Technology advancements in computing capabilities and miniaturization of radiation detectors have led to lower cost radiation imaging that is a rapidly developing area for safeguards and treaty verification. Early in the post-9/11 era, much interest focused on remote detection and identification of threats for interdiction efforts at ports of entry, highways, and high attendance events; however, this is inherently difficult with radiation and lent reason for a second look at the concept of utilizing

Fig. 3 Canberra active well neutron coincidence counter. Source: [11]

the Compton scattering effect for standoff radiation detection. This concept has been deployed in space applications with silicon-based detectors for some time as deep space imagers, while terrestrial imagers have had limited success as they are prone to phantom sources from reflections and weak signals due to scatter through air or other materials and naturally occurring radioactive material (NORM) background. Development efforts focused on detector and electronics design, background reduction, and algorithm improvements have led to a new generation of standoff gamma imagers capable of performing tasks associated with safeguards and arms control.

Standoff imagers are generally of two types, semiconductor and scintillation based, where sensitivity and detector resolution are the prime differentiators. Semiconductor types consist of one or more detectors with multiple electronic readouts, effectively pixellating the device into many, small detectors. These have high directional resolution along with high energy resolution; however, given the size of the detectors, the sensitivities are rather low. Scintillator types consist of many detectors combined to form a high sensitivity array, but with medium energy resolution. Often times, either method may be combined with a collimator for a pinhole camera effect or a coded aperture to improve positioning information. In most cases, the gamma images are 2-D heat maps, similar to thermal images, which provide little depth information and rely on overlaid visible camera images for position identification. For safeguards and arms control purposes, these detectors could be left in situ to monitor consistent presence of material or brought into an area to locate hidden or missing material.

4 RDE Implementation Challenges

Implementation of RDE for use in transparency and arms control regimes faces many barriers from basic shipping and receiving of the RDE to simultaneously ensuring both host and monitor confidence in methods, information integrity and security. In many cases, these challenges have led to cooperation between instrument manufacturers, national laboratories, countries, and supranational agencies (e.g., IAEA and EURATOM) to address these challenges through RDE design instrumentation methods and/or changes in policies and procedures.

4.1 Arms Control Implementation Challenges

Unique challenges associated with utilizing radiation detection equipment (RDE) in an arms control framework include the balancing required between: (1) the ability of the monitoring country to confirm the declared information related to the treaty accountable item (TAI) and to develop and maintain confidence in the measurement results (i.e., authentication); and (2) the host country's ability to approve the RDE for deployment in the monitored facility and to prevent unauthorized disclosure of sensitive information (i.e., certification). Authentication and certification are complementary processes and trade-offs between authentication and certification concerns must be balanced such that both the host and monitoring country are able to establish an acceptable level of confidence. This confidence is based upon the combination of RDE design features as well as the associated RDE procedures and administrative controls that ensure that the equipment is operated correctly.

Successful authentication of a RDE system would assure the monitor that when the RDE is used as designed and intended, accurate and reliable information is collected. Additionally, the monitor must ensure that the results produced by the RDE are representative of the true state of the measured TAI such that irregularities, including hidden features, are detected. A wide range of tools and methods exist to support authentication, including: detailed RDE protocols and procedures for use of the RDE for all monitoring activities, inspection/evaluation of the RDE hardware and software, designing the RDE to facilitate inspection/evaluation, functional testing with calibration sources, random selection of equipment, and the use of chain of custody tools such as tamper-indicating tags and seals. An authentication protocol can vary significantly in the complexity of equipment used, the intrusiveness of the measurement techniques (i.e., how much information they reveal about the TAI being measured), and cost to implement.

Successful certification of a RDE system would ensure the host that the RDE can be: (1) operated securely such that operation will not release any unauthorized sensitive information regarding the TAI being measured, the facility, or ongoing operations, and (2) safely such that the RDE meets all applicable host facility safety and quality requirements to protect people, the facility, and the environment. Several of the same tools identified to support authentication may also be used to support certification, including detailed RDE protocols and procedures for use of the RDE for all monitoring activities and designing the RDE to facilitate inspection/evaluation. Additional certification tools, such as security and vulnerability testing/evaluation of RDE hardware and software, information barriers, and facility acceptance reviews may also be used to ensure that safety and security concerns are addressed for the host facility.

4.2 General Implementation Challenges

A common difficulty in implementation of a field measurement is calibration of the instrument in the field. It is common practice to fully vet an instrument when it leaves a clearinghouse and when it returns to give confidence to the collected data; however, some instruments have special requirements upon startup or during operation that may yet cause concern with data fidelity. For photomultiplier tube (PMT) based scintillation instruments, calibration upon startup is generally required to offset the temperature effects to both the PMT and scintillator that affect the peak position. This calibration is most effectively performed with a radiation source that effects a similar detection path as the gamma sources to be measured. In most cases, reliance on ambient potassium-40 background is tenuous due to a low count rate or conflicting gamma signatures, and reliance on an internal or relocatable radiation source is difficult due to shipping, import, or licensing regulations in the host location. A similar difficulty affects active well coincidence counters that require neutron sources for operation. For monitoring operations relying on short notice inspections, a multi-week waiting period for shipments in customs is counterproductive to the nature of the agreement. Workarounds for these issues have often relied on agreements for shipping and receiving of instruments, special courier possession licenses, or sometimes the host location procuring instrumentation that is then cross-calibrated to readily available standards. Other options have been the development of equipment that do not require an external calibration source, including CZT-based instruments and newly developed silicon photomultiplier devices utilizing scintillators for gamma detection.

As previously stated, maintaining confidence in devices shipped through common carrier with single measurements presents a difficult situation for ensuring data integrity should a device fail upon return or is suspected of having been tampered with. Tampering is often prevented with tamper-proof seals that will indicate whether an instrument was disassembled during transit in an attempt to gain access to data. Inspectors will often travel with instruments when possible to maintain physical control. Instrument failure in the field from wear and tear, accidental breakage, or due to shipping presents an instance where data must be recollected if possible, resulting in a significant cost to an inspection agency. It is generally good practice to subject instruments to a battery of tests with calibrated standard sources with laboratory-quality precision and accuracy in order to determine variance from ideal conditions. Instruments that are approaching a calibration boundary may not be deployed but rather serviced or replaced, and instruments that pass and return will be subjected to the same tests and criteria upon return. If a device performs identically upon exit and return, then data acquired in the interim can be accepted with a high confidence.

5 Conclusions

The efficacy of a monitoring regime relying upon radiation-based measurements is heavily reliant upon the reliability, accuracy, and cost of deployment of the radiation detection methods utilized. Methods and instruments are conservatively chosen through vigorous tests and prolonged parallel trials to maintain high integrity of the measurements. As a result, an addition or modification to an inspections method or instrument is a deliberate and costly process that focuses on areas of greatest potential gains. In many cases, existing technology is suitable and amenable to agreements, but lacks the corresponding authentication and/or certification procedures that would be used to justify host and monitoring country confidence for the proposed use of the RDE. Further development of the tools, techniques, and procedures to demonstrate authentication and certification capabilities should be further explored to better understand how confidence is built and maintained for potential arms control regimes utilizing existing RDE technologies.

Future developments in safeguards technologies are aiming to shorten detection time through improved algorithms and detectors, and lower the cost of high performance instruments to increase deployment. Improvements in detector methods are typically at odds with the low cost aspect, which has led to development efforts towards higher performance scintillators and semiconductors with lower price points that meet or closely approach the baseline criteria for ruggedness, stability, and performance for the materials they intend to replace or complement. In many cases, the methods most widely deployed are derived from a common method of gamma or neutron detection that has a significant cost-effective advantage, as evidenced by the ubiquity of sodium iodide based spectrometers. Gamma detection utilizing $<4\%$ resolution (at 662 keV) scintillation detectors have focused on reducing cost of lanthanum bromide or achieving comparable resolution with lower cost materials as a means to improve detection capabilities and algorithm performance. Cryogenically cooled germanium for semiconducting gamma detection is still unrivalled in terms of spectral resolution, but the negative aspects (e.g., operational limitations and cost of operation) are sufficiently high that it is increasingly supplanted by CZT based spectrometers in applicable usage scenarios, and HPGe is utilized as a secondary method as needed to reduce the number of detectors deployed and associated costs. Neutron detection with helium-3 based neutron capture detectors are increasingly being replaced due to decreasing supply. Alternative methods have focused on lithium- or boron-based scintillator and semiconductor detectors, while complementary methods for neutron scattering detection with liquid xenon and loaded plastic scintillators have shown significant promise. New detection materials suitable for safeguards usage overlap with security markets, providing a synergistic effect for R&D funding.

This work of authorship and those incorporated herein were prepared by Consolidated Nuclear Security, LLC (CNS) as accounts of work sponsored by an agency of the United States Government under Contract DE NA0001942. Neither the United States Government nor any agency thereof, nor CNS, nor any of their employees, makes any warranty, express or implied, or assumes any legal liability or responsibility to any non governmental recipient hereof for the accuracy, completeness, use made, or usefulness of any information, apparatus, product, or process disclosed, or represents that its use would not infringe privately owned rights. Reference herein to any specific commercial product, process, or service by trade name, trademark, manufacturer, or otherwise, does not necessarily constitute or imply its endorsement, recommendation, or favoring by the United States Government or any agency or contractor thereof. The views and opinions of authors expressed herein do not necessarily state or reflect those of the United States Government or any agency or contractor (other than the authors) thereof.

References

1. IAEA (2011) Safeguards Techniques and Equipment: 2011 Edition. IAEA. Vienna
2. Forsburg CW, Lewis LC (1999) ORNL-6952: Uses for Uranium-233: What Should Be Kept for Future Needs? Oak Ridge National Laboratory. Oak Ridge
3. National Nuclear Data Center (2016) Interactive Chart of Nuclides. Brookhaven National Laboratory. http://www.nndc.bnl.gov/chart/. Accessed 28 Feb 2019
4. Duderstadt J, Hamilton LJ (1976) Nuclear Reactor Analysis. Wiley, New York
5. T-2 Nuclear Information Service (2011) ENDF/B-VII.1 Incident-Neutron Data. Los Alamos National Laboratory. http://t2.lanl.gov/nis/data/endf/endfvii.1-n.html. Accessed 28 Feb 2019
6. Online Spectrum Catalogs for Ge and Si(Li) (2009)
7. Knoll GF (1999) Radiation Detection and Measurement, 3rd) ed. Wiley, New York
8. Saint-Gobain (2018) Properties of Common Inorganic Scintillators. Saint-Gobain Crystals. https://www.crystals.saint-gobain.com/sites/imdf.crystals.com/files/documents/physical-properties-of-inorganic-scintillators_0.pdf. Accessed 28 Feb 2019
9. Cantrell JA (2012) Uranium Enrichment Standards of the Y-12 Nuclear Detection Sensor Testing Center. In: Proceedings of the 53rd annual meeting of the Institute of Nuclear Material Management, Orlando.
10. Reuter Stokes ^{3}He Neutron Detectors for Homeland Security Radiation Portal Monitors. Fact Sheet. https://www.gemeasurement.com/sites/gemc.dev/files/reuter_stokes_he-3_neutron_detectors_for_homeland_security_radiation_portal_monitors_english_0.pdf. Accessed 28 Feb 2019
11. Canberra (2009) Model JCC-51 Active Well Neutron Coincidence Counter. Canberra Industries, Inc. http://www.canberra.com/products/waste_safeguard_systems/pdf/JCC-51-SS-C36907.pdf. Accessed 28 Feb 2019

Environmental Signatures and Forensics

Klaus Mayer, Maria Wallenius, Yetunde Aregbe, and Magnus Hedberg

Abstract Locard's principle "every contact leaves a trace" is the overarching foundation for any kind of forensic investigations, including nuclear forensics. In some instances, it doesn't even take a contact to leave a trace, as we know from emissions of gases or particles (e.g. from industrial facilities) which can then be found in the environment and interpreted as footprint of the facility. Such trace can be understood as "signature" of the process they originate from. In the present chapter we want to illuminate the power of "signatures" (i.e. characteristic parameters that point at the provenance, history or intended use of nuclear material) and their application in nuclear forensics and in nuclear safeguards. Particular attention will be given to environmental particle analysis as applied for safeguards purposes but also for nuclear forensic investigations. Moreover, nuclear forensic signatures of bulk samples will be discussed with emphasis on uranium and plutonium age dating, isotopic signatures in Pu and chemical impurities in natural uranium. Ensuring measurement quality by providing appropriate reference materials and establishing state-of-the-practice through interlaboratory comparison exercises are essentially contributing to the credibility of results. Synergies between nuclear material analysis for safeguards purposes, for nuclear security applications and for non-proliferation are discussed.

K. Mayer (✉) · M. Wallenius
EC Directorate General Joint Research Centre, Directorate G - Nuclear Safety and Security, Karlsruhe, Germany
e-mail: klaus.mayer@ec.europa.eu; maria-s.wallenius@ec.europa.eu

Y. Aregbe
EC Directorate General Joint Research Centre, Directorate G - Nuclear Safety and Security, Geel, Belgium
e-mail: yetunde.aregbe@ec.europa.eu

M. Hedberg
EC Directorate General Joint Research Centre, Directorate G - Nuclear Safety and Security, Karlsruhe, Germany

IAEA, Vienna, Austria
e-mail: magnus.hedberg@ec.europa.eu; p.m.hedberg@iaea.org

I. Niemeyer et al. (eds.), *Nuclear Non-proliferation and Arms Control Verification*,
https://doi.org/10.1007/978-3-030-29537-0_18

1 "Signatures" as Source of Information

Nuclear non-proliferation and nuclear security are largely information driven areas. Information can be obtained from a variety of sources such as open literature, scientific publications, satellite images or the internet. The nuclear material itself or samples containing nuclear material may also provide relevant information. The concept is based on one of the fundamentals of (classical) forensic science, Locard's principle, which makes use of physical evidence. Kirk [1] put it straight: "Physical evidence cannot be wrong, ... Only human failure to find it, study and understand it, can diminish its value." The hints and information contained in the evidence, however, is not directly accessible but needs to be inferred from measurable material characteristics, referred to as "signatures". The isotopic composition of the nuclear material, the isotopic composition of trace elements, chemical composition, chemical impurities, macroscopic appearance or microscopic morphology are examples of such signatures. A careful evaluation of these signatures provides information or hints on the intended use of the material, on its processing history, on its place of production and on the date of production. This evaluation may require comparison information e.g. data from literature, from databases in order to match the material under investigation against an individual reference sample or against a population of samples sharing a common feature; in consequence the signature is called "comparative signature". Only few signatures can be understood without comparison data; these are referred to as "predictive signatures". The age of sample (i.e. the time elapsed since its last chemical processing) is a typical example. Signatures have proven to be highly useful in nuclear non-proliferation when checking consistency of material characteristics against declared facility operations or for verifying the absence of undeclared nuclear activities. Another area of application addresses investigations related to nuclear material found out of regulatory control, e.g. material intercepted from illicit trafficking. The efforts for establishing linkages between nuclear material, nuclear security events and the origin and intended use of the material are referred to as "nuclear forensics".

2 Environmental Particle Signatures and Measurements

Particle analysis has proven to be a very powerful tool for establishing signatures of the history of nuclear installations, in particular uranium enrichment facilities. The determination of the isotopic composition of a reasonably large number of individual uranium particles allows establishing a picture of the range of isotopic enrichments in the uranium materials handled or produced in the facility under investigation. The techniques typically applied for analysing (sub-)micrometer sized particles are fission-track TIMS (thermal ionization mass spectrometry) and secondary ion mass spectrometry. The latter one is gaining increased attention due to the much simpler sample preparation, the capability of analysing a large number of particles and (using the latest generation of instruments) comparable measurement uncertainty.

2.1 New Instruments Opening New Avenues: The LG-SIMS

A SIMS (Secondary Ionisation Mass Spectrometry) instrument sputters and ionises a sample by an ion bombardment from a primary ion beam [2]. The ionised material is extracted and analysed in a mass spectrometer. The produced secondary ions typically show an energy spread in the region of several tens of eV, thus a single magnetic sector will not provide a good mass dispersion. A double focusing SIMS has in its secondary ion optics both a magnetic and an Electrostatic Sector Analyser (ESA). A magnet is dispersive both in mass and in energy while an ESA is dispersive only in energy. The principle of double focusing mass spectrometers is to cancel as much as possible of the magnet energy dispersion by a compensated ESA energy dispersion.

A SIMS instrument is capable of analysing all elements in the periodic table, at trace levels. Some SIMS instruments can both be operated as an isotope mass spectrometer and as an isotope microscope, thus adding also isotope imaging capabilities over the surface area. The latter is very important in the case of environmental particle analysis as this provides a tool to find the particles of interest in a matrix of other material.

Particle analyses have predominantly been performed in the past using the small geometry CAMECA IMS xF instrument series [3]. The SIMS analytical work consists of two main sub tasks; the first one is to find the particles of interest in a matrix of other materials, the second is to perform accurate and precise isotopic measurements of the individual particles. In summary, the SIMS technique features numerous advantages for small uranium particle analysis:

1. Sample preparation is fast and requires little effort using the vacuum impactor technique [4].
2. High ion yield for uranium, at around 1%, which allows for analysis of particles down to sub micron size range.
3. SIMS instruments with ion imaging provide the exact location of uranium particles within a sample matrix.
4. The screening measurements can be followed immediately by precise and accurate isotopic micro beam measurements of individual particles using the same instrument.
5. Timely analysis compared to other techniques like the fission track TIMS method [5]. A typical throughput is about four samples per week.

A breakthrough to improve the SIMS measurements has been the implementation of Large Geometry SIMS, the CAMECA IMS 1280/1300 models, for this type of analysis [6, 7]. Basically, LG-SIMS instruments were developed in response to a demand for high mass resolution—high transmission instruments in geosciences and cosmochemistry. These instruments are, like the small geometry SIMS, based on a double focusing mass spectrometer, but with the implementation of a large-radius magnetic sector and improved secondary ion optics. LG-SIMS instruments feature numerous instrumental advantages, in particular removal of background interferences with minimal loss of transmission. The LG-SIMS instrument is equipped with a multi-collection detector system, which allows simultaneous isotope measurements for elements from lithium up to uranium. This system consists of five moveable collector units (L2, L1, C, H1, H2). Each unit is equipped with a set of three slits for adjusting the mass resolution conditions, and with either a discrete dynode electron multiplier (EM) or a Faraday cup (FC) detector for higher signal intensities. The instrument is also equipped with a mono-collection detector system with two FCs and one EM detector.

In addition, the productivity has significantly increased with the implementation of an automated particle measurement (APM) software that performs fast screening of the sample to determine the exact location of the particles of interest and a first estimate of their isotopic composition [8]. In short, the LG-SIMS today provides isotopic data for particle analysis that are at a state of the art level, combining highest quality with high throughput. The latter is important for the safeguards application, where timely analysis can be critical, especially in facilities with a high throughput. Figures 1 and 2 show an example of implementation of a LG-SIMS instrument in a clean-room facility. The high sensitivity of the technique requires great care in the sample preparation in order to avoid contamination of the sample.

2.2 Expanding the Set of Established Signatures in Bulk and Particle Samples

Particle analysis is used in both nuclear safeguards inspection work and in nuclear forensic investigations for trying to establish the origin of confiscated or found nuclear material. A classic case of applying nuclear forensics methodology with particle analysis in Safeguards was IAEA's finding of highly enriched uranium particles at different locations in Iran in 2003 [9]. The uranium isotopic composition of particles found in Iran was compared with particles found from equipment that had been sold by a large clandestine nuclear trading network [10, 11] and contaminated centrifuge equipment previously used in an Asian country for the making of weapon grade uranium. The investigation showed that the material found in Iran could be a contamination originating from an enrichment facility used for high enriched uranium production outside Iran [12, 13]. Other, more recent cases where particle analysis has been part of nuclear forensic investigations are the finding of uranium contaminated scrap metal in a few European scrap yards. The contaminated scrap was very likely imported from outside the EU. As the isotopic composition is one of the key signatures telling about the intended use of the material as well as in narrowing down the possible production facilities, isotopic

Fig. 1 The control room is separated from the instrument room to minimise the noise level for the operator. This also minimizes the contact with the instrument and reduces the particle levels in the instrument room. A window allows the operator to have a clear view of the instrument. There is a small changing area between the control room and the instrument room for the operators to suit up in clean room clothes before entering the instrument room. (Picture © European Commission)

composition measurements are always performed on nuclear forensic samples. In the case of the contaminated scrap metal it was observed that the ^{235}U enrichment varied between 10 and 33 wt% in different samples. The particle analysis revealed, however, that the material was inhomogeneous and the ^{235}U enrichment ranged from natural to weapon grade level as seen in Figs. 3 and 4.

3 Nuclear Forensic Signatures

Whenever nuclear material is encountered out of regulatory control, questions on its origin and intended use arise. Nuclear forensic science, or shortly "nuclear forensics", is a discipline which aims at answering these questions. In essence, nuclear forensics shares the basis of all forensic investigations, known as Locard's Principle: "every contact leaves a trace". According to this principle, nuclear material, be it of natural origin such as natural uranium or of anthropogenic origin such as plutonium or enriched uranium, will carry traces of the environment it

Fig. 2 The instrument room with the IMS 1280-HR is maintained as a class 100 clean room level. Eight H14 HEPA filter modules continuously clean the air. The room has been verified to fulfil all CAMECA requirements. (Picture © European Commission)

has been in contact with. Environment in this context may be understood as the geological formation from which uranium has been mined or the process the material has gone through. As a consequence, nuclear forensic scientists need first to identify those measurable parameters in nuclear material or any combination thereof, which can be related to the material history. These parameters are referred to as "nuclear forensic signatures". Therefore, nuclear forensics can assist in the identification of the materials, as well as how, when, and where the materials were made, and what was their intended lawful use [14].

Parameters such as the elemental composition, isotopic composition of major constituents (i.e., uranium or plutonium), isotopic composition of minor constituents (e.g., Pb, Sr, or Nd), metallic and anionic impurities, radioactive decay products, microstructure, or molecular structure can serve as nuclear forensic signatures [15]. These signatures may be source material inherited (e.g., natural uranium carrying impurities originating from the uranium ore) or process inherited (e.g., through neutron irradiation in a reactor). The following chapter will review nuclear forensic signatures that are associated to arms control materials, i.e. highly enriched uranium and plutonium.

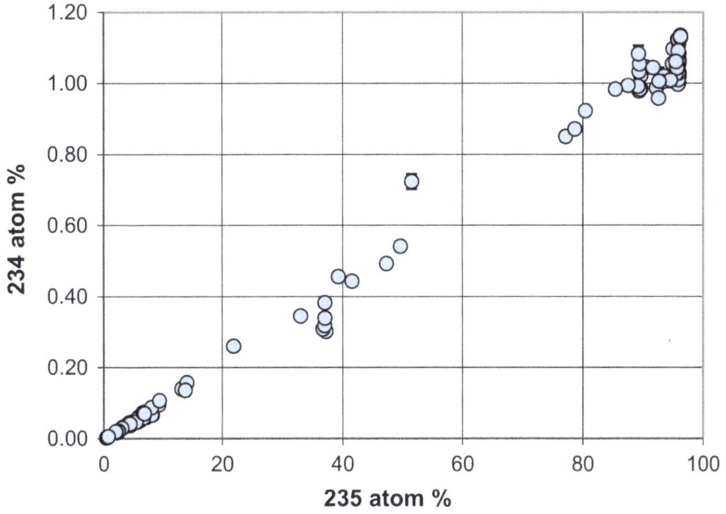

Fig. 3 Plot showing SIMS measurements of ^{234}U versus ^{235}U in particles from a set of samples taken from contaminated scrap metal objects

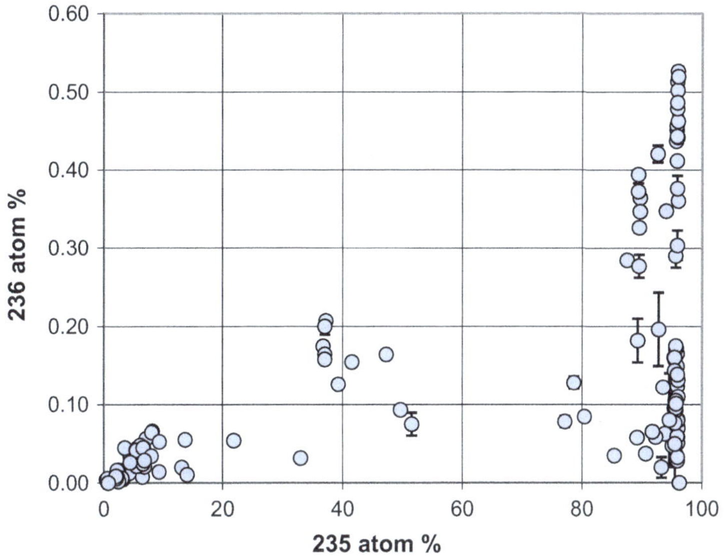

Fig. 4 Plot showing SIMS measurements of ^{236}U versus ^{235}U in particles from a set of samples taken from contaminated scrap metal objects

3.1 Radiochronometry

Nuclear material, like any other radioactive material, is subject to radioactive decay. While a radioactive isotope decays over time, its progeny, also referred to as daughter nuclide, is built up. This decay is mathematically described by an exponential law with the decay time being the main variable. In consequence, by measuring the number of parent atoms and the number of daughter atoms the decay time can be determined. The method is referred to as "radiochronometry" or "age dating".

In nuclear forensics this unique property of radioactive decay can be utilised to determine the time elapsed since the last chemical processing (separation of parent and daughter nuclide) of the material. This time is generally referred to as the "age" or "production date" of the nuclear material. This information is important when limiting the number of operational production facilities of seized nuclear material or, in best case, one can even identify the exact production batch. For accurate age dating we have to assume a complete separation of parent and daughter nuclides and that the material behaved as "closed system", i.e. no parent or daughter nuclides were added or removed during the decay period.

There are several parent/daughter ratios that can be used for the age dating of U and Pu as shown in Fig. 5. The use of multiple parent/daughter ratios will increase confidence in the result, if two or more ratios yield consistent ages and it serves as proof that the assumptions made (see paragraph above) were indeed correct.

The chronometer used for the age determination usually defines which analytical techniques can be employed for the analysis (i.e. radiometric vs. mass spectrometric

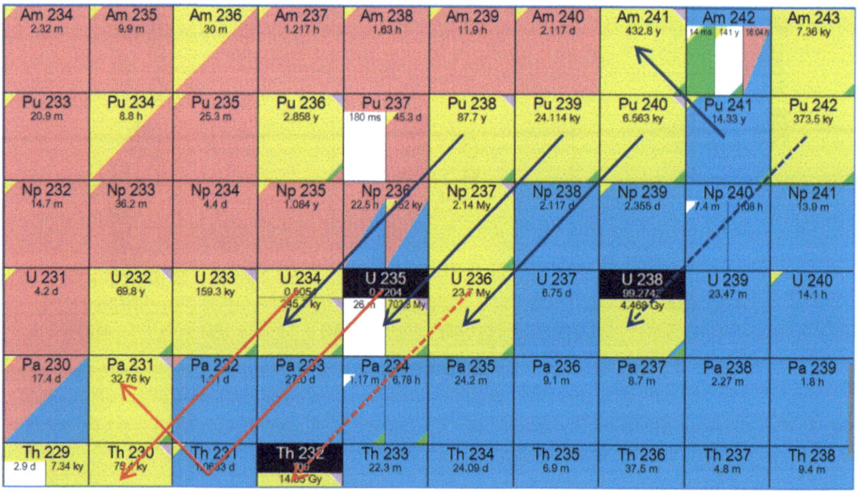

Fig. 5 Direct parent/daughter relations of Pu (blue lines) and U (red lines). The use of parent/daughter ratios with dashed lines is commonly not advantageous

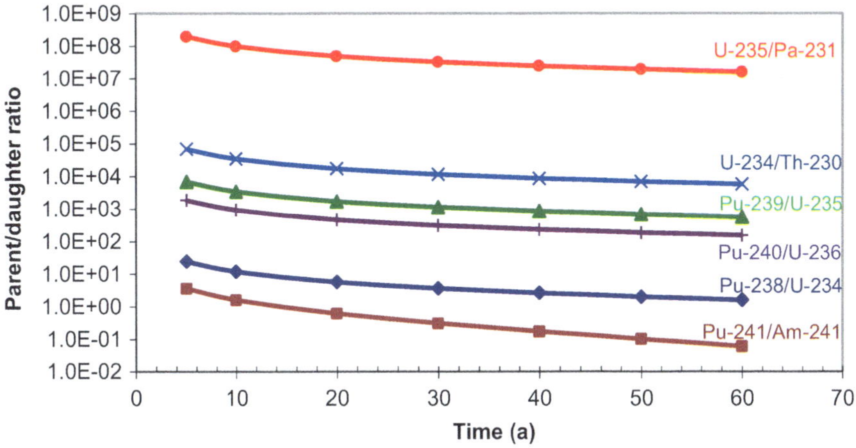

Fig. 6 Parent/daughter ratio vs. age of commonly used U and Pu chronometers

technique). Figure 6 shows the order of magnitude of the parent/daughter ratios of the main chronometers as a function of decay time (age). The following chapters describe in more detail specifics of uranium and plutonium age determination, respectively.

Age Dating of Plutonium

The separation of decay products from plutonium could have taken place either when the material was reprocessed or, for reprocessed plutonium, after a long storage period, when the material was purified from the ingrown ^{241}Am. Thus, the determined age does not necessarily reflect the original date of chemical processing of the plutonium, but the last purification process.

The age of plutonium can be determined, in principle, using five different parent/daughter relations: ^{238}Pu/^{234}U, ^{239}Pu/^{235}U, ^{240}Pu/^{236}U, ^{241}Pu/^{241}Am, and ^{242}Pu/^{238}U (Fig. 1, blue arrows). However, in practice, only the first four ratios are used. The reason is that ^{242}Pu has a very long half-life ($T_{1/2} = 3.75 \cdot 10^5$ a), which results in only a minute amount of ingrown ^{238}U. In addition, as the daughter nuclide ^{238}U is the most abundant isotope in natural uranium, even a small amount of cross contamination would cause positively biased results.

Because of the two different parent–daughter elemental ratios available (i.e., Pu/U and Pu/Am) and the isotopes of Pu having orders of magnitude differing half-lives, there are several measurement techniques on hand for the age determination of plutonium. A frequently used method utilises high resolution gamma spectrometry (HRGS) for determining the age based on the ^{241}Pu/^{241}Am ratio. This technique is obviously non-destructive, quick and does not require any sample preparation. The downside, however, is that it relies on one parent-daughter ratio only. Therefore, incomplete separation of Am (which is known to be the case for certain reprocessing methods), will result in a positively biased "age". On the other hand, if this bias is noted, it can be also used as an indication of the reprocessing method.

Age determination methods based on the other parent/ daughter ratios of plutonium utilise typically mass spectrometry techniques (TIMS, ICP-MS), which are combined with spiking and chemical separation of the elements[16, 17]. Although such methods are relatively time-consuming, the precision is typically superior ($\leq$0.3%) to HRGS. In addition, it provides increased confidence in the result if all the Pu/U ratios yield consistent age values.

Due to the moderately short half-lives of Pu isotopes ^{238}Pu, ^{239}Pu, and ^{240}Pu (87.74, 24110, and 6563 a, respectively), the age can be determined also using direct measurement by ICP-MS or it can be performed even for μm-sized plutonium particles by SIMS. To compensate for the ionisation efficiency difference of U and Pu in SIMS measurements, a reference sample of known age needs to be available to determine so-called relative sensitivity factor (RSF) for the Pu/U ratio beforehand.

Age Dating of Uranium

The number of available U chronometers is lower than for plutonium. The most frequently reported methods are using the following parent/daughter pairs: ^{230}Th/^{234}U, ^{231}Pa/^{235}U and ^{214}Bi/^{234}U.

The ^{230}Th/^{234}U ratio is the most commonly applied method due to several reasons: ^{234}U has three orders of magnitude shorter half-life than ^{235}U, i.e. more daughter nuclide grows in; many spike isotopes are available for Th (e.g., ^{228}Th, ^{229}Th, ^{232}Th); both radiometric (by alpha spectrometry) and mass spectrometric measurements are possible [18]. Direct measurement (i.e. without chemical separation of U and Th) from the ^{230}Th/^{234}U ratio by ICP-MS has also been studied. The precision of the method can be improved by using spectral deconvolution to decrease the effects of the spectral interference (i.e., tailing) of the neighbouring highly abundant uranium peaks on the ^{230}Th signal [19].

Uranium age determination can be performed also quasi non-destructively from the ^{230}Th/ ^{234}U ratio using LA-ICP-MS. However, similar to the Pu age dating by SIMS, the quantification of the chronometer requires the application of proper calibration using a standard material with known ^{230}Th content and a chemical composition similar to the sample [20].

Age dating of uranium using the ^{231}Pa/^{235}U chronometer can be performed either by mass spectrometry or alpha spectrometry. Due to the very long half-life of ^{235}U ($7.038 \cdot 10^8$ a), only minute amount of daughter nuclide grows in, and therefore it is more difficult for natural or low enriched uranium samples. In addition to that, the chemical recovery estimation of the ^{231}Pa is hampered by the lack of a longer-lived Pa tracer. This can be overcome by using gamma spectrometry for the ^{233}Pa chemical recovery measurement and alpha spectrometry for the ^{231}Pa/^{235}U ratio determination [21].

High resolution gamma spectrometry (HRGS) has been found to be applicable for the age measurements of uranium using the $^{234}Bi/^{234}U$ ratio. A big advantage of this technique is that the quantification can be performed using so-called intrinsic efficiency calibration, i.e. no calibrator or standard is required [22]. A disadvantage of this chronometer is the longer ingrowth time of ^{214}Bi compared to other chronometers: as ^{214}Bi is several steps away from the parent ^{234}U nuclide with a few longer-lived intermediate progenies in the decay chain (such as ^{230}Th and ^{226}Ra), longer time is necessary for ^{214}Bi to be formed in measurable quantities. Therefore, the method is more appropriate for older highly enriched uranium samples.

3.2 *Isotope Correlation for Pu Reactor Type Determination*

Plutonium is formed as a by-product in nuclear (fission) reactors through consecutive neutron capture reactions of uranium and subsequent decay of short-lived intermediate products. Different reactor types (e.g. heavy water reactor, fast breeder reactor) show different neutron energy distributions, and since the neutron capture cross sections vary with neutron energy, the relative yield of plutonium isotopes generated by neutron capture will vary. Additionally, the isotopic composition of plutonium is a function of the initial ^{235}U enrichment of the fuel and of the irradiation conditions (e.g. neutron flux, irradiation cycles). In consequence, the isotopic composition of plutonium present in the nuclear fuel after irradiation may serve as a “signature” of the type of reactor it was produced from [23, 24]. Obviously, longer irradiation times and higher neutron fluxes will result in an increased build-up of the heavier plutonium isotopes. In other words, the difference in isotopic composition between different reactor types will increase with increasing irradiation time. Weapons grade Pu (WGPu) is typically produced in heavy-water reactors (e.g. Candu) or graphite-moderated reactors (e.g. RBMK), where on-load refuelling is possible or in fast breeder reactors (FBR) as blankets. WGPu is a low burn-up Pu, where the difference in Pu isotopic composition between different reactor types is rather small. However, it is still possible to draw conclusions on the type of reactor in which the plutonium was produced as shown in Fig. 7. Similar isotopic correlations can be used to estimate other parameters, such as the date of discharge (^{241}Pu relationship to ^{240}Pu and ^{242}Pu) or the time elapsed between reprocessing and subsequent purification of ^{241}Am (combined with the age dating using $^{241}Am/^{241}Pu$) [25].

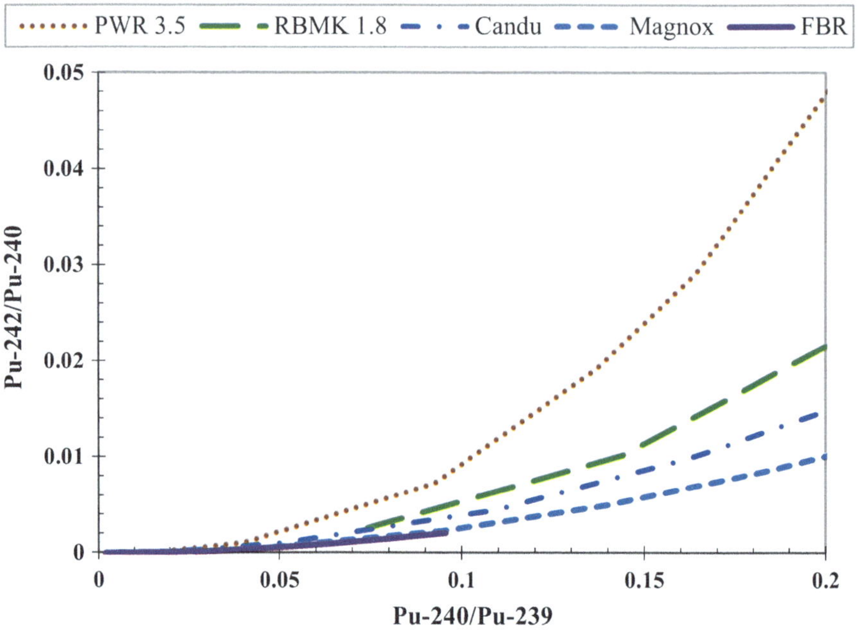

Fig. 7 Isotope correlation for Pu reactor type determination

4 Confidence in Results

In traditional nuclear safeguards conclusions are largely based on paired comparisons of operator's and inspector's measurement results. More recently, safeguards have become increasingly investigative and deals with consistency of measurable material properties (e.g. chemical impurities) with declared processes. In nuclear forensics, characteristic parameters (also referred to as "signatures") are used for re-establishing the history of nuclear material found out of regulatory control. Signatures can either be "predictive" (i.e. self-explaining) or "comparative" in nature. The latter require external data of known nuclear material for supporting data interpretation. In either case, accurate measurement results need to be performed with a validated method and expressed as a measured quantity value with uncertainty and traceability of a measurand (analyte in the investigated matrix). Accreditation under ISO17025 or similar international/national standards [26–28] strongly supports the confidence in results obtained by a measurement laboratory. The increased use of the Guide to the expression of uncertainty in measurement (GUM), the respective EURACHEM guides and the GUM-VIM terminology contributed also in the nuclear security and safeguards area [5–7] to a harmonized way of evaluating and expressing uncertainties. Quality control (QC) is a means to demonstrate the fitness for purpose of measurement laboratories and to establish confidence in analytical measurement results. The proper application of certified reference materials (CRMs) establishes traceability of a measurement result to the International System of Units (SI) and inter-laboratory comparisons give evidence of laboratory capabilities. The European Commission-Joint Research Centre (EC-JRC) has therefore put effort into the development of CRMs and the provision of conformity assessment tools of importance to both, the nuclear forensics and nuclear safeguards community.

(a) Tailor Made Reference Materials and Certified Test Samples A large number of nuclear reference materials has been produced by specialized laboratories, traditionally focusing on element and isotope assay. However, new parameters such as impurities or "age" (i.e. the time elapsed since the last chemical separation of the parent from the daughter nuclides) have gained increasing interest in nuclear safeguards and nuclear forensic applications., To that end, the EC-JRC-Geel and the EC-JRC-Karlesruhe jointly produced IRMM-1000a and IRMM-1000b (see Fig. 8) in accordance with ISO Guide 34. These are uranium reference materials are certified for the production date based on the ^{230}Th/^{234}U radio-chronometer [29, 30]. In addition, the US Department of Energy New Brunswick Laboratory (US-DOE-NBL) also provides CRM-125A uranium dioxide-pellets and CRM U630 triuranium octoxide certified for 'model purification date' [31].

Currently JRC-Geel and the CEA's (French Alternative Energies and Atomic Energy Commission) Committee for the establishment of analysis methods (CETAMA) have released IRMM-0243, the first spike CRM certified for the ^{243}Am and ^{241}Am amount contents and Am isotope amount ratios by Isotope Dilution-Thermal Ionisation Mass Spectrometry (ID-TIMS) and TIMS to be used for accurate measurements of the age of plutonium materials using the ^{241}Pu/^{241}Am radio-chronometer [32].

(b) Quality Control Laboratories performing measurements in nuclear safeguards and nuclear security area have to demonstrate their competence on an interna-

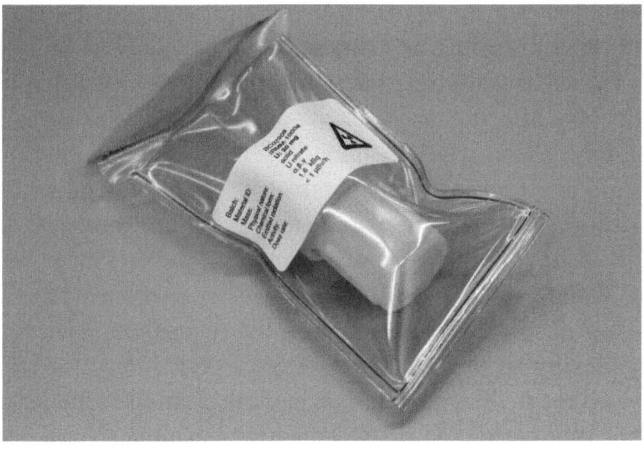

Fig. 8 IRMM-1000a (20 mg)—certified uranium reference material for the production date based on the ^{230}Th/^{234}U radiochronometer. (© European Commission)

tional level and through internal and external quality control measures [33, 34]. The EC-JRC-Geel's Nuclear Signatures Inter-laboratory Measurement Evaluation Programme (NUSIMEP) is an example of an external quality control program in the field of environmental sample analysis involving element and isotopic analysis of trace amounts of nuclear material in the environment [35]. Recent NUSIMEP campaigns included major and minor isotope amount ratios in uranium particles and uranium and plutonium isotope amount ratios in low-level synthetic nitrate solution, [36–38]. With REIMEP-22 the first ever inter-laboratory comparison on 'U Age Dating' was organised. It offered a unique opportunity to participating laboratories to demonstrate their capability for measuring the age of (relatively young) uranium materials [39].

Similar technical and instrumental methods and laboratory network approaches can be applied in safeguarding of nuclear material stemming from the military fuel cycle to be included in the civil nuclear fuel cycle as disarmament proceed.

5 Synergies Between Safeguards, Non-proliferation and Nuclear Forensics

The legal bases for nuclear safeguards and for nuclear security (including nuclear forensics) are largely different. International organizations (such as the IAEA) have a strong mandate for verifying treaty compliance of the signatory States, sanctioned by the Non-proliferation treaty. In contrast to that, nuclear security is strictly national responsibility with the international organizations essentially limited to providing guidance. Another notable difference is the fact that Safeguards addresses nuclear material under regulatory control while nuclear security deals with nuclear material out of regulatory control. In spite of these differences, however, there are a number of common features at technical level and also synergies from which both areas can benefit. IAEA Safeguards have evolved significantly, from verification of known amounts of material and criteria driven conclusions to a comprehensive view of a state's nuclear activities and information driven safeguards conclusions. Accordingly, nuclear material analysis in safeguards is increasingly making use of the information inherent to the material. Thus, reading the "signatures" of nuclear material is developing into a source of information feeding into safeguards conclusions. The identification of parameters providing hints on the history of nuclear material (signature development) and the development, adaptation and optimization of analytical methods (method development) have been core activities in nuclear forensic science. In consequence, the measurement needs and tools in both areas are converging and enable developing synergies at scientific/technical level. The measurement of impurities in natural uranium samples may serve as example. Chemical impurities in uranium material originate from various sources. On the one hand already by the starting materials feeding into the process carried a number of impurities (source material inherited); on the other hand they may have been introduced intentionally or accidentally through chemical reagents, solvents or corrosion of vessel or pipework (process inherited). In uranium ore concentrate samples (yellow cake) the pattern of the rare earth elements has been proven to be highly characteristic of the geological environment from the uranium ore was mined [40, 41]. It could also be shown that this signature remains unchanged through chemical processing of natural uranium. While the concentration of the impurities is going down with progressing purification of the uranium, the relative distribution of the rare-earth elements remains unaffected. This is due to the very similar chemical behaviour of these elements, all being trivalent and all having similar ionic radii.

Figure 9 shows the rare-earth element pattern as measured in uranium samples obtained from the same facility. Samples were taken at different stages of the process: uranium ore, the ore leachate, the eluate of the purification (by ion exchange –IX), the raffinate, the ammonium diuranate (ADU) and the uranium oxide (U_3O_8) after calcination. As can be seen, the shape of the curve (i.e. the REE signature) remains unaffected by the various chemical and thermal treatments. We observe a curve slightly decreasing towards the heavier rare-earth elements with a pronounced "Europium anomaly". This signature is source material inherited and persists throughout the process. In consequence, it can be used for nuclear forensic purposes for provenancing of natural uranium (note that only the type of geological deposit can be identified using this signature, not the individual mine) and it can be used for safeguards purposes for verifying consistency of materials present in a facility with declared operations [33, 41].

Similarly, also impurities in enriched uranium serve as source of information on the history of the material. Metallic impurities typically arise from corrosion of the pipework and the vessel material. Isotopic signatures in plutonium have been used for decades in nuclear safeguards to verify consistency of declarations with measurable material properties.

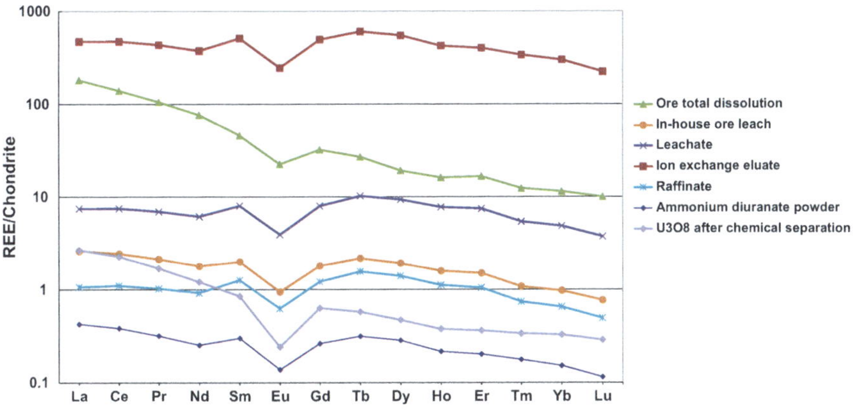

Fig. 9 Relative distribution of the rare-earth elements (chondrite normalized) in materials related to uranium ore processing (For more details, see text)

Overall, we notice that—despite differences in the legal basis—at technical level there is common interest in making use of information inherent to nuclear material. Cross-fertilization between these areas arises from applying the same analytical toolset, from re-purposing analytical methods and from transferring useful signatures.

6 Conclusions

In nuclear security, nuclear safeguards and in non-proliferation we observe an increasing use of information inferred from measurable material properties. Investigative analytical and interpretational techniques have been developed that are capable of exploiting signatures from samples of nuclear material. Examples of such signatures are the isotopic composition of the nuclear material, the isotopic composition of trace elements, the concentration or pattern of chemical impurities or the morphology of the material. Advanced instrumental techniques have become available that enable the accurate analysis of minute amounts of sample (e.g. LG-SIMS offers very small measurement uncertainties on isotope ratios obtained from individual particles of less than one micrometer in diameter, equivalent to less than 2 pg of uranium). Even from odd matrices (like uranium deposited as contamination on the surface of scrap metal) material signatures can be usefully exploited. In summary, signatures have become an important source of information on the provenance and the processing history of nuclear material for safeguards authorities, for investigating agencies and for the scientific community.

References

1. Kirk PL (1953) Crime investigation: physical evidence and the police laboratory. Interscience Publishers, Inc.: New York
2. Vickerman JC, Brown A and Reed NM (1989) Secondary Ion Mass Spectrometry; Principles and Applications. Oxford University Press, New York
3. Tamborini G, Betti M, Forcina V, Hiernaut T, Giovannone B and Koch L (1998) Application of secondary ion mass spectrometry to the identification of single particles of uranium and their isotopic measurement. Spectrochim. Acta, Part B 53:1289–1302
4. Esaka F, Watanabe K, Fukuyama H, Onodera T, Esaka KT, Magara M, Sakurai S, Usuda S (2004) Journal Nucl Sci Technol 41:1027
5. Esaka F, Lee CG, Magara M, Kimura T (2012) Fission track–secondary ion mass spectrometry as a tool for detecting the isotopic signature of individual uranium containing particles. Analytica Chimica Acta 721:122–128
6. Ranebo Y, Hedberg PML, Whitehouse MJ, Ingeneri K, Littmann S (2009) Improved isotopic SIMS measurements of uranium particles for nuclear safeguard purposes. J. Anal. At. Spectrom. 24:277–287
7. Hedberg PML, Peres P, Fernandes F, and Renaud L (2015) Multiple ion counting measurement strategies by SIMS – a case study from nuclear safeguards and forensics. J. Anal. At. Spectrom.30:2516

8. Hedberg PML, Peres P, Cliff JB, Rabemananjara F, Littmann S, Thiele H, Vincent C, Albert N (2011) Improved Particle Location and Isotopic Screening Measurements of Sub Micron Sized Particles by Secondary Ion Mass Spectrometry. J. Anal. At. Spectrom.26:406–413
9. International Atomic Energy Agency (IAEA) (2004) Implementation of the NPT Safeguards Agreement in the Islamic Republic of Iran. IAEA Board of Governors reports GOV/2004/34
10. IAEA (2004) Implementation of the NPT Safeguards Agreement of the Socialist People's Libyan Arab Jamahiriya. IAEA Board of Governors reports GOV/2004/12
11. IAEA (2008) Implementation of the NPT Safeguards Agreement of the Socialist People's Libyan Arab Jamahiriya. IAEA Board of Governors reports GOV/2008/39
12. IAEA (2005) Implementation of the NPT Safeguards Agreement in the Islamic Republic of Iran. IAEA Board of Governors reports GOV/2005/67
13. IAEA (2006) Implementation of the NPT Safeguards Agreement in the Islamic Republic of Iran. IAEA Board of Governors reports GOV/2006/15
14. Mayer K, Wallenius M, Varga Z (2013) Chem Rev 113(2):884–900
15. Mayer K, Wallenius M, Ray I (2005) Analyst 130:433–441
16. Wallenius M, Mayer K, (2000) Fresenius J Anal Chem 366:234–238
17. Varga, Z, Nicholl, A, Wallenius, M et al. (2016) J Radioanal Nucl Chem 307:1919. doi:10.1007/s10967-015-4418-5
18. M. Wallenius, A. Morgenstern, C, Apostolidis, K. Mayer, Anal Bioanal Chem 374 (2002): 379–384
19. Vargo Z, Wallenius M, Mayer K (2010) J Anal At Spectrom 25:1958–1962
20. Varga Z, Suranyi G (2007) Anal Chim Acta 599:16–23
21. Morgenstern A, Apostolidis C, Mayer K (2002) Anal Chem 74:5513–5516
22. Nguyen C, Zsigrai J (2006) Nucl Instr Meth Phys Res B 243:187–192
23. Wallenius M, Peerani P, Koch L (2000) J Radioanal Nucl Chem 246:317–321
24. Luksic AT, Collins BA, Friese JI, Schwantes JM, Starner JR, Wacker JF (2010) Proceedings of the INMM 2010 meeting. PNNL-SA-73572
25. Alamelu, D, Aggarwal, SK (2016) J Radioanal Nucl Chem 307:277. doi:10.1007/s10967-015-4426-5
26. International Organization for Standardization (2017) General requirements for the competence of testing and calibration laboratories. ISO/TEC 17025:2017, Geneva
27. GUM anniversary issue (2014) Metrologia 51
28. International Vocabulary of Metrology (2012) Basic and General Concepts and Associated Terms (VIM 3rd edition) JCGM 200:2012
29. Varga Z, Venchiarutti C, Nicholl A, Krajko J, Jakopic R, Mayer K, Richter S, Aregbe Y, (2016) Journal of Radioanalytical and Nuclear Chemistry 307:1077–1085. doi:10.1007/s10967-015-4227-x
30. Venchiarutti C, Varga Z, Richter S, Nicholl A, Krajko J, Jakopic R, Mayer K, Aregbe Y (2016) Journal of Radioanalytical and Nuclear Chemistry 308:105–111. http://doi:10.1007/s10967-015-4368-y
31. U.S. Department of Energy NBL Office homepage. http://science.energy.gov/nbl. Accessed 28 Feb 2019
32. Venchiarutti C, Fankhauser A, Varga Z, Jakopič R, Richter S, Aregbe Y, Mayer K, Maillard C, Rivier C, Roudil R (2015) Quality Control Tools for Age Dating in Nuclear Safeguards and Forensics, Proceedings of 37th ESARDA Symposium on Safeguards and Nuclear Non-Proliferation, Manchester, UK, 19–21 May 2015
33. Roudil D, Rigaux C, Rivier C, Hubinois J.C, Aufore L (2012) Procedia Chemistry 7:709–715
34. International Atomic Energy Agency (2010) International Target Values 2010 for Measurement Uncertainties in Safeguarding Nuclear Materials. IAEA-STR-368
35. Aregbe Y, Jakopič R, Richter S (2014) Venchiarutti C, Conformity assessment in nuclear material and environmental sample analysis, Proceedings Symposium on International Safeguards Linking Strategy, Implementation and People. October 2014, S13-01
36. European Commission (2018) EU Science Hub.https://ec.europa.eu/jrc/en/interlaboratory-comparisons/NUSIMEP. Accessed 28 Feb 2019

37. Truyens J, Stefaniak EA, Aregbe Y (2013) NUSIMEP-7: uranium isotope amount ratios in uranium particles. Journal of Environmental Radioactivity 125:50–55
38. Jakopič R, Bujak R, Aregbe Y, Richter S, Buda R, Zuleger E (2014) NUSIMEP-8: Uranium and plutonium isotope amount ratios in low-level synthetic nitrate solution. Report EUR 26671 EN
39. Venchiarutti C, Varga Z, Richter S, Jakopič R, Mayer K, and Aregbe Y (2015) REIMEP-22 inter-laboratory comparison: U Age Dating – determination of the production date of a uranium certified test sample. Radiochim. Acta 103(12):825–834
40. Varga Z, Wallenius M, Mayer K (2010) Radiochim. Acta 98, 771–778
41. Varga, Z, Krajkó, J., Peńkin, M. et al. J Radioanal Nucl Chem (2017) 312: 639. doi:10.1007/s10967-017-5247-5

Part V
Information Analysis

Verification of compliance is dependent on technical data and other supporting information. It can be argued that the current information revolution has changed the way we can collect, process and analyze a wide variety of information. The virtual explosion of available data provides opportunities and challenges. All data does not necessarily have with the same relevance or quality, so it must be managed and evaluated with new techniques. For the credibility of the verification judgments, discrimination of relevant information from background noise (signal-to-noise) must be optimized. New computational capabilities may move us beyond what we thought was possible.

Information Management: The Role of Information Technology in Treaty Verification

Nicholas Kyriakopoulos

Abstract Arms control treaties rely on extensive collection, processing and analysis of data to ascertain compliance with treaty obligations. Monitoring and evaluation relies on a broad range of technologies for sensing, communications and processing to collect, transmit and analyze the data necessary to detect non-compliance. The effectiveness of a treaty verification system depends on the ability of the information management system to collect and evaluate the information necessary to ascertain compliance with the treaty obligations. The information management systems for treaty verification have the same architecture because they are similar in concept and operation. They collect, transmit, store, process and analyze data. The effectiveness of each of these operations is a function of the capabilities of the associated technology. This chapter presents an overview of the technologies available for use in each of the elements of a treaty information management system. In sensing, these range from low frequency seismic sensors to multispectral and hyperspectral imaging. Data can be transmitted in the order of hundreds of Gigabits per second through fiber optic cables for terrestrial communications and tens of Megabits per second for satellite channels, while available technologies allow for storing petabytes of data and processing speeds of tens of Gigabytes per second. These capabilities are more than sufficient to collect the necessary data for use in sophisticated analysis algorithms for detecting treaty violations.

1 Introduction

Verification of treaties is based on information derived from measurements. The term "information" gained prominence when Claude Shannon first used it to measure the performance of noisy communications channels [1]. Subsequently,

N. Kyriakopoulos (✉)
The George Washington University, Washington, DC, USA
e-mail: kyriak@gwu.edu

I. Niemeyer et al. (eds.), *Nuclear Non-proliferation and Arms Control Verification*,
https://doi.org/10.1007/978-3-030-29537-0_19

with the advent of computer science, the trend has been to distinguish between the measurable concept of *data* and the terms *information* and *knowledge*. In this chapter we will use the term "information" in the context of the obligations of the States to comply with the terms of specific international treaties, and for consistency, the term "technical organization" will be used to denote the international authority responsible for the information management specified by a treaty [2, 3].

The technologies used in the generation, collection, transmission storage and analysis of data constitute what is collectively referred to as information technology; information management comprises the functions performed by the various components of this technology. For treaty verification, "information management" is the manipulation of treaty-relevant data for the purpose of detecting non-compliance. In contrast to other categories of information management systems, the expected outputs of treaty verification systems present unique challenges that arise from the specification of the verification regimes through international negotiations.

This chapter presents an overview of the functions and capabilities of technologies used in the collection and processing of data associated with treaty verification systems. The focus is on multilateral treaties that establish technical organizations to operate such systems. To provide a perspective for the information technologies associated with treaty verification, the chapter begins with a brief review of the monitoring and verification systems of the Nuclear Nonproliferation Treaty (NPT), Chemical Weapons Convention (CWC) and the Comprehensive Nuclear Test Ban Treaty (CTBT) in order to establish the expected output of an information management system. The flow of data from the generation, transmission, storage and analysis to the detection of non-compliance is presented, concluding with a discussion of the requirement for data integrity.

2 Establishing the System Requirements

The design requirements of a verification system are derived from the text of the treaty and have evolved over time starting with the international monitoring system by the International Atomic Energy Agency (IAEA) Statute in 1956. With the advent of the NPT, monitoring and verification became a treaty obligation. The information managed by the verification system consists of data files provided by the States, measurement data collected by in situ instruments. and data collected by inspectors. The latter belong to three categories: visual observations, reading data files, and instrument measurements. Under INFCIRC/153 monitoring and verification are applied to a closed system that consists of specified and declared processes and locations. The expected output of the processing of information for that system is also well defined, namely, detection of diversion.

With the Additional Protocol (INFCIRC/540) the scope of safeguards is expanded from declared facilities to the entire State. While INFCIRC/153 defines the process (specific facilities associated with the nuclear fuel cycle), the monitoring mechanism (auditing), the data (measurements, counting) and the expected result (detection of diversion), INFCIRC/540 does not define a mechanism for quantifying the objectives of effectiveness and efficiency. Thus, the information relevant to the broad aim of the NPT to prevent the spread of nuclear weapons falls into two categories: quantitative (measurable) and qualitative (unmeasurable) [4].

A similar partition exists for the information relevant to the verification of the CWC. The types and quality of information collected by the monitoring process varies according to the categories of chemicals under the purview of the Convention. For those designated as Schedule 1 chemicals, monitoring is done through the evaluation of declarations submitted by the State, inspections at declared facilities and instruments operating continuously in situ. Monitoring of chemicals designated as Schedule 2 is through the evaluation of declarations by inspections for the purpose of detecting production of Schedule 1 chemicals, discrepancies in the declarations, and diversion of Schedule 2 chemicals. For Schedule 3 chemicals, the declarations are in ranges rather than actual quantities. Monitoring consists of inspections conducted for the purpose of detecting the presence of Schedule 1 chemicals and verifying that "activities are consistent with the information to be provided in declarations." For other chemical production facilities covered by the Convention, the quantities of produced chemicals are declared in ranges and monitoring consists of random inspections for the purpose of detecting the presence of Schedule 1 chemicals and verifying consistency with the declarations [5].

The information used for the detection of non-compliance may range from secure sensor data to open source data of unknown reliability. Consequently, the information collected and processed by a treaty information system needs to be categorized on the basis of its impact on the level of confidence in the probability of detecting non-compliance. As in the case of the NPT, the inputs to the information processing system are partitioned into quantitative and qualitative. The first includes data generated by sensors and instruments, computer files and humans. Qualitative information is either verbal or in text form, or as images. The inputs may be generated by inspectors under the control of a Technical Secretariat, States Parties, or it may exist in public sources.

A similar partition exists for the verification objectives of the CTBT. Detection of nuclear explosions is a measurable objective. To this end the International Monitoring System, with a global communications infrastructure, in conjunction with the International Data Centre form the information infrastructure of the CTBT. The system is designed to detect all events that generate acoustic waves and to discriminate between benign events and those that could possibly be nuclear tests. The information processing requirements for this type of obligation are straightforward. Signals from acoustic waves are detected by sensors, transmitted through communications and processed by computer systems that use elaborate signal analysis algorithms. The treaty, however, imposes some additional obligations on the States that are difficult if not impossible to quantify, such as "refrain from causing, encouraging, or in any way participating in the carrying out of any nuclear weapon test explosion..." Thus, the information processing requirements of the CTBT as with those of the Additional Protocol and the CWC are partitioned into quantitative and qualitative [6, 7].

Another partition of the information may be on the basis of the integrity and reliability of the data, as they impact the level of confidence in the detection probability. These characteristics can be maximized only by an authority that has control of the design and operation of a system. For example, the confidence level derived from the data collected by a monitoring system operated by the technical organization is the highest, because the Secretariat has complete control of the sources whether they are sensors, instruments or inspectors. Data entering the verification system may be deemed less reliable if they originate in sources outside the control of the organization. Such is the case for data supplied by States Parties and even more so data from open sources. Another partition may be on the basis of sufficiency for maximizing the detection probability. Thus, the performance of the information processing technologies is heavily influenced by the quantity and quality of the data entering the verification system.

3 Sources of Information

Quantitative data lead to rigorous analysis and evaluation, while the use of qualitative information introduces subjective judgment that is not amenable to independent evaluation. Nevertheless, qualitative information can have a role as a trigger for the monitoring system to generate supplementary inputs. For example, observations of an inspector en route to a facility may be used as an input to the monitoring system to perform additional measurements. On the other hand, comments of a political leader or in the news media may not have sufficient credibility to justify activation of a treaty verification system. The distinction in the two sources of information lies in the relation of the source of information to the defined treaty verification system. The inspector is a well-defined component of the monitoring system, while the media are not. The reliability of the open source data is even more questionable, because of the lack of an objective mechanism for assessing the integrity of the information. It is a challenge to formulate an approach for utilizing all available qualitative information.

The major sources of the quantitative information are inspectors performing measurements, instruments performing measurements either automatically or under the control of an operator and sensors detecting signals remotely or in situ. Monitoring systems collect data either from specified locations with in situ sensors or remotely with sensors observing phenomena that are not in physical contact with the sensor. Security sensors and instruments measuring process variables at an enrichment facility are examples of in situ sensing, while seismometers, hydrophones and infrasound detectors sense signals at a distance from events or processes through the propagation of waves. Practically all sensors that are components of a monitoring system translate physical phenomena into electrical signals that are sampled and digitized.

Remote sensing typically involves signals in all regions of the electromagnetic spectrum ranging from extremely low frequencies in the order of 1 Hz or less for seismometers to Gamma rays in the Exahertz frequency range for Gamma ray detectors. Sensors are designed to detect signals within specified frequency ranges and power levels corresponding to specific sources. At the low end of the spectrum the different types of seismometers used in various applications from oil exploration to seismology are capable of detecting signals in the range form 0.01 Hz up to 600 Hz. Detection of signals at the upper range of the electromagnetic spectrum is done with radiation detectors using various technologies such as gas-filled, scintillation or solid state. Radiation detectors for non-destructive assay of nuclear materials are an essential tool of nuclear safeguards [8, 9].

Remote sensing provides the capability of identifying materials (solids, liquids and gases) by creating spectral images in the bands of the electromagnetic spectrum from the visible, near infra-red and short-wave infrared. Multispectral or broad band imaging such as that used on Landsat and Spot satellites is applicable for characterizing landscapes by obtaining imaging data in discrete bands over a wide range of the spectrum. Current satellite imaging technology is capable of multispectral resolution in the order of 1 m and panchromatic resolution in the order of centimeters ground sample distance. Hyperspectral imaging is applicable to detecting anomalies and distinguishing between natural and man-made, detecting and recognizing target and characterizing natural features associated with atmosphere, land surface or ocean. The principles of multispectral imaging are equally applicable to the design of detectors for special nuclear materials and gasses [10–13].

4 Data Transmission

The architecture of the treaty networks for data collection and distribution is a star network, with the technical organization being the root node of the network. However, at the physical layer, the communications channels composing the application layer star network are a complex network of interconnected nodes with channels that utilize different elements of the latest available communications technologies.

For data sources that are instruments, signals are digitized, formatted and converted into electrical signals at the source for transmission, while the data from computer data bases, are already in digital format. Depending on the requirements for timeliness, these may be transmitted to the central node in a storage medium that may be paper, magnetic, optical or electronic, or through the communications network. Each data source is a logical node of the monitoring system and a physical node of the communications network.

Data may be transmitted to the root node continuously in real time or near-real time, as is the case for the primary seismic stations of the CTBT's International Monitoring System, or in batch format from the auxiliary stations on demand. From the perspective of the communications network the impact of this distinction is whether the outbound communications channel is dedicated or available on demand, and on the expected input data rate that the network should be able to accommodate at each node; the critical parameter is the output data rate of a given node. These data rates can vary from kilobits/second as in the case of seismic signals to gigabits/second for high definition video. Thus, depending on the verification regime, the demand for capacity upon the communications network may be anywhere in that range.

The technology for accommodating such a wide range of data communications requirements has been available for some time. Copper wiring is sufficient to accommodate the low data rates. Coaxial cable can handle data rates in the order of 10 Gbps depending on distance, but the distances over which coaxial cable may be used without signal boosters is in the order of meters. Fiber optic cables have data rates typically in the order of 30 Gbps and can accommodate rates up to 100 Gbps. The distance over which fiber optic cables are used without signal amplification is in the order of kilometers. Radio frequency communications links are either terrestrial or satellite and use frequencies in the range 2–40 GHz. Terrestrial microwave links provide point-to-point communications and use directional antennas mounted on towers on a line of sight. Due to the curvature of the earth maximum distance between towers is in the order of 40–60 km. Satellite links for civilian communications are in the C (4–8 GHz), Ku (12–18 GHz) and Ka (26.5–30 GHz) bands. As satellite channels become widely available, they replace microwave links, except for specialized uses, due to cost considerations [14, 15].

Historically, the transmission of data between two points has been through dedicated channels. Such is the case for the continuous real-time transmission of the data from the monitoring stations of the International Monitoring System. The advent of packet switching has given rise to a radically difference notion of communications. The data are converted into packets. Each packet is propagated through a network of interconnected computer nodes. At the destination the packets are re-assembled into their proper order although they may arrive through different paths and not necessarily in sequence in which they were transmitted.

Regardless how the data are transmitted, from the perspective of the technical organization, the main considerations are preservation of the integrity of the data and timeliness of arrival, when necessary. The requirements are satisfied by using dedicated communications channels that are designed to meet specifications for capacity and speed with a high level of confidence. In packet-switched networks, it is not theoretically possible to specify performance requirements for the availability and timeliness of the data transmission from a sensor to the central node.

Nevertheless, the increase in the capacity of the backbone network, with the laying of fiber optic cables between continents and the placement of more satellites on geostationary orbits, may make this theoretical constraint purely academic. From the perspective of treaty monitoring, the available communications technology is capable of satisfying any foreseeable data communications requirements.

5 Data Storage and Processing

The distinguishing characteristics of the different storage media are the capacity to store data and the speed of recording and retrieving the data. Storage densities in the order of 500 gigabits/in^2 with the potential to increase to 1000 gigabits/in^2 in the near future, make it possible to store 3 terabytes of data on a 3.5 in Hard Disk Drive. Current storage platforms have capacities in the order of 30 petabytes and internal bandwidth in the order of 80 gigabytes/s. For comparison the entire print collection in the U.S. Library of Congress is about 10 terabytes and could be stored in three to four 3.5 in Hard Disk Drives or in a fraction of the storage space of one large storage platform. In terms of processing speeds, in 2009, Google was processing more than 20 petabytes of data per day [16, 17].

In the processing of information, speed of access to the stored data determines the role of the available technologies. Volatile memories using semiconductor technologies, managed by the computer system hardware, are used as cache and main memory, because they provide the fastest access to the data. Larger volumes of data available on-line for easy access at relatively high access speed are stored primarily in magnetic disks. Optical devices and magnetic tapes are the principal technologies for off-line storage. While tapes are relatively inexpensive as a storage medium for large quantities of data, reading of the data is done in a serial manner resulting in a slower access time compared to disk storage. Flash memory is another type of electronic device that is non-volatile that can be erased and re-written multiple times. Flash memories are widely used as memory cards in various electronic devices such as digital cameras, mobile telephones, USB flash drives, etc. If speed of access is not essential to a processing task, quantities of data in the order of petabytes can be stored in systems consisting of drives connected in parallel. Connection of multiple disk drives in parallel as a redundant array of independent disks (RAID) is used to provide for redundancy of the data in addition to the improvement of performance through parallelism. As a general rule for storage media, there is an inverse relationship between speed of access and cost; cache and main memories, in addition to their volatility, have the highest cost of storage.

In addition to the devices for storing and processing information, efficient processing relies on how the data are organized in a computer system. The manner in which the data are stored constitutes the physical record, while the description of the data and the relationships among the various types of data form a logical record, consisting of a set of entities. These fall into two categories: formatted or structured and unstructured. The former, such as signals generated in real time from sensors, reside in a fixed field within a record, while the latter, such as text, etc., do not reside in fixed locations.

The structured data are characterized by their attributes and the relationships among them. For example, information about the source of the segment, the time period when the signal was generated, spikes in the signal, and the spectrum of the signal are a few of the attributes that are associated with a file corresponding to a segment of a seismic signal. The spectrum calculated from a segment of the signal is an example of a derived attribute. For a given seismic event and a given time interval, signals are recorded at more than one stations resulting in multiple files with related characteristics. The collection of the records and the relationships among the respective attributes form a relational data base that allows the manipulation of the data and the extraction of information from the files and their interrelationships [18, 19].

Structured data reside in relational databases which define the relationships among different forms of information that allow their manipulation. The management of structured data in a relational database management system (RDBMS) is by a Structured Query Language (SQL) that provides a mechanism for relating different elements in a database on the basis of specified attributes that could be simple or composite, single-valued or multi-valued. Conversely, there is no mechanism for relating distinct elements of unstructured data, such as images stored in different locations [20, 21].

Computer systems may be configured into different architectures depending on the services they are expected to provide. The major configurations are: centralized, parallel, and distributed systems. Centralized systems are the simplest consisting of a central memory and CPU that serve a single user. For fast simultaneous performance of multiple tasks, computer systems are configured into parallel structures that are interconnected by high-speed data networks. In distributed systems processors at multiple sites interconnected through high-speed data networks have access to the databases at all the sites. The interconnection allows the integration of multiple databases into a virtual single database from the perspective of each site. The combination of continuing progress in data storage density, transaction speed and data communications speed drive the development computer system architectures that make it possible to process ever increasing volumes of data at faster speeds for increasing number of clients. In summary, the existing data storage and processing capabilities are more than sufficient to process the data of current and future verification systems.

6 Verification System Information Processing

The requirements of a verification system for communications systems (connectivity and bandwidth) and computer systems (processing speed and storage capacity) are determined by the type and quantity of data that are necessary to detect violations of treaty obligations as well as the timeliness of detection. While the available information technology is more than sufficient to accommodate any conceivable data processing need for a treaty verification system, the extraction of information relevant to the detection of non-compliance presents major challenges. These may be classified as economic/political and scientific/technological. The first category pertains to technically solvable verification problems that arise from economic/political considerations. Two examples illustrate the impact of economic/political considerations on the design and use of information processing for treaty verification purposes.

- The seismic network of the International Monitoring System of the CTBT comprises 50 primary and 120 auxiliary stations. According to the treaty, both categories have the same technical and operational requirements; in other words they have similar signal detection capabilities. While the primary seismic stations transmit data continuously in near-real time, the auxiliary stations provide segments of data only upon demand. Although the combined network of 170 stations would have an improved detection capability, the 120 stations were designated as auxiliary during the negotiations to minimize the cost of data transmission. With current technologies, the cost of transmission is negligible. Nevertheless, the configuration cannot be changed because it is enshrined in the treaty language!
- A more complicated example of the interplay between technology and economic/political considerations is the monitoring and verification system for Schedule 3 chemicals in the CWC. It covers toxic chemicals that are either precursors to highly toxic agents, or have been used or may be used as chemical weapons. The purpose of the verification regime is to detect diversion into activities prohibited by the Convention, namely, chemical weapons. The nuclear safeguards system considers the nuclear fuel cycle as a closed system and applies material accountancy and collects all information necessary for the detection of diversion. In contrast, the information collected by the corresponding monitoring system for Schedule 3 chemicals is incomplete for the detection of diversion, akin to the information collected under the Additional Protocol. The extensive use of these chemicals in commercial applications and the large quantities involved make it economically and politically difficult to implement material accountancy on such a large scale.

Another challenge to the design of information processing systems may arise from the definition of the obligations under a treaty. If the obligations are quantifiable, it would be technically feasible to collect sufficient data and design a system to detect violations. For example, the monitoring system for the CTBT, as specified in the treaty, was designed on the basis of achieving a system-wide detection threshold in the order 1.0 kiloton. As the system has been deployed, the threshold for some regions is below 0.1 kiloton. The threshold could be further lowered through the use of advanced sensing technologies such as phased arrays and incorporating into the treaty-specified system additional regional networks. In addition, integration of the sensor network with a position detection system such the GPS would help decrease the region of uncertainty about the location of a detected event. However, since the monitoring system is specified in the treaty, any substantive modification, however technologically sound, would require re-negotiation of the treaty, a nearly impossible task. Thus, technological improvements may also be constrained by political considerations [22].

Perhaps, the best example of the challenge presented by unquantified or incompletely specified objectives is the design of a data processing system to detect violations of the NPT. With the State-level approach, the Additional Protocol aims to augment the capability of safeguards to detect undeclared activities in addition to the detection of diversion based on the materials balance model. Implicitly, the State-level approach assumes that an undeclared fuel cycle has existed prior to the application of safeguards or it is started while safeguards are in place. Detection of such potential diversion paths requires the extension or modification of the conventional nuclear fuel cycle model to incorporate additional models of proliferation.

The critical difference between the State-level approach and the materials balance approach lies in the lack of completeness of the former in contrast to that of the latter. The absence of a "mechanistic or systematic" manner for verifying declarations makes it nearly impossible to detect potential discrepancies. It is difficult to search data for unknown or hidden patterns. The tools for such searches fall under the category of data mining, which is a generic term for a collection of search and statistical analysis techniques. These include techniques for identifying anomalies, relationships and structures in a record, classifying data into categories, as well as using regression analysis to fit the data into a model. In effect, this approach is based on using the measurement data to compute the a *posteriori* probability that a violation has occurred. Although these statistical analysis techniques are computationally intensive, the information processing requirements are relatively small compared to, e.g., weather prediction models. While the data processing capabilities are more than sufficient to accommodate the computational demands, the challenge lies in modeling vague and unquantifiable obligations.

7 Data Integrity

Decisions about compliance or non-compliance which are based on the stored and processed data can have significant global consequences. Therefore it is important that the integrity of the data be preserved, i.e., the data may not be altered by any unauthorized means. The integrity of the data may be threatened at their source, or during transmission, storage and processing. Due to the nature of the treaties, processing is centralized under the control of an international authority with responsibility to ensure the integrity of the data. At the other end of the system, the sources are sensors, as in the case of the CTBT's International Monitoring System, process measuring instruments as in the case of IAEA safeguards, and databases as in the cases of the CWC and safeguards. The integrity of the sensor data and the measurements is the responsibility of the international authority that ensures that the data represent what they purport to represent. The use of seals and intrusion detection sensors ensure that attempted violations of the integrity of the data at the source are detected. For databases originating at States, protection starts at the point of entry into the verification information system.

The main approaches for protecting data during transmission are authentication and encryption. The integrity of an authenticated message is ensured by appending a signature that is unique to that message. It is generated by the transmitter using a secure key and the message itself. The receiver is assured of the integrity of the message by generating a signature using the received message and matching it to the received signature. Authentication is used in treaties such as the CTBT to provide transparency of the monitoring data while ensuring their integrity during transmission. For those treaties that might require confidentiality and assurance of integrity during transmission the data are scrambled using encryption into a sequence that approximates the characteristics of white noise.

The threats to the verification data at the central processing facility may be internal or external. Internal threats may materialize due to the composition of the personnel of the technical organization. As an international organization it acquires its staff either through nominations by member States or in the open market. Under such circumstances it would be impossible to devise procedures for ensuring that the loyalty of the staff would be solely to the technical organization.

The threats to the verification data at the central processing facility may be internal or external. Internal threats may materialize due to the composition of the personnel of the technical organization. As an international organization it acquires its staff either through nominations by member States or in the open market. Under such circumstances it would be impossible to devise procedures for ensuring that the loyalty of the staff would be solely to the technical organization. Thus the potential exists for the integrity of the data to be compromised internally. Access control by authorized personnel is used to minimize the internal threat.

External threats may take the form of downloading or modifying sensitive verification data by breaching the security of the central processing facility. Computers may be infected with viruses and worms. Trojan horses execute unauthorized operations in infected computers and spyware may execute unauthorized operations in infected computers. Access to the verification database can also be gained through rootkits which are programs that, when installed in a computer, allow unauthorized persons to conceal malicious activities by obtaining administrator-level access.

There is a broad range of defense tools against cyber threats ranging from anti-virus patches to firewalls, front-end hardware and private virtual networks that are designed the information-processing system from the external environment. Due to the very broad range of potential threats, selection of the appropriate defense tools must be based on a detailed analysis of the specific risks associated with the treaty verification information system. Perpetrators of the threats may range from lone wolves motivated by the challenge of disrupting any system to threats that have the verification information system as primary target. Risk analysis that prioritizes the significance of the various categories of threats is a prerequisite to the development of appropriate strategies for mitigating potential risks.

8 Conclusions

The demands of treaty verification systems on the communications and computer technologies are relatively benign compared to the capabilities of these technologies both in terms of speed of communications, data storage capacity and processing speed. The principal challenges lie in the development of models that describe the obligations under a treaty, the identification of the parameters to be measured in order to design proper detectors of violations, and mechanisms for extracting usable information from incomplete and noisy data.

References

1. Shannon C E (1948) A Mathematical Theory of Communication. The Bell System Technical Journal 227(July, October): 379–423, 623–656
2. Meadow CT Yuan W (1997) Measuring the Impact of Information: Defining the Concepts. Information Processing & Management 33(6):697–714
3. Järvelin K, Wilson TD (2003) On conceptual models for information seeking and retrieval research. Information Research 9(1)
4. Cooley JN (2006) International Atomic Agency Safeguards under the Treaty on the Non-Proliferation of Nuclear Weapons: Challenges in Implementation. In: Avenhaus R et al (eds) Verifying Treaty Compliance: Limiting Weapons of Mass Destruction and Monitoring Kyoto Protocol Provisions, Springer
5. Convention on the Prohibition of the Development, Production, Stockpiling and Use of Chemical Weapons and on their Destruction. http://www.opcw.org/chemical-weapons-convention/. Accessed 29 Feb 2019
6. Comprehensive Nuclear-Test-BanTreaty. http://www.ctbto.org/fileadmin/content/treaty/treaty_text.pdf. Accessed 29 Feb 2019
7. IRIS PASSCAL Sensors. https://www.passcal.nmt.edu/content/instrumentation/sensors. Accessed 29 Feb 2019
8. Tsoulfanidis S, Landsberger S (2011) Measurement and Detection of Radiation, 3rd edn. CRC Press, Boca Raton, FL
9. Hastings A, Smith Jr HA, Lucas M (1991) Gamma-Ray Detectors. In: Reilly D, Eneslin N, Smith Jr. HA, Kreiner S (eds) Passive Nondestructive Assay of Nuclear Materials, U. S. Government Printing Office, Report NUREG/CR-5550
10. IKONOS Stereo Satellite Imagery. http://www.satimagingcorp.com/satellite-sensors/ikonos/ikonos-stereo-satellite-images/
11. Shaw GA, Burke HK (2003) Spectral Imaging for Remote Sensing. Lincoln Laboratory Journal 14(1):3–28
12. Farley V, Vallières A, et al (2006) Performance of the FIRST: a long-wave infrared hyperspectral imaging sensor. Proc. SPIE 6398, Optically Based Biological and Chemical Detection for Defence III, 63980T. https://doi.org/10.1117/12.689487
13. Satellite Footprints. http://www.sadoun.com/Sat/Installation/Dish-Size.htm. Accessed 29 Feb 2019
14. Frequency Bands. http://www.idirect.net/Company/Resource-Center/Satellite-Basics/Frequency-Bands.aspx. Accessed 29 Feb 2019
15. Seagate HAMR hard drive technology. https://newatlas.com/seagate-hamr-hard-drive-technology/21880/. Accessed 29 Feb 2019
16. Hitachi TagmaStore® Universal Storage Platform and Network Storage Controller. http://www.spectra.com/wp-content/uploads/tagmastore_usp.pdf. Accessed 29 Feb 2019
17. How many bytes for… https://searchstorage.techtarget.com/definition/How-many-bytes-for. Accessed 29 Feb 2019
18. Dean J, Ghemawat S (2008) MapReduce: Simplified Data Processing on Large Clusters. Communications of the ACM 51(1):107–113
19. Lee KH, et al (2011) Parallel Data Processing with MapReduce: A Survey. SIGMOD Record 40(4):11–20

20. Silberschatz A, Korth HF, Sudarshan S, (2010) Database System Concepts, 6th edn. http://codex.cs.yale.edu/avi/db-book/db6/slide-dir/. Accessed 29 Feb 2019
21. Agrawal R, et al, (2009) The Claremont Report on Database Research. Communications of the ACM 52(6):56–65. http://db.cs.berkeley.edu/claremont/claremontreport08.pdf. Accessed 29 Feb 2019
22. National Research Council (2012) The Comprehensive Nuclear Test Ban Treaty: Technical Issues for the United States. The National Academies Press, Washington, DC. https://www.nap.edu/catalog/12849/the-comprehensive-nuclear-test-ban-treaty-technical-issues-for-the. Accessed 29 Feb 2019

Enhancing Verification with High-Performance Computing and Data Analytics

James M. Brase, Eric G. McKinzie, and John J. Zucca

Abstract Hidden within our rapidly growing global streams of business, scientific, and communications data, is information that can reduce the global danger of nuclear weapons proliferation and use. Significant advances in machine learning, the explosion of available data and the ability of high performance computing that can be used to process massive amounts of data has the potential to provide a leap-ahead in the ability to detect proliferation and verify treaty commitments.

1 Introduction

The timely and accurate detection of the proliferation of nuclear weapons-related materials, technology, and activities remains one of the key challenges in maintaining an effective global nonproliferation regime. Domestic monitoring for national security purposes and verification of compliance with treaties and agreements relies on detection, understanding (discrimination) and, ideally, early warning of activities in very complex, interconnected national and international systems. It becomes even more complicated when one tries to take a more systems approach towards integrating domestic and cooperative regimes across nonproliferation and arms control communities. The access to data related to a wide range of science and technology development, technology-focused databases, material and equipment procurement and transportation and financing have exploded in the last decade and continues to grow exponentially. Automated systems for organizing and mapping massive data streams in the context of models of proliferation activities will be more and more critical for guiding human decision makers to the important threads of information in the sea of available data. New methods in machine learning and data analytics (sometimes called Big Data) are emerging to enable the integration of massive amounts of complex data. These new technologies and methods have the potential

J. M. Brase (✉) · E. G. McKinzie · J. J. Zucca
Lawrence Livermore National Laboratory, Livermore, CA, USA
e-mail: brase1@llnl.gov; mckinzie1@llnl.gov; zucca2@llnl.gov

I. Niemeyer et al. (eds.), *Nuclear Non-proliferation and Arms Control Verification*,
https://doi.org/10.1007/978-3-030-29537-0_20

to link related concepts, even when they come from disparate data modalities, such as news reports, material shipping transactions, and remote sensor data. To effectively integrate information and extract actionable information, development of new capabilities in machine learning, big data analytics, and the underlying data-intensive high-performance computing systems will be needed.

For non-proliferation and arms control monitoring and verification, large-scale integrated data models will allow analysts to answer state-specific questions and better track the evolution of potential proliferation activities to answer questions, such as:

- Is the nuclear technology activity typical? Has there been significant change in the last year? The last month?
- Are there coherent threads of activity consistent with a proliferation threat? What are the data elements that lead to that conclusion? What is the uncertainty?
- Is there an efficient way to integrate analysis over monitoring and verification data from a broad range of existing verification regimes?
- Can greater insight into national policies and intent be seen and how can that increase confidence and international security?

The exploration of the application of High Performance Computing (HPC) and data analytics in support of nuclear security is just beginning. In this chapter, the current status of capabilities and the central supporting technologies will be described at a conceptual level, and some potential application or use cases for such information mapping systems will be considered.

2 Underlying Technologies: From High Performance Computing to Machine Learning

2.1 High Performance and Data Intensive Computing

Bringing state-of-the-art high-performance computing capability to data analytics will enable new approaches to integrating science-based models and simulations and enable analytics to scale to increasing data volumes and model complexities. At the time of this writing, 'traditional' high performance computing (HPC) has come to mean massively parallel computers that are optimized for solving partial differential equations. Parallel computing breaks up a problem into sub problems and runs the sub problems at the same time (i.e., in parallel). When a problem can be broken down and run on a computer that contains thousands to millions of compute nodes, the problem is said to be massively parallel.

HPC performance is measured in Floating Point Operations per Second (flops). Current high performance computers are in the mega, tera or peta flop range (i.e. 10^9, $10^1 2$ or 10^{15} flops). The fastest computers at the time of this writing are in

Fig. 1 Sequoia supercomputer located at Lawrence Livermore National Laboratory containing over 1.5M cores

the 100 petaflop range. Computers in the mega to teraflop range are in common use (Fig. 1).

A well-known application of HPC is to solve partial differential equations to, for example, run predictions of the weather or simulate seismic wave propagation in the earth. The solution to such problems may use finite difference methods that model the earth or the atmosphere in a large number of cells. A sample problem may have on the order of 10^{10} cells and require 10^{15} flops to complete the simulation. Depending on the class of the computer and the degree to which the computer algorithm takes advantage of parallelism, these simulations can run in tens of seconds to a few weeks. General overviews of high performance computing can be found in [1, 2].

On-going integration of data analytics and simulation requirements are driving new directions in high-performance computing architectures. New approaches to heterogeneous architectures provide new types of processing units and innovative memory systems, which allow for tighter memory-processor integration to improved access to ever-growing data streams in analytic applications. Data intensive computers, such as Catalyst (Fig. 2), are designed with large amounts of computer memory at the compute nodes. This architecture facilitates the rapid processing of the data that is stored near the compute node. Specialized software is available that facilitates the distribution of the data to the computer architecture and enables the calculations on the data.

Fig. 2 The 150 teraFLOP/s Catalyst, a unique high-performance computing (HPC) cluster, serves as a proving ground for new HPC and big data technologies, architectures, and applications

Fig. 3 A circuit board with 16 of the new brain-inspired chips in a 4×4 array along with interface hardware—one whose architecture is inspired by the neuronal structure of the brain and requires only a fraction of the electrical power of conventional chips [7]

In the longer term, R&D for new computing technologies will focus on data analytics applications. Neuromorphic computing architectures (Fig. 3) inspired by the structure of the mammalian brain will emerge as an extremely efficient and low-power approach to applications in machine leaning and other analytics, allowing analytics to process data at the point of collection. Looking further out into the future, quantum information systems may enhance the performance of selected analytic algorithms. These technologies will be integrated with existing Von Neumann architectures and will serve as efficient co-processors for specific parts of the calculations. Research is underway in these areas and initial machine learning algorithms, particularly for neuromorphic systems, are being developed and evaluated for predictive analytics applications.

2.2 *Bringing the Data Together: Machine Learning and Multimodal Semantic Feature Spaces*

Machine learning has grown into a critical tool for large-scale data analytics. It is the ability to discover patterns in data, to learn models of activities, and to infer unobserved entities and connections. A machine learning approach currently gaining increased attention called Deep Learning is creating unprecedented pattern discovery capabilities. Deep Learning refers to the latest generation of artificial neural networks that, unlike previous generations, are able to train many (>2) hidden-layers deep. In some domains, Deep Learning is so effective that automated systems built from deep networks are surpassing human-level performance [3]. Learning algorithms for these multi-layer networks are extremely computationally intensive. High-performance computing is essential to efficient training on large sets of data.

A central concept in predictive analytics is the multimodal semantic feature space that brings together multiple classes of data in a common framework for analysis. These semantic feature spaces then support a variety of analytic methods to map information structure, discover activities, and detect changes. Machine learning algorithms map entities (e.g. places, organizations, specific objects) and concepts extracted from sources such as text streams, videos, transportation and financial transactions, or sensor networks into a common multimodal feature space. In this space, distance is directly related to semantic meaning, that is, items that are close together in the feature space are related to each other in the application context (Fig. 4).

The primary driver of the common feature space will be massive collections of structured and unstructured natural text (e.g. Wikipedia, open source web crawls, news articles, other databases, etc.) that provide the semantic "backbone" of the feature space. The text backbone will also prove highly useful for intuitive analyst interaction with the system via natural language. The semantic feature space creates the basis for information mapping pipelines that can create geospatial, temporal, and relational views of large, dynamic data streams for the classes of analysis described above.

Network analysis algorithms and methods are a second pillar of predictive analytics. Network models provide representations for relational data that describe large communication and transactional data sets and provide essential elements

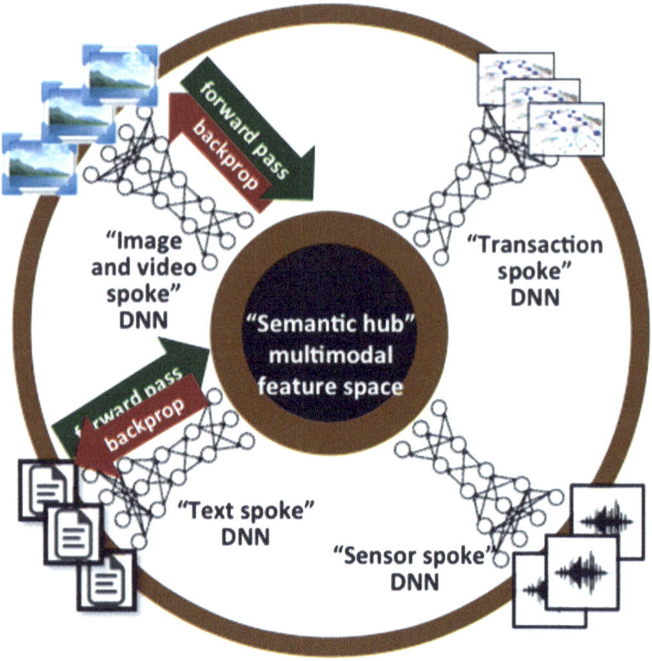

Fig. 4 Conceptual diagram showing how different data type or 'modes' can be brought together to a semantic feature space

of semantic feature spaces. Networks are analyzed using graph analytics which provide methods for finding connections between multiple data elements, discovering communities of related entities in data, and detecting temporal changes in relationships among entities and concepts. Graph analytics have been extended to very large networks with trillions of connections and can incorporate machine-learning algorithms to discover structure and relationships in large networks [4, 5].

Deep learning is beginning to open new approaches to predictive modeling of complex systems. In many cases relevant to the detection of proliferation activities, we must bring together physics-based models for some components of the activity with models based on observed data for other activity components. Multi-target and transfer learning approaches provide a systematic framework for integrating these different modeling approaches. With these tools physics-based simulations are used to learn feature spaces that "respect the physics". Machine learning algorithms then build predictors using these features with significantly fewer training examples than would be needed to learn directly from the data. Initial applications of these approaches to predictive models for inertial confinement fusion performance [8] illustrate the potential for application to complex physical systems.

Deep learning has been demonstrated in national security applications ranging from counter-WMD analysis to cybersecurity. For a general discussion of data science and machine learning refer to [6].

3 Possible Applications

It is easy to see applications for such capability in the context of national monitoring of proliferation activities or verification of existing treaties and commitments. For example, when monitoring countries with latent nuclear capabilities, it is valuable to objectively estimate the length of time it would take from the (hypothetical) decision to develop capability to possession of a nuclear weapon. An integrated analysis of information about the state's scientific R&D and the strength of its underlying technical-industrial infrastructure, in related products and disciplines, would help inform such an estimate.

There are also needs for such techniques in the international cooperative arena. This new capability could aid the International Atomic Energy Agency (IAEA). Annually, the IAEA is required to draw nuclear proliferation-relevant inferences from a State's safeguards declarations, the IAEA verification of those declarations and integrate this information with multiple sources of voluminous information about states' scientific, technical, and industrial activities and capabilities. The range of relevant data includes, *inter alia*, scientific and technical publications, data from specialized databases on patents, trade, and company registers, financial data, and data from open sources including new media, as well as data from various in-house IAEA databases.

Data analytic capabilities could assist by alerting IAEA country analysts to trends that should be explored further. By creating a dynamic map of a state's existing capabilities and/or underlying potential in a set of various fuel-cycle and weaponization-relevant technologies that evolve over time and in light of new information, could assist in flagging significant changes in capability that might be too subtle for an analyst to have noticed unaided. This could increase the timeliness of follow up actions, ranging from increased analytical attention to more frequent inspection activities in the field or even focused investigation of potential undeclared activities.

Just as data analytic capabilities could aid IAEA country analysts, it could also work hand-in-hand with a systems approach and identify possible signatures or indicators that would improve arms control verification regimes. It may provide a credible way to incorporate analysis of information collected from highly uncertain sources, such as social media. Furthermore, these approaches are potentially useful for future arms control agreements that could be multilateral. Such techniques could allow a country without extensive national technical means a capability to contribute substantially to the verification of the agreement.

4 Next Steps

As the use of machine learning and graph analytics evolve in a high-performance computing environment it will need to be coupled with subject matter experts in proliferation activities to constrain the problem domain. Additional research on what data is relevant to today's proliferation activities and the specific challenges the nonproliferation and arms control communities are facing will help drive the use of machine learning and graph analytics to address these challenges. Today's machine learning methods for complex pattern discovery are largely black boxes, providing little information on the evidence supporting an inference of activity. Nonproliferation applications will require an understanding of how an inference was computed and to estimate the confidence levels of the inference. This will require additional research as these two disciplines of data science and security come together, to provide a chain-of-reasoning explanation and robust confidence bounds.

This document was prepared as an account of work sponsored by an agency of the United States government. Neither the United States government nor Lawrence Livermore National Security, LLC, nor any of their employees makes any warranty, expressed or implied, or assumes any legal liability or responsibility for the accuracy, completeness, or usefulness of any information, apparatus, product, or process disclosed, or represents that its use would not infringe privately owned rights. Reference herein to any specific commercial product, process, or service by trade name, trademark, manufacturer, or otherwise does not necessarily constitute or imply its endorsement, recommendation, or favoring by the United States government or Lawrence Livermore National Security, LLC. The views and opinions of authors expressed herein do not necessarily state or reflect those of the United States government or Lawrence Livermore National Security, LLC, and shall not be used for advertising or product endorsement purposes. LLNL-BOOK-753657

References

1. Hager G, Wellein G (2011) Introduction to high performance computing for scientists and engineers. In: Chapman and Hall (eds) CRC, Computational Science Series, pp 1–32
2. Vetter J (2013) Contemporary high performance computing from petascale to exascale. In: Chapman and Hall (eds) CRC, Computational Science Series, pp 1–13
3. He K, Zhang X, Ren, S, Sun J (2015) Delving deep into rectifiers: Surpassing human-level performance on imagenet classification. In: Proceedings of the IEEE International Conference on Computer Vision, pp 1026–1034
4. Pearce R, Gokhale M, Amato N (2013) Scaling techniques for massive scale-free graphs in distributed (external) memory. In: Parallel & Distributed Processing (IPDPS), IEEE 27th International Symposium, IEEE, pp 825–836
5. Pearce R, Gokhale M, Amato N (2010) Multithreaded asynchronous graph traversal for in-memory and semi-external memory. In: Proceedings of the 2010 ACM/IEEE International Conference for High Performance Computing, Networking, Storage and Analysis, SC'10, IEEE Computer Society, Washington, DC, pp 1–11

6. Alpaydin E (2016) Machine Learning: the new AI. In The MIT Press Essential Knowledge Series, Cambridge, MA
7. Modha, D (2015) Introducing a brain-inspired computer: TrueNorth's neurons to revolutionize system architecture. http://www.research.ibm.com. Accessed 28 Feb 2019
8. Peterson JL, Humbird KD, Field JE, Brandon ST, Langer SH, Nora RC, Spears BK, Springer PT (2017) Zonal flow generation in inertial confinement fusion implosions. Physics of Plasmas 24(3):032702

Open Source Analysis in Support to Non-proliferation: A Systems Thinking Perspective

Guido Renda, Giacomo G. M. Cojazzi, and Lance K. Kim

Abstract Open source information, here defined as "publicly available information that anyone can lawfully obtain by request, purchase, or observation" is playing an increasing role in treaty monitoring, compliance verification and control. The increasing availability of data from an increasing number of sources on a vast range of topics has the potential to provide cues about complex programmes subject to international treaties such as the *Treaty on the Non-Proliferation of Nuclear Weapons* (NPT).

Indeed, within its State Level Concept, the International Atomic Energy Agency (IAEA) envisions an objectives-based and information-driven approach for designing and implementing State Level Approaches (SLAs). To achieve its objectives, the IAEA foresees a holistic approach, in which all the information gathered is analyzed and assessed as a whole.

This chapter will take a journey in the world of open source analysis and, through a holistic systems thinking approach, will highlight the potential and the challenges that open source analysis can offer to gain insights about various aspects of a complex engineering programme such as the nuclear fuel cycle.

1 Introduction

Nuclear energy programmes are complex endeavors lasting several years, involving multiple sites and a substantial number of personnel with a wide range of expertise. States willing to embark on such an enterprise would either sign the *Treaty on the Non-Proliferation of Nuclear Weapons* (NPT) and benefit from the transfer of technology enabled by its membership or develop the necessary science, technology and industrial capability indigenously. State signatories of the NPT

G. Renda (✉) · G. G. M. Cojazzi · L. K. Kim
EC Directorate General Joint Research Centre, Directorate G - Nuclear Safety and Security, Ispra, Italy
e-mail: guido.renda@ec.europa.eu; giacomo.cojazzi@ec.europa.eu

I. Niemeyer et al. (eds.), *Nuclear Non-proliferation and Arms Control Verification*,
https://doi.org/10.1007/978-3-030-29537-0_21

renounce the possibility to develop a nuclear military programme and are subject to international nuclear safeguards implemented by the *International Atomic Energy Agency* (IAEA).

Within its State Level Concept [1], the IAEA envisions an objectives-based and information-driven approach for designing and implementing State Level Approaches (SLAs). To achieve its objectives, the Agency foresees the adoption of a holistic approach, in which all the information gathered is analyzed and assessed as a whole "... to draw and maintain a conclusion of the absence of undeclared nuclear material and activities in that State." ([2], p. 20)

While the Agency, in its assessment, performs an *all-source analysis* making use of all the information and knowledge available to it,[1] either closed-source (official State's declarations, inspection data, third parties information) or open source, other treaty monitoring regimes often rely on fewer potential sources. No matter the domain in which open source analysis is employed to inform international treaty monitoring, the IAEA's intuition about the need for holism to be effective warrants serious consideration.

This chapter provides an overview of *what* is open source analysis (Sect. 2) from a holistic perspective and *how* it could be used to detect and monitor a complex engineering programme such as nuclear energy, here described from a *systems thinking*, *holistic* [3] point of view (Sect. 3). In the process, some applications to nuclear non proliferation are provided. Finally, Sect. 4 identifies open source analysis scenarios that can be encountered when analysing complex programmes for treaty monitoring purposes. This chapter takes the point of view of an open source analyst, having access only to openly available information and knowledge.

The objective of this chapter is not to provide complete coverage of all possible uses of open sources of information. Therefore, the examples in this overview are intended to clarify concepts and are not meant to be exhaustive.

2 *What* Is Open Source Analysis?

There is no normative definition of open source information that is universally accepted. For this exposition, open source information is defined as "publicly available material that anyone can lawfully obtain by request, purchase, or observation." ([4], p. 5) This definition is very broad, and covers a vast range of potential sources. For instance, the IAEA defines it to include "information generally available from external sources, such as scientific literature, official information, information issued by public organizations, commercial companies and the news media, and commercial satellite imagery" ([5], p.10), and also includes trade data [6]. While classified information is certainly outside the open source information envelope, the

[1] In this respect, although they do collect and make use of open source information, the IAEA analysts are here considered not to be open source analysts.

exact boundaries of this envelope are not well defined. For example, gray literature is considered to be open source [7], even if technically not reachable by "anyone".

Within a systems thinking approach, open source analysis (OS) could be seen as a process of "getting the right information (what) to the right people (who) at the right time (when) for the right purpose (why) in the right forum (where) and in the right way (how)" ([8], p. 190) by merging openly available data and information from a wide variety of accessible sources into an overall comprehensive and cohesive picture. Usually the process involves the gathering and analysis of a large amount of data and information, a very small percentage of which is relevant. A common scenario involves filtering an enormous amount of data to end up with a sparse and incomplete set of information not all of which contributes to knowledge [9]. When investigating a covert military engineering programme, the analyst will have to deal with low quality data and is always exposed to deliberate deception [10, 11]. Nonetheless, the analyst may be able to obtain valuable insights about what a State might be pursuing.

3 *How* Can Open Source Analysis Monitor an Engineering Programme?

There are many ways of formalizing a complex engineering programme, and depending on the final goal, one might work better than another. This discussion will propose a formalization of a complex system adapted from the "Doing it Differently" (DID) approach [12], originally conceived as a holistic approach in the construction industry and then applied in several other fields, including nuclear safeguards [14] and the proliferation resistance of nuclear energy systems [13].

3.1 Defining the System to be Analyzed: A Systems Thinking Approach

A central idea in the DID approach is that complex issues involve many players with diverse needs, wants and objectives which must be taken into account. In the DID approach, everything is modeled as a *process*. A process is defined by establishing the need for its existence and how it develops over time to achieve its objectives. The first process is that of *being*. The very fact of existence involves changes that might have considerable impact on the eventual success of the project. Every process involves an interaction between "*hard*" technical systems (which might range from basic equipment/products or very complex technical systems) and "*soft*" social systems (the behavior of people who play various roles in the process and who interact with the hard system). All of this is set within a given *context* [8].

Each process is considered to be a *holon* (i.e. both a whole and a part) [3]. Every process is a whole made up of the *hard* and *soft* layers that interact to produce change, but it is also a part of other processes to which it contributes. For example a nuclear reactor is a whole made up of primary and secondary technical systems operated by the personnel and according to the regulatory framework which, taken together, constitute the process of operating a nuclear power plant. The reactor is also a part of the overall nuclear fuel cycle process in which it is embedded [13].

The complexity of a programme is described as a process view where the traditional separation of the *hard* technical layer from the *soft* social one and the *context* in which they take place is replaced by a systemic comprehensive view, where the processes of a complex system are organised into a hierarchical structure of the scope of levels of definition.

All three layers (hard, soft and context) of a process will produce several observables, some of which could be detected and monitored through open source analysis. The following paragraphs will give an overview of the three layers, together with the characteristics of the related observables.

3.2 **Hard** *Layers' Observables*

The *hard* layer of an engineering programme refers to its technical infrastructure. The observables typically relate to the existence and functioning of the technical infrastructure of the system. They provide information about the topology, layout, shape and size of the infrastructure. When particular shapes are recognizable (such as the reactor dome of a nuclear power plant), they give the analyst the opportunity to gain insights about the technical layer's function.

Usually these observables are continuously available (e.g., the visual signature of a building is always available for collection), but—at least in the field of open source analysis—signals collection is discontinuous to varying degrees (e.g., the visual signature of a building will be captured only during an overhead passage of a commercial satellite). Since the technical infrastructure is generally geo-localized, the data gathered can be processed and analyzed with geospatial tools.

Examples of open source information related to hard layers' observables include those obtained through satellite/overhead imagery and remote sensing [15]. These techniques already demonstrated their effectiveness in analyzing and monitoring nuclear-related sites to gain knowledge of their status and activities, both in nuclear safeguards and in non-proliferation. For instance, the interest in monitoring proliferation sensitive fuel cycle steps, such as irradiation in reactors and enrichment via remote sensing, is not new [16, 17]. For example, nuclear-related sites in Iran have been extensively monitored by NGOs active in the field of nuclear non-proliferation (see e.g., [18]). The DPRK nuclear test site has been the subject of several geospatial analyses, published both in scientific journals (see e.g., [19, 20]) and non-proliferation blogs (see e.g., [21, 22]). Open source analysis has also investigated sites that underwent nuclear accidents [23, 24]. Figure 1 provides an

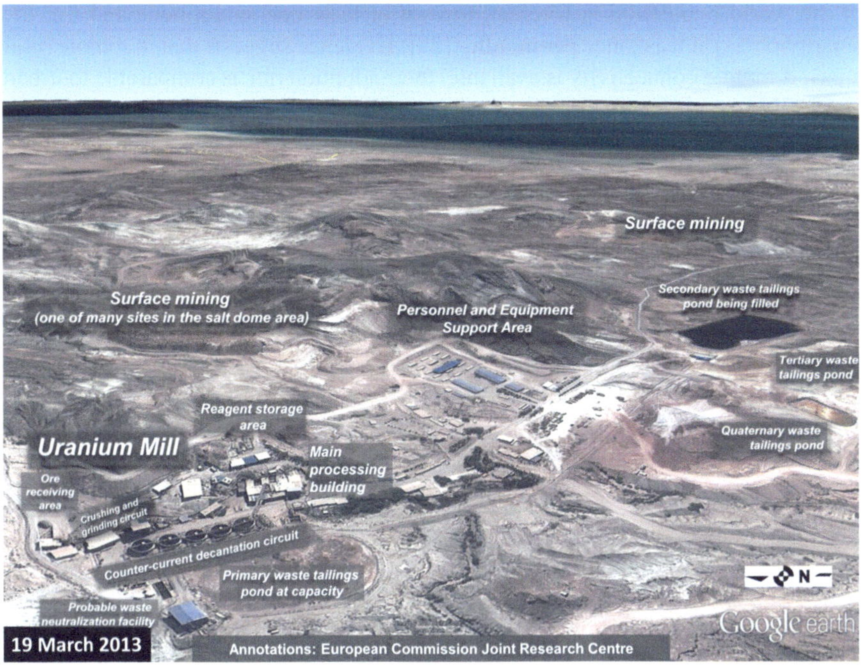

Fig. 1 Annotated satellite image of a uranium mine, where the function of some of the infrastructure is identified. Courtesy of F. Pabian. Image credits as in the figure

example of open source analysis of technical infrastructure through the use of satellite imagery.

3.3 Soft *Layers' Observables*

The *soft* layer of an engineering programme relates to the people that conceive, design, build and operate the *hard* layer under investigation. This soft layer opens possibilities to examine the social aspects of the programme under analysis, and gives the analyst the possibility to gain insights about the function of the various parts of the infrastructure, the people working in and operating it, the working security rules and policies of the system, together with hints about the context to which the process belongs.

Signal emitters of a soft layer include the people themselves. In addition to the scientific literature and "traditional" media sources, the advent of the internet and social networks [25, 26], coupled with technological advancements in the field of mobile connectivity, led to the fact that nowadays many people own and regularly use internet-enabled camera phones. The penetration of smartphones worldwide grew globally from 15% in 2012 to 24% in 2014 ([27], p. 1) and in 2016, 31.2%

of the world population is foreseen to own at least one smartphone and use the smartphone(s) at least once per month. ([28]) Evidence suggests that in remote rural areas, if mobile connectivity is available, their adoption rate is comparable to urban areas [27].

Where people have access to reliable connectivity, such as in the European Union and in the Unites States of America, a high percentage of them regularly uses one or more social networks. In the US in 2014, 52% of online adults used two or more social media sites. ([29], p. 2) Having a camera-phone in one's pocket increases the likelihood that people capture and share pictures of the places and events they frequent. In 2014, an average of one million photos were shared each day on the social network Flickr.

Compared to the observables for the *hard* layer described in Sect. 3.2, *soft* layer observables are discontinuous in nature (a single person publishes or posts only intermittently), but the enormous number of potential emitters provides a remarkable signal—albeit most of it is, from a treaty monitoring perspective, just noise. Contrary to the *hard* layer's observables, some *soft* observables can be *continuously collected* posts on social networks such as Twitter or Instagram can be monitored near-real time, and the same is technically possible with (micro-)blogs and online media sources [30].

The information collected can either be geo-referenced (as in the case of some social network posts) or not (as in the case of scientific and official literature, online blogs, online and traditional media). Geo-localized information can be processed and analyzed with geospatial tools. For example, when social networks data are analysed, the analyst is potentially able to derive information from much more than just the actual content shared by the user. In particular, the following aspects of social networks' posts can be of potential interest:

- *Layout* patterns: when geo-referenced, the pattern of the posts can reveal useful information about, for example, the security rules of a nuclear site. When analyzed over a long enough period of time, the pattern could reveal, for example, whether the site allows smart-phone use, and in which areas and buildings their use is restricted. Usually civilian nuclear research sites are very active, but military nuclear sites and research centres are virtually silent.
- *Temporal* patterns: sequential posts from one source can provide insights of the system's evolution (people and related activities moving from one building to another, dismissal of activities); concentration of posts from a given building in a given period of the day can provide insights of the function of a given building (e.g., a building active between 12H00 and 14H00 is likely the canteen/cafeteria), providing hints of its function and level of sensitivity. Sharp increases or decreases at certain periods of the day from the entire site might reveal working habits, such as usual working hours.
- *Content*, although most of the times irrelevant, can occasionally provide insights on the type of activities and equipment involved, for example, in laboratories, or whether renovation and refurbishment of the equipment is ongoing.

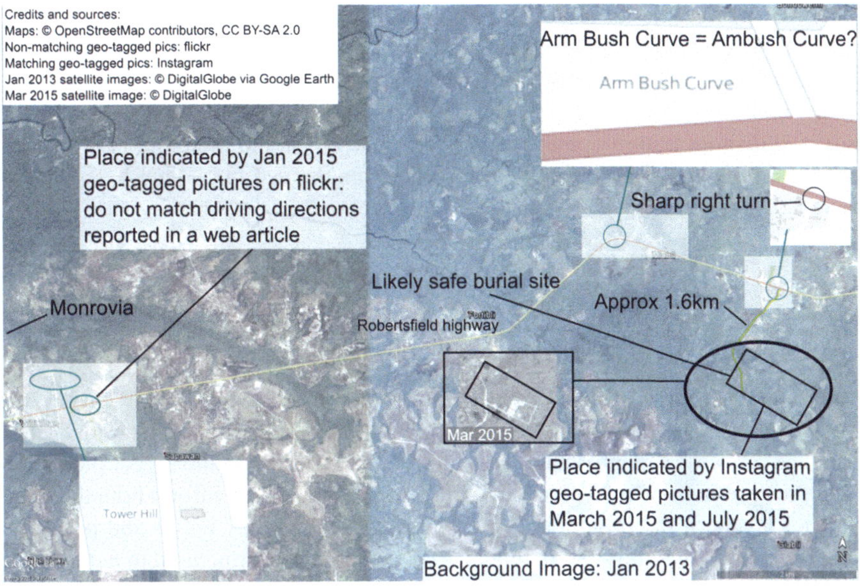

Fig. 2 Effort to use social media information to locate the Ebola cemetery opened by the Liberian authorities in Monrovia at the beginning of 2015, starting with the last high-resolution satellite image of the area available for free dating back to 2013. Following the description of a blog post on openstreetmap and Google Earth, it was possible to identify a candidate area. Low resolution images from DigitalGlobe ImageFinder confirmed the presence of new activities after the cemetery was established but were not sufficient to link the activities to the cemetery. Images uploaded on social networks taken on *Decoration Day* (a Liberian national holiday in which the people remember their dead relatives by decorating their graves) in the area under investigation corroborated the location of the site. Another Instagram image uploaded in July 2015 further increased confidence that the site had been located. Credits as in the figure

Monitoring news and social media allows the open source analyst to follow the development and the evolution of the system under analysis in an almost dynamic way, potentially filling the gap between the availability of signals collected from the *hard* layer. In particular, the geo-referenced information coming from social networks posts and online collaboration efforts can allow the follow-up and analysis of events such as radiological releases [31, 32], accidents, riots and protests [33].

As an example of synergy between *soft* and *hard* layers' observables, Fig. 2 shows an effort to locate the Ebola cemetery opened by the Liberian authorities in Monrovia at the beginning of 2015. At that time, the last high-resolution satellite image of the area available for free dated back to 2013. Social networks, collaborative tools and online blogs filled the gap between the last available satellite image (January 2013) and a given point in time (July 2015). Finally, in early 2016, a new high resolution satellite image of the area became available in Google Earth, which confirmed the candidate site as the actual burial site.

While it might be an issue for any type of open source [34], the *soft* layer's observables, especially those related to social network posts, face the risks of being unreliable, inaccurate, or even subject to manipulation to misinform and deceive [11]. There have been cases of States having allegedly set up "troll farms" as a part of a disinformation operation [35]. The analyst should therefore be aware of this possibility and look for possible corroboration derived from other independent sources.

3.4 Taking the Context *into Account*

Any human activity is shaped by and carried out within a particular *context*, the knowledge of which might be extremely important for the understanding of the observed process. The information about the *context* relates to the bigger picture of which the programme under investigation is part. Context can inform both the *hard* and *soft* layers of the programme, but is usually coloured by biases and points of view. Typically, the information related to the context is unstructured in nature, and comes from a wide variety of sources, including news and media reports, international relationships and obligations, political and government statements and acts.

The start of complex programmes such as a nuclear proliferation effort is often influenced by the context perceived by the State initiating it, and influences how the programme could be structured and carried out. For instance "two programs, those of the United States and Soviet Union, were unquestionably initiated in response to what were perceived as dire outside threats and, in the rush to develop weapons, both the enriched uranium and plutonium routes were pursued simultaneously." ([36], p. 11)

Although there is no definitive answer to why states decide to start a nuclear weapons programme and how influential the context may be, [37, 38] sometimes knowledge of the external influencing factors is the only means to explain the observed evolution of a nuclear programme. An example of how the context might be critical to understand the evolution of a programme is provided in Fig. 3. The figure shows two snapshots of the Iranian nuclear fuel cycle: one in 2013 [39] and another in 2014 [40]. While in 2013 the fuel cycle covered all the steps of an open fuel cycle and presented a coherent nuclear fuel cycle process 1 year later the same fuel cycle apparently evolved into a much less understandable process flow. The explanation for this evolution is the agreement between Iran and the EU/E3+3 (France, Germany, United Kingdom, China, the Russian Federation, the United States of America, the European Union) on a Joint Plan of Action [41] that shaped its nuclear fuel cycle as depicted.

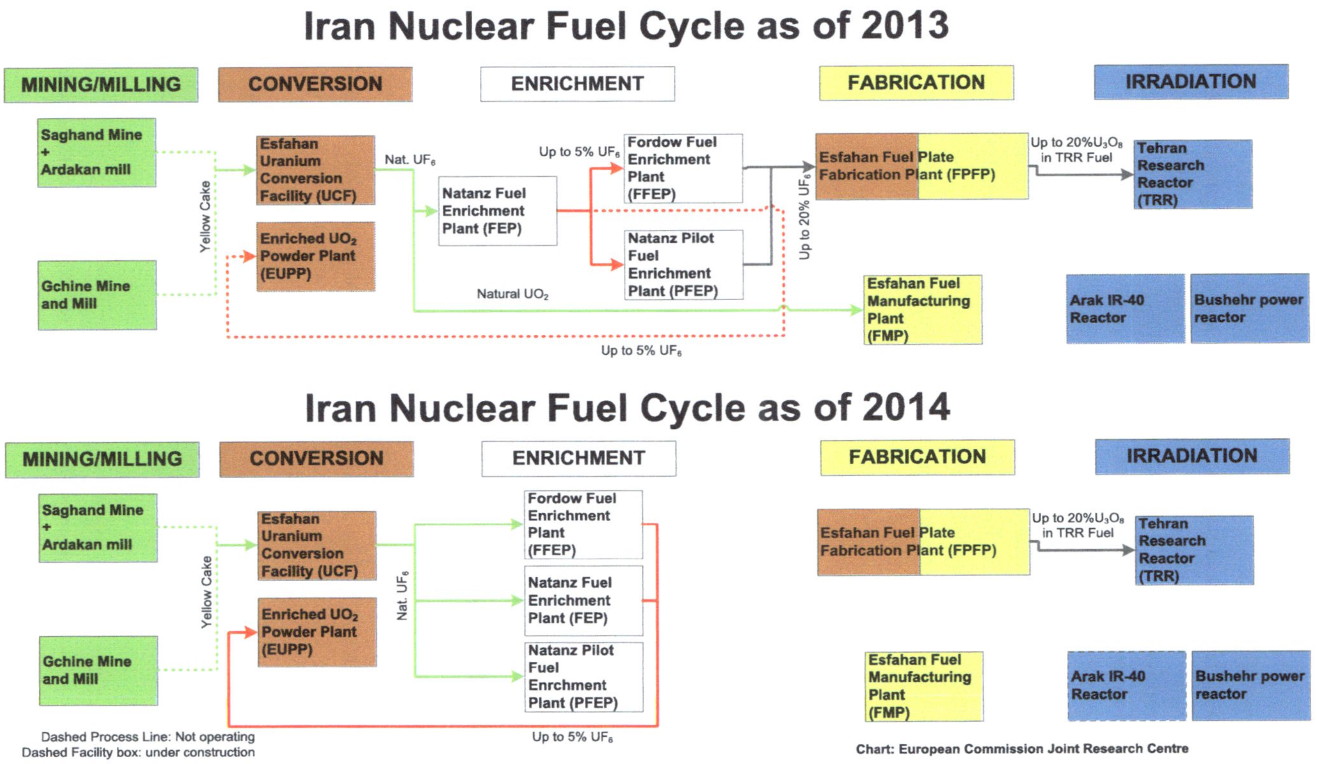

Fig. 3 Snapshot of the Iranian nuclear fuel cycle in 2013 and in 2014 derived from the IAEA Director General reports to the IAEA Board of Governors [39, 40]. The evolution of the programme would not be understandable without the knowledge of the geopolitical context (signature of the EU/E3+3 and Iran Joint Plan of Action related voluntary measures [41] foreseeing changes in the scope and operation of some Iranian nuclear fuel cycle's facilities)

4 Types of OS Analysis Scenarios for Nuclear Non-proliferation

As with any type of analysis, an open source analysis in support to nuclear non-proliferation comprises a set of data to be analysed and a paradigm[2] to make sense of those data. Depending on the quality and quantity of the data and the quality of the models[3] at the analyst's disposal, four different analysis scenarios can be identified [44], illustrated in Table 1. To be able to effectively perform OS analysis in support to nuclear non-proliferation, these scenarios are to be understood and assessed.

The following paragraphs will briefly discuss the four categories of Table 1. It has to be noted that the classification of the analysis as being part of one of the four scenario types depends on both the analysis and the analyst: an analysis could fall into the *data foraging* category for one analyst and into the *model building* one for another. For this reason, the presented examples are to be considered as illustrative and not normative.

4.1 *Puzzle Solving*

When high quality data and a well-established paradigm with dependable models are available to the analyst, the outcome of the analysis is usually dependable and the analyst's role is that of gathering the data, processing them according to the need and visualizing the outcome in the most suitable way. An example of puzzle solving analysis is the description of the declared Iranian production nuclear fuel cycle depicted in Fig. 3. The information used to build the diagram comes mainly from the IAEA Director General reports to the IAEA Board of Governors [39, 40]. The nuclear fuel cycle production processes are very well known in their general terms and make for a high quality model for the purpose of the analysis. The outcome can be considered as a dependable picture of the status of the Iranian nuclear fuel cycle's *declared* processes.

While the Iran nuclear fuel cycle is a good example of puzzle solving, it is also a very particular one. Usually the amount of information available on a nuclear fuel cycle is much more limited and attributable to sources that are much less reliable than the IAEA. Thus, it is not expected to see many puzzle solving problems in performing OS analysis in support to nuclear non-proliferation.

[2]Following Kuhn, a *paradigm* is here defined as "universally recognized scientific achievements that for a time provide model problems and solutions to a community of practitioners." ([42], p. viii).

[3]A model is here defined as "a physical, mathematical or logical representation of a system entity, phenomenon, or process." ([43], p. 105).

Table 1 Research dimensions for OS analysis applied to nuclear non-proliferation

		Paradigm strength	
		Low	High
Data quality/quantity	High	Model building	Puzzle solving
		What could work for NP?	How likely in NP?
	Low	Mystery framing	Data foraging
		How to find "unknown unknowns"?	What and how in NP?

General framework adapted from [44]

4.2 *Data Foraging*

A more likely analysis scenario is one in which the paradigm is well established but dependable information is lacking. In this case, the main task and challenge for the OS analyst is to find and collect data as dependably as possible, and the dependability of the analysis will be mainly driven by the quality of the gathered data. This scenario also opens a research dimension, i.e. to look into innovative ways of finding and collecting data that may not have been generated/intended for non-proliferation analyses but can adapted to the task.

An example of data foraging for a nuclear non-proliferation analysis is the analysis of the technical details of Iranian enrichment centrifuges provided by Jungwirth et al. [45]. While the knowledge of gas centrifuge enrichment technologies is well established in open literature and a lot of information is available on the Iranian nuclear fuel cycle, virtually no official information was available about the technical details of the various types of Iranian gas centrifuges. By merging different sources of open information, including measurements derived from public photographs, the report shows an attempt to derive the main technical characteristics of some centrifuges types developed by Iran.

In some areas data foraging can be supported by *ad hoc* tools. For example, in the domain of online news media monitoring tools to monitor, screen and collect information in multiple languages [46, 47] can inform the open source analyst and contribute to gaining early awareness of nuclear-related activities [30]. The use of trade analysis to support the non-proliferation regime is discussed in the chapter *Strategic Trade Analysis for Non-Proliferation* by Cristina Versino and Giacomo G.M. Cojazzi in this book.

As with any data gathering campaign, a critical point is being able to assess "how much information is enough". Having more information does not necessarily increase the accuracy of an analysis: studies show that "once an experienced analyst has the minimum information necessary to make an informed judgement, obtaining additional information generally does not improve the accuracy of his or her estimates. Additional information does, however, lead the analyst to become more confident in the judgement, to the point of overconfidence." ([48], p. 52)

4.3 Model Building

Sometimes reliable data could be gathered, but the available paradigms are not adequate for making sense of the available information. Where models are known, they are often classified and not publicly available. There are therefore areas of investigation where a model building activity is necessary for open source analysis. As mentioned, sometimes this activity is aimed at "rediscovering" in the public domain what is already known in classified environments, as in the case of the suitability of the nuclear material available in a commercial nuclear fuel cycle for a military nuclear programme [49–52]. Sometimes there is a need to investigate the signature of nuclear facilities to compare with open source data and signals [16, 17].

Even when a well established and dependable paradigm is available, identifying the right model for a given analysis is a complicated task. For example, in an analysis of the technical difficulty for a given State to set up and operate a gas centrifuge enrichment facility—typical of an acquisition pathways analysis [53], it might be very difficult to select the most appropriate model. Indeed, "...some analysts have claimed that '... all enrichment techniques demand sophisticated technology in large and expensive facilities' ([55], p. 31), suggesting that enrichment is out of reach of all except the most capable States. In contrast, others suggest that a small centrifuge plant is 'feasible for countries with no prior experience, that possess relatively little technical skills and which have relatively little industrial activity' ([56], p. 47)." ([54], p. 6)

In this scenario, the dependability of the analysis will mainly depend on the quality of the model and on the final goal of the analysis.

4.4 Mystery Framing

In a nuclear proliferation effort, the information related to the programme is usually a closely guarded secret, and both data and possible related models are unknown or highly uncertain. When both the quality of the data and the model are poor, the analyst faces a mystery framing scenario. This is typical of looking for clandestine nuclear weaponization activities and is usually addressed by national intelligence services. Open source analysis could theoretically play a role but faces formidable challenges in both data gathering and modelling. Moreover, analysts cannot conclude the absence of a nuclear weaponization activity from the absence of a signal.

Occasionally, the analyst stumbles upon ambiguous data that do not fit neatly into any known signature but still cannot be dismissed as innocuous (or when there are multiple competing hypothesis that seem plausible). These are sometimes referred to as "enigmas", and are another form of mystery framing. An example of an enigma in the nuclear non-proliferation community was the Dair Alzour site in Syria, which was evidently subject to an air strike in September 2007.[4] In the aftermath of the air strike, the claims that the destroyed site was hosting a nuclear reactor [58] were disputed by some analysts on the grounds that many of the usual signatures associated to this type of facility, such as "barbed wire or air defences that would normally ring a sensitive military facility" ([59]) were not evident. Others interpreted this absence as being the outcome of a complex concealment effort [60, 61]. In 2011, the IAEA assessed that "... the destroyed building was very likely a nuclear reactor..." ([62], para. 24).

5 Conclusions

Proliferation efforts are complex engineering programmes lasting several years, involving multiple sites and a lot of people with a wide range of expertise. As such, they are prone to leave behind several traces and potential cues that can be identified and monitored via open source analysis with varying degrees of success.

With a systems thinking approach, open source analysts can gain insights into a given programme. The process usually involves the gathering and analysis of a large amount of data and information, much of which is irrelevant. The most common scenario entails filtering an enormous amount of data to end up with a sparse and incomplete set of information. When after a covert military engineering programme, the analyst will likely encounter low quality data and is always exposed to deliberate deception. Nonetheless, analysts can gain valuable insights into what a State's pursuits.

This chapter suggests a system's thinking view of open source analysis in support of non-proliferation, identifying the dimensions involved and discussing the scenarios an open source analyst might face. A system's thinking approach to open source analysis has the potential to increase the international community's confidence in its ability to detect an undeclared proliferation programme.

Acknowledgements The reflections here presented stem from the activities performed within the project *Information Analysis and Methodologies for Nuclear Non-proliferation and Security* (IANNP), funded within the European Commission (EC) Euratom Horizon 2020 Research and Training Programme. The authors are grateful to all the researchers that contributed to these projects, In particular Dr. Rainer Jungwirth, Mr. Frank Pabian and Mr. Erik Wolfart. The discussions with them stimulated the ideas that are here expressed. The views expressed in this chapter are those of the authors and do not necessarily represent the official views of the European Commission. This chapter is based and adapted from the JRC Report: G. Renda et al, "Open Source Information Analysis in Support to Non-Proliferation: A Systems Thinking Approach", EC Joint Research Centre, EUR29515 EN.

---

[4]For a reconstruction of the events, for example see [57].

References

1. International Atomic Energy Agency (2013) The Conceptualization and Development of Safeguards Implementation at the State Level - Report by the Director General (GOV/2013/38). http://www.isisnucleariran.org/assets/pdf/GOV201338.pdf. Accessed November 2019
2. International Atomic Energy Agency (2002) IAEA Safeguards Glossary. No. 3 in International Nuclear Verification Series, IAEA
3. Smuts JC (1936) Holism and Evolution, 3rd edn. MacMillan and Co., Limited
4. Best RA Jr, Cumming A (2007) Open source intelligence (OSINT): Issues for Congress. Tech. Rep. RL34270. Congressional Research Service
5. International Atomic Energy Agency (2009) The safeguards system of the International Atomic Energy Agency
6. Versino C, Cojazzi GGM, Contini F, Tsois A, Chatelus R, Tarvainen M (2010) Global trade data to support IAEA safeguards. ESARDA Bulletin 45:14–21
7. Holland BR (2012) Enabling open source intelligence (OSINT) in private social networks. Tech. rep., Iowa State University
8. Blockley DI (2010) The importance of being process. Civil Engineering and Environmental Systems 27(3):189–199
9. Few S (2015) Signal: Understanding What Matters in a World of Noise. Analytics Press
10. Katz JI (2006) Deception and denial in Iraq: The limits of secret intelligence and the intelligent adversary corollary. Techn. rep., Washington University
11. Tsikerdekis M, Zeadally S (2014) Online deception in social media. Communications of the ACM 57(9):72–80
12. Blockley DI, Godfrey P (2000) Doing it differently: systems for rethinking construction. Thomas Telford
13. Renda G (2008) Resisting nuclear proliferation through design — a systems approach to nuclear proliferation resistance assessment. PhD thesis, University of Bristol
14. van Wijk LGA (2005) A process model for nuclear safeguarding. PhD thesis, University of Bristol
15. Pabian F, Renda G, Jungwirth R, Kim LK, Wolfart E, Cojazzi GGM (2015) Recent developments promoting open-source geospatial synergy: Emerging trends and their impact for nuclear nonproliferation analysis. In: Proc. 56th INMM Annual Meeting, Indian Wells, USA
16. Zhang H, von Hippel FN (2000) Using commercial imaging satellites to detect the operation of plutonium-production reactors and gaseous-diffusion plants. Science & Global Security 8(3):261–313
17. Bernstein A (2001) Monitoring large enrichment plants using thermal imagery from commercial satellites: A case study. Science & Global Security 9(2):143–163
18. Albright D, Hinderstein C (2003) The Iranian gas centrifuge uranium enrichment plant at Natanz: Drawing from commercial satellite images. http://www.isis-online.org/publications/iran/natanz03_02.html. Accessed 28 Feb 2019
19. Schlittenhardt J, Canty M, Grünberg I (2010) Satellite earth observations support CTBT monitoring: A case study of the nuclear test in North Korea of Oct. 9, 2006 and comparison with seismic results. Pure and Applied Geophysics 167(4–5):601–618
20. Pabian FV, Hecker SS (2012) Contemplating a third nuclear test in North Korea. Bulletin of the Atomic Scientists 6
21. Lewis J (2014) The tunnels at Punggye-ri: An alternative view. 38 North, http://38north.org/2014/03/jlewis032014/. Accessed 28 Feb 2019 http://38north.org/2013/12/punggye122013/figure6-17/. Accessed 28 Feb 2019
22. Liu J (2013) North Korea's Punggye-ri nuclear test site: no indication of nuclear test preparations. https://www.38north.org/2013/12/punggye122013/. Accessed 15 Nov 2019
23. Sadowski FG, Covington SJ (1987) Processing and analysis of commercial satellite image data of the nuclear accident near Chernobyl, USSR. Tech. rep., US Geological Survey

24. Albright D, Brannan P, Walrond C (2011) Fukushima crisis: Unmonitored releases - preliminary assessment of accident sequences and potential atmospheric radiation releases. http://isis-online.org/uploads/isis-reports/documents/Accident_Sequence_Fukushima_31March2011.pdf. Accessed 28 Feb 2019
25. NTI (2014) Innovating verification: New tools & new actors to reduce nuclear risks - redefining societal verification. Techn. rep., Nuclear Threat Initiative. http://tinyurl.com/pjk9eb8. Accessed 28 Feb 2019
26. Crosbie V (2002) What is New Media? Sociology Central. http://www.sociology.org.uk/as4mm3a.doc. Accessed 12 Jan 2017
27. Heimerl K, Menon A, Hasan S, Ali K, Brewer E, Parikh T (2015) Analysis of smartphone adoption and usage in a rural community cellular network. In: Proc. Seventh International Conference on Information and Communication Technologies and Development, ACM
28. eMarketer (2014) Smartphone users worldwide will total 1.75 billion in 2014. eMarketer.com, http://tinyurl.com/qcjkkj8. Accessed 28 Feb 2019
29. Duggan M, Ellison NB, Lampe C, Lenhart A, Madden M (2014) Social media update 2014. Techn. rep., Pew Research Centre
30. Cojazzi GGM, van Der Goot E, Verile M, Wolfart E, Rutan Fowler M, Feldman Y, Hammond W, Schweighardt J, Ferguson M (2013) Collection and analysis of open source news for information awareness and early warning in nuclear safeguards. ESARDA Bulletin 50:94–105
31. Plantin JC (2011) The map is the debate: Radiation webmapping and public involvement during the Fukushima issue. Techn. rep., Université de Technologie de Compigne
32. Glassman M, Kang MJ (2012) Intelligence in the internet age: The emergence and evolution of open source intelligence (OSINT). Computers in Human Behavior 28(2):673–682
33. Stefanidis A, Crooks A, Radzikowski J, Croitoru A, Rice M (2014) Social media and the emergence of open-source geospatial intelligence. Human Geography: Socio-Cultural Dynamics and Global Security, US Geospatial Intelligence Foundation, Herndon, VA 1:109–123
34. Renda G, Kim LK, Jungwirth R, Pabian FV, Cojazzi GGM (2014) The potential of open source information in supporting acquisition pathway analysis to design IAEA state level approaches. In: Proc. IAEA International Safeguards Symposium: Linking Strategy, Implementation and People, Vienna
35. Joseph R (2015) Open source indicators and asymmetric advantage in security planning. http://www.css.ethz.ch/en/services/digital-library/articles/article.html/192425/pdf. Accessed 28 Feb 2019
36. Ullom J (1994) Enriched uranium versus plutonium: Proliferant preferences in the choice of fissile material. The Nonproliferation Review 2(1):1–15
37. Erickson SA (2001) Economic and technological trends affecting nuclear nonproliferation. The Nonproliferation Review 8(2):40–54
38. Ogilvie-White T (1996) Is there a theory of nuclear proliferation? An analysis of the contemporary debate. The Nonproliferation Review 4(1):43–60
39. International Atomic Energy Agency (2013) Implementation of the NPT Safeguards Agreement and relevant provisions of Security Council resolutions in the Islamic Republic of Iran (GOV/2013/56). GOV/2013/56, IAEA
40. International Atomic Energy Agency (2014) Implementation of the NPT Safeguards Agreement and relevant provisions of Security Council resolutions in the Islamic Republic of Iran (GOV/2014/58). GOV/2014/58, IAEA
41. EU/E3+3 and The Islamic Republic of Iran (2013) Joint statement by EU High Representative Catherine Ashton and Iran foreign minister Zarif. http://www.eeas.europa.eu/statements/docs/2013/131124_02_en.pdf. Accessed 28 Feb 2019
42. Kuhn TS (1970) The structure of scientific revolutions, 2nd edn. University of Chicago Press
43. Defense Systems Management College (1999) Systems Engineering Fundamentals. Defense Systems Management College, Fort Belvoir, Virginia
44. Bachastow TS (2015) Geospatial Intelligence & the Geospatial Revolution. Penn State University

45. Jungwirth R, Cojazzi GGM, Kim L, Pabian F, Renda G (2015) Technical details of Iranian fuel cycle assets - an example of an open-source analysis. Technical Report - Limited JRC 95228, Joint Research Centre
46. Atkinson M, Van der Goot E (2009) Near real time information mining in multilingual news. In: Proc. 18th international conference on World wide web, ACM, pp 1153–1154
47. Steinberger R, Pouliquen B, Van Der Goot E (2013) An introduction to the Europe media monitor family of applications. In: Proc. SIGIR 2009 Workshop - Information Access in a Multilingual World, Boston, Massachussetts, USA
48. Heuer RJ (1999) Psychology of intelligence analysis. Center for the Study of Intelligence
49. Lovins AB (1980) Nuclear weapons and power-reactor plutonium. Nature 283(5750):817–823
50. Pellaud B (2002) Proliferation aspects of plutonium recycling. Comptes Rendus Physique 3(7):1067–1079
51. Kessler G, Höbel W, Goel B, Seifritz W (2008) Potential nuclear explosive yield of reactor-grade plutonium using the disassembly theory of early reactor safety analysis. Nuclear Engineering and Design 238(12):3475–3499
52. Bathke CG, Ebbinghaus BB, Collins BA, Sleaford BW, Hase KR, Robel M, Wallace RK, Bradley KS, Ireland JR, Jarvinen GD, Bradley KS, Ireland JR, Johnson MW, Prichard AW, Smith BW (2012) The attractiveness of materials in advanced nuclear fuel cycles for various proliferation and theft scenarios. Nuclear Technology 179(1):5–30
53. Renis T, Yudin Y, Hori M (2014) Conducting acquisition path analysis for developing a state-level safeguards approach. In: Proceedings of the 55th INMM Annual Meeting
54. Kim LK, Renda G, Cojazzi GGM (2015) Methodological aspects on the State Level Concept and related Acquisition Path Analysis. In: Proceedings of the 37th ESARDA Annual Meeting
55. Panel on Reactor-Related Options for the Disposition of Excess Weapons Plutonium, Committee on International Security and Arms Control (1995) Management and disposition of excess weapons plutonium - reactor related options. Techn. rep., National Academy of Science
56. Kemp RS (2014) The nonproliferation emperor has no clothes. International Security 38:39–78. http://hdl.handle.net/1721.1/89182. Accessed 28 Feb 2019
57. Follath E, Stark H (2009) The story of "Operation Orchard": How Israel destroyed Syria's Al Kibar nuclear reactor. Spiegel Online International, http://tinyurl.com/kcdsjbh. Accessed 28 Feb 2019
58. AAVV (2008) Background briefing with senior U.S. officials on Syria's covert nuclear reactor and North Korea's involvement. http://www.dni.gov/files/documents/Newsroom/SpeechesandInterviews/20080424_interview.pdf. Accessed 28 Feb 2019
59. Broad WJ (2008) Syria rebuilds on site destroyed by Israeli bombs. http://www.nytimes.com/2008/01/12/world/middleeast/12syria.html?_r=0. Accessed 28 Feb 2019
60. Albright D, Brannan P (2008) The Al Kibar reactor: Extraordinary camouflage, troubling implications. Techn. rep., Institute for Science and International Security
61. Pabian FV (2012) Strengthened IAEA safeguards-imagery analysis: Geospatial tools for nonproliferation analysis. In: Nuclear Nonproliferation Safeguards and Security in the 21st Century, 2012-06-11/2012-06-29 (Brookhaven, New York, United States), LA-UR-12-24104
62. International Atomic Energy Agency (2011) Implementation of the NPT Safeguards Agreement in the Syrian Arab Republic (GOV/2011/30). GOV/2011/30, IAEA

Geospatial Information and Technologies for International Safeguards Verification

Erik Wolfart

Abstract International safeguards verification is based on the analysis of a large variety of information from different sources including state-declared information, inspection data, open source and third-party information. Geographic Information Systems (GIS) are used for the management and analysis of spatial information, such as satellite imagery and site maps declared under the Additional Protocol. They can also be utilized for the management of non-spatial safeguards information which often relates to a specific site, facility or location. Mobile geospatial technologies can facilitate on-site inspector tools such as positioning and navigation; location-based information retrieval; location-tagging of information for efficient post-inspection analysis; and Augmented Reality (AR) applications. These capabilities can put the inspector in a more pro-active and investigative role and can further integrate safeguards activities in the field and at headquarters. This chapter discusses the current and possible future use of geospatial information and technologies for safeguards analysis at the inspectorates' headquarters and illustrates the potential of mobile geospatial tools for on-site inspections.

1 Introduction

Geospatial information includes a reference to a coordinate system on the earth's surface (geographic information) or any other spatial component such as an address or a position in a local coordinate system. Geospatial information can be represented as vector data (e.g. a point location or the outline of an area) or as raster data (e.g. a geo-referenced satellite image).

Geographic Information Systems (GIS) allow the user to manage, analyze and visualize geographic information, e.g. create and plot maps, make spatial queries

E. Wolfart (✉)
EC Directorate General Joint Research Centre, Directorate G - Nuclear Safety and Security, Ispra, Italy
e-mail: erik.wolfart@ec.europa.eu

I. Niemeyer et al. (eds.), *Nuclear Non-proliferation and Arms Control Verification*,
https://doi.org/10.1007/978-3-030-29537-0_22

and conduct spatial analysis. They have been used for many years for professional mapping and analysis purposes in a wide variety of application areas [1]. *Geospatial technologies* is a term that emerged more recently with the popularization of geospatial information. It has a broader scope than GIS and includes any tool that can be used to acquire, visualize, manage and analyze geospatial information. Examples include inter alia: remote sensing analysis software; GPS and other sensors for acquiring location information; location-based services on mobile devices and web-based mapping and navigation tools.

Nuclear safeguards aims to verify that nuclear materials are not diverted from their declared peaceful use [2]. In order to verify the correctness and completeness of a State's safeguards declaration, the IAEA analyses all available safeguards-relevant information including State-declared information; information collected by the IAEA during on-site inspections; open source information including commercial satellite imagery and information received from third parties [3]. Integrating and connecting all disparate information pieces is essential for creating a complete picture of a State's nuclear activities and identifying any potential inconsistency in the available information set.

Over the last two decades, the adoption of the Additional Protocol and the advances of commercial satellite imagery have significantly increased the importance of geospatial information in nuclear safeguards verification. Also most non-spatial information related to nuclear activities refers to a specific country, site, facility or building. Therefore Geographic Information Systems are valuable tools for the integrated analysis of all safeguards-relevant information, spatial and non-spatial.

Whereas safeguards analysts at inspectorates' headquarters use geospatial information and technologies on a routine basis, they are less commonly applied during on-site inspections. Today's mobile consumer devices offer geospatial capabilities (such as localization; navigation; retrieval of location-based information; and geo-tagging), which could considerably increase the effectiveness and efficiency of on-site inspections. However, information security concerns need to be addressed before mobile, geospatial technologies can be adopted more widely during nuclear safeguards inspections.

Section 2 describes how safeguards analysts use geospatial information and technologies at inspectorates' headquarters and how it supports the integration of disparate information for an all-source information analysis. Section 3 shows the potential of mobile, location-based applications for supporting on-site inspections and Sect. 4 illustrates the challenges related to using such tools in nuclear safeguards. Section 5 provides a summary and draws conclusions.

2 Geospatial Information and Technologies for Nuclear Safeguards Analysis

Under the State-level concept (SLC), the IAEA implements safeguards in a manner that considers a State's nuclear activities and capabilities as a whole, within the scope of the State's safeguards agreement [2]. For the evaluation of a State's nuclear activities, the IAEA integrates and assesses all available information including State declarations, inspection results and open source information [3]. Geospatial information is an essential component of all-source safeguards analysis. In particular, the Additional Protocol and the availability of commercial satellite imagery have significantly increased the importance of geospatial information. The remainder of this section provides some examples for the use of geospatial information at IAEA headquarters.

2.1 Additional Protocol

The Additional Protocol (AP) was designed in order to strengthen the effectiveness and efficiency of the safeguards system for States having a Safeguards agreement with the IAEA [4]. It was negotiated in the late 1990s and is now being implemented by a majority of IAEA Member States. The AP requires that States declare additional information about their nuclear fuel cycle; inter alia they should provide a detailed map for each nuclear site including the layout and description of each building. The analysis of the site declaration significantly improves IAEA's understanding of the facility operation and is an essential part of the overall evaluation of a State's nuclear activities. However, the format of the site maps is not standardized and they are typically submitted to the IAEA as hardcopy or pdf format and therefore cannot be ingested and analyzed in a GIS environment directly. In order to maximize the benefit from AP site declarations and related geospatial analysis tools, the IAEA is currently proposing that the States submit the AP site declarations digitally in a standardized format [5].

2.2 Satellite Imagery

Since the early 2000s, satellite imagery has advanced dramatically in terms of capabilities and commercial availability. It is now an integral part of safeguards verification at the IAEA [6], inter alia for the verification of state declared information (e.g. of the site declarations submitted under the Additional Protocol); for preparing on-site visits; and for remotely monitoring sites that are difficult to access. It is also becoming increasingly relevant for the identification and verification of possible undeclared activities [7, 8]. See also Chaps. 23 and 24 for more details.

2.3 Open Source Information

Open source information is routinely used for nuclear safeguards analysis to support the state evaluation process. It comes in many different types and from different sources, such as news media; information from governmental sources or NGO's; scientific and technical literature; and trade data. Increasingly, open source information includes a geographic location, this is particularly the case for social media, e.g. social networks, photo sharing sites and microblogging, which typically is posted from a mobile device and often enriched with the corresponding GPS location. Although social media is not part of the core set of information sources for nuclear safeguards, it can prove useful in specific cases. For example, analyzing the spatial and temporal patterns of social network posts originating from a (nuclear) facility can help to detect and understand the activities on that site. See Chaps. 20 and 24 for more details.

2.4 Inspection Results

Inspection results remain the central source of information for verifying safeguards declarations. Analyzing the inspection results is often complex since large amounts of measurements of different types and from different periods have to be considered. In order to facilitate the management and analysis, the IAEA has started using geospatial tools, such as the Visual Environmental Sampling (Visual ES) tool [9]. Environmental sampling was introduced at the IAEA in the 1990s. It enables the IAEA to better understand the processes at nuclear facilities, e.g. by verifying the uranium enrichment levels in swipe samples coming from enrichment facilities. The samples are analyzed in specialized laboratories and the results are then evaluated at the IAEA headquarters. The number of samples can be very high for large facilities and the amount of data that accumulates over the years can become difficult to manage and understand. The IAEA developed the Visual ES in order to improve the effectiveness of environmental sample planning and the evaluation of results. Using Visual ES, the analyst can interactively create maps of the inspected facilities and plot the environmental samples according to their locations. The spatial mapping and visualization of the environmental samples facilitates the understanding of the facility operation and material flows, the sampling history and analytical results.

2.5 GIS-Based Information Integration and Analysis

As described above, Geographic Information Systems (GIS) are suitable tools for the integration of spatial and non-spatial information from multiple sources, multiple time-frames and resolutions. The concept of GIS-based information integration for nuclear safeguards was proposed and demonstrated in several research

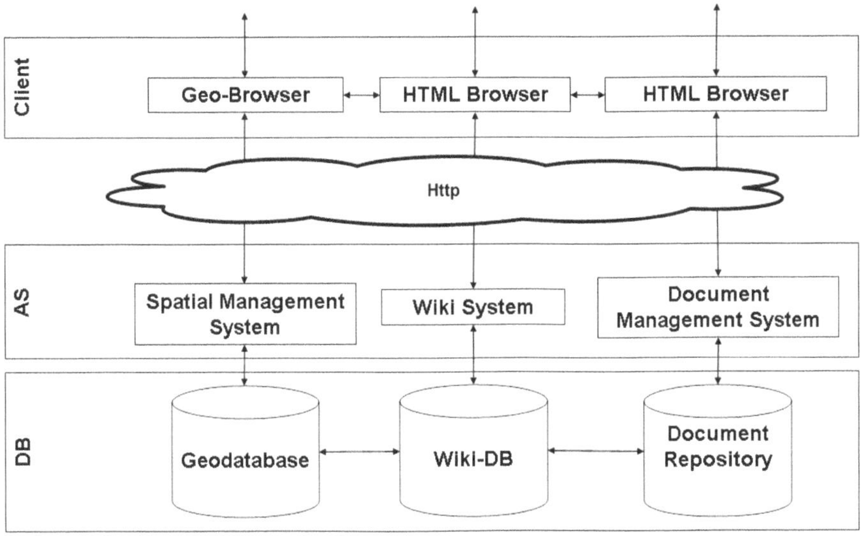

Fig. 1 High-level three-tier architecture of the integration platform proposed in the G-SEXTANT project: database (DB), application server (AS) and client. The system integrates a Geographic Information System (left), a Wiki (centre) and a document repository in order to support spatial and non-spatial information. Chart: European Commission Joint Research Centre

projects [10, 11]. The G-SEXTANT project proposed an integration platform where a GIS provides a central point of access to retrieve, view and analyze all available (spatial and non-spatial) information for a given nuclear site, including satellite imagery, GIS information, external databases and other information. It is based on a standard three-tier architecture (database, application server and web client) and incorporates three independent pillars each serving a particular purpose: a geographic information system, a Wiki system and a document repository. Figure 1 shows the high-level architecture of the system and Fig. 2 illustrates the spatial and non-spatial information that can be provided for a given site.

The European Commission explored the use of GIS-based information integration for Euratom safeguards and developed SIT-ES (Site Information Tool for European Safeguards), a Geo-Portal integrating information from diverse Safeguards-related internal databases and providing a single point of access to all relevant information. SIT-ES was mainly designed to support the management, analysis and verification of AP declarations, but also links to other types of information e.g. material accountancy, inspection planning and reports [12, 13].

The Geospatial Exploitation System (GES) is a GIS-based information platform developed and deployed at the IAEA. It initially focused on the dissemination of geospatial information (site maps, satellite imagery and analytical products) to internal users which are not GIS-experts, for example nuclear inspectors. The IAEA intends to widen the scope and make the GES a "*central element of an effective analysis environment, providing access to the majority of data within the*

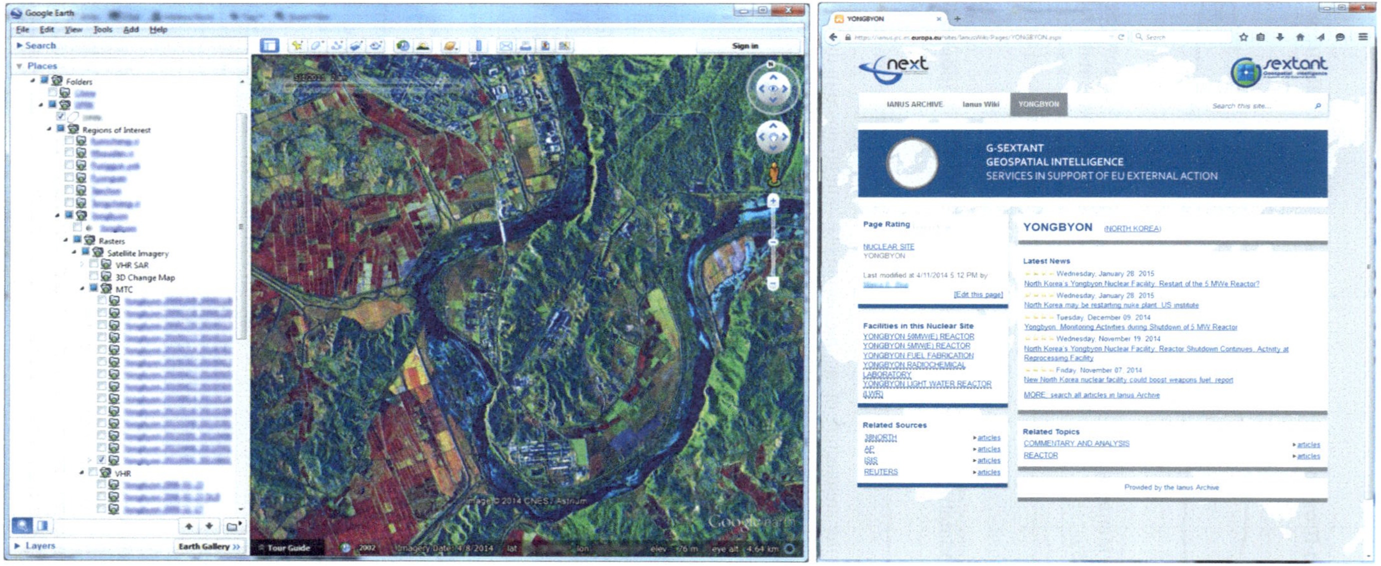

Fig. 2 Left: Geo-browser (Google Earth) which allows loading and viewing all spatial data (satellite images, vectors, processing results) available in the local database for a given nuclear site. Right: HTML browser to view the non-spatial data for the same nuclear site. The two interfaces are inter-linked to provide immediate access from one to the other. Reproduced from [11]

Department of Safeguards that have a spatial component" [14]. Visual ES has been integrated into the GES and can be launched directly from the browser. The concept is now being extended to the representation of destructive assay (DA), non-destructive assay (NDA), and material balance evaluation (MBE) results. Other examples of possible extensions include the seamless integration of site maps which are digitally submitted under the AP. Furthermore, the GES facilitates information sharing and collaborative information analysis inside the IAEA thus improving the effectiveness and efficiency in drawing safeguards conclusions [3].

3 Geospatial Information and Technologies for On-Site Inspections

This section describes the use of 3D spatial information for verifying design information and the absence of undeclared changes in nuclear facilities. It then illustrates how advanced localization technologies and location-based services could be deployed to increase the effectiveness and efficiency of on-site inspections.

3.1 3D Laser Scanning

3D laser scanners have been used in nuclear safeguards for several years in order to create as-built 3D models of nuclear facilities [15, 16]. The models are used for documenting the facility; for Design Information Verification (i.e. to verify that the facility layout corresponds to the design information provided by the operator); and for monitoring changes over time.

Traditional systems are based on high-accuracy, high-resolution 3D laser scanners mounted on a tripod during data acquisition. In order to completely cover a given area of interest, several scans are acquired in a so-called 'stop-and-go' mode, which are then registered into a single coordinate frame in an offline post-processing phase. The resulting 3D model is used to verify the correctness and completeness of the design drawings provided by the operator and is stored as a reference for subsequent visits. On return, the inspector re-scans the area of interest to verify that no undeclared modifications to the facility have occurred.

The recent availability of smaller and faster 3D laser scanners enabled the development of mobile scanning systems, for example mounted on a backpack. It is now possible to generate 3D maps on-the-fly while walking through a nuclear facility (Fig. 3). Mobile laser scanning systems significantly increase the efficiency of 3D laser scanning applied to nuclear safeguards verification. They also allow computing the position of the sensor with respect to the acquired map [17]. This technique, known as Simultaneous Localisation and Mapping (SLAM), can be used for location-based services in indoor environments as described below.

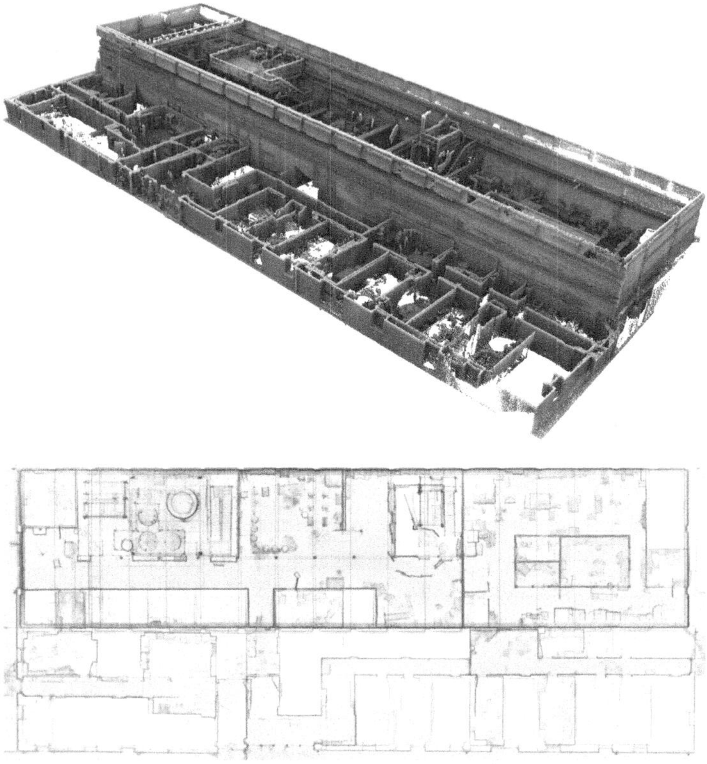

Fig. 3 3D model and corresponding floor plan of an industrial facility created with a mobile laser scanning system. The acquisition time was about 20 min and the processing was carried out automatically. The 3D model can be used for design information verification, localization and change detection. Image: European Commission Joint Research Centre

A combination of stop-and-go and mobile laser scanning has been used since 2014 for verifying the design information in the geological repository for spent nuclear fuel currently under construction in Finland. The repository was completely scanned with the stop-and-go approach to generate a highly accurate 3D model for the initial verification in 2014 and mobile laser scanning was used in subsequent DIVs to efficiently verify the absence of undeclared changes [18]. Figure 4 illustrates the use of mobile laser scanning for design information verification

3.2 Location-Based Services

The proliferation of mobile consumer devices equipped with GPS, photo cameras and other sensors has led to the wide-spread use of mobile geospatial applications, e.g. for navigation or location-based information retrieval [19]. Location-based

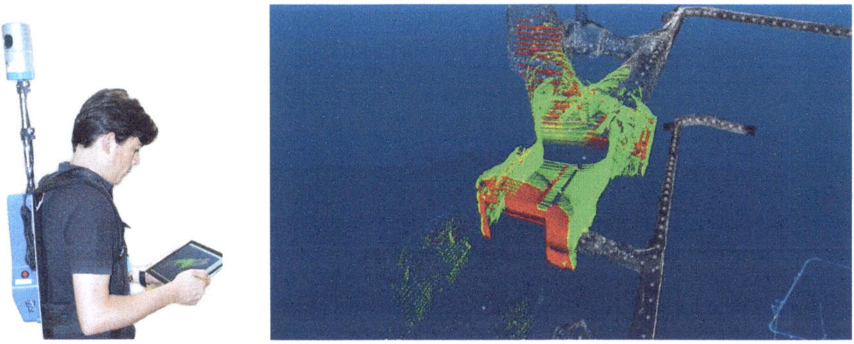

Fig. 4 Left: Mobile laser scanner mounted on a backpack. Right: Snapshot of the user interface as it is provided to the inspector in real-time. The reference model acquired a-priori is shown in grey. The real-time data acquired by the mobile scanner is color-coded as follows: green corresponds to objects that already existed in the reference model (i.e. the main tunnel excavation); red corresponds to changes (e.g. the fire door that was constructed later). The data was acquired during technology demonstrations in 2007 and 2014. Image: European Commission Joint Research Centre

services are not yet applied commonly in the context of nuclear safeguards because of information security concerns. However, nuclear inspectors could benefit from location-based services in several ways:

- The inspector can independently verify the current position and navigate within nuclear facilities.
- Location-based services can provide access to information that is contextual to the current position, for example to measurements, notes and observations acquired during previous inspections.
- Inspectors can be prompted to carry out specific tasks according to the current location.
- The inspector can tag measurements and observations taken during the inspection with the current position to facilitate later analysis of the data and increase the efficiency of follow-up inspections.

Current mobile devices typically use GPS sensors to acquire position information, i.e. they cannot be used for inspection activities inside nuclear facilities. Providing accurate indoor position information has been a challenging research topic in recent years and there is not yet a large-scale commercial deployment in this field. Different technologies are being explored, including inertial sensors, triangulation using radio signals and approaches based on optical or 3D sensors [20, 21]. Choosing between different technologies is a trade-off between accuracy, portability and cost of the system. In order to evaluate the state-of-the-art, IAEA's Technology Foresight programme hosted a workshop on indoor positioning systems in 2014 [22]. Following the positive outcome, the IAEA developed an indoor position system that helps the inspectors to document their activities and integrate records from different instruments used during in-field inspections [23].

4 Challenges for Applying Geospatial Technologies to Nuclear Safeguards

The use of geospatial information and technologies has the potential to further enhance safeguards effectiveness and efficiency in many ways. Nevertheless, there are several challenges that need to be addressed:

4.1 Information Security

Information security is essential to nuclear safeguards and the protection of state-declared and inspection data is of paramount importance. This imposes stringent security requirements on the information infrastructure and software solutions deployed by the inspectorates. At IAEA headquarters, sensitive information is stored on the isolated, highly secure Integrated Safeguards Environment (ISE) which excludes any access to the Internet, and additionally strict role-based access rules are applied on a need-to-know basis.

On the other hand, many modern geospatial tools rely on Internet access and cloud-based services and commercial GIS packages might not provide the fine-grained access control that is required by safeguards inspectorates. As a consequence, it is often difficult to use open source or commercial off-the-shelf software and the introduction of new information systems involve costly and time consuming customizations.

4.2 Legacy Infrastructure

The information infrastructure within the IAEA Department of Safeguards has evolved over decades and consists of a collection of many disparate systems, often based on different technologies. The improvement of the safeguards information infrastructures has been a continuous effort for several years [24]. Currently, IAEA's Modernization of the Safeguards Information Technology (MOSAIC) project aims to enhance the existing capabilities while strengthening the protection of safeguards information [25]. The integration of all safeguards information originating from many different applications and sources into a single GIS-based system remains challenging.

4.3 Safeguards Culture

Traditionally, nuclear safeguards is characterized by a high degree of compartmentalization and a lack of information sharing. Hence, all-source information analysis using GIS-based information integration requires not only the development of appropriate technical systems, but also a shift in working culture and working procedures.

4.4 *Operator Acceptance*

The use of advanced technologies during on-site visits needs to be accepted by the State Authority or site operator. For example, 3D mapping applications acquire detailed geometric information which the operator might not want to reveal and location-based services might transmit sensitive information to and from the headquarters' information systems. The operator will typically only accept advanced technologies if the information security can be ensured and if the introduction is of mutual benefit, e.g. if the inspector presence on site can be reduced.

5 Conclusions

Geospatial information plays an important role for international safeguards verification. Different types of geospatial information are being used, such as satellite imagery and site maps declared under the Additional Protocol. Also, most of the non-spatial safeguards information relates to a specific site, facility or building. Therefore, geospatial technologies play a central role for all-source safeguards analysis as it facilitates information management, integration and sharing. The importance for safeguards analysis will further increase in the future as more and better geospatial information will become available. The information infrastructures at the inspectorates will be modernized over time and allow to maximize the benefit that can be achieved from geospatial information and related technologies.

The use of geospatial technologies for in-field inspections is still limited. Recent developments in the area of location-based services provide an enormous potential in increasing the effectiveness and efficiency of safeguards inspections. They can (1) provide the inspector with all relevant information according to the current location, (2) enable close collaboration between the inspectors in the field and the analysts at the headquarters and (3) facilitate subsequent data analysis by associating all measurements and observations with the related location. The required indoor localization techniques are becoming more mature and are being introduced for nuclear safeguards inspections. However, the operational use does not only require the development of the technical systems, but also needs to address the issue of information security and confidentiality.

Current use of geospatial technologies for safeguards is focused on the site or facility level, i.e. it is used to map, integrate and analyse all available information of a specific nuclear site. However, modern geospatial tools work at all scales, i.e. the user can zoom out from the building level to the entire country or region. Therefore, geospatial technologies can support state-level analysis by mapping, visualizing and analyzing all nuclear facilities and activities within a country or relationships between countries (e.g. trade flows). Advanced analysis of geospatial information and satellite imagery might help to detect anomalies or undeclared nuclear activities.

This chapter illustrates the use of geospatial information and technologies for nuclear safeguards applications. The underlying concepts can also apply to the verification of other non-proliferation or disarmament agreements: facilities and activities can be mapped geographically; different types of spatial and non-spatial information need to be evaluated and often can be associated with a location; geographically visualizing all information at different scales supports integrated information analysis; and the use of geospatial technologies during on-site inspections increases the effectiveness and efficiency and facilitates the cooperation between on-site inspectors and headquarter analysts.

Acknowledgements The author would like to thank the colleagues that supported the writing of this chapter, in particular Giacomo Cojazzi, Guido Renda, Vitor Sequeira, Simone Ceriani, Carlos Sanchez and Pierluigi Taddei. Their work and discussions with them contributed significantly to the chapter. Any incorrectness or mistake is the authors' sole responsibility. The activity is funded within the European Commission (EC) Euratom Horizon 2020 Research and Training Programme. The views expressed in this chapter are those of the author and do not necessarily represent the official views of the European Commission.

References

1. Burrough PA, McDonnell RA, Lloyd CD (2015) Principles of Geographical Information Systems - Third Edition, Oxford: Oxford University Press
2. IAEA (2015) IAEA Safeguards Serving Nuclear Non-Proliferation, IAEA, Vienna
3. Ferguson M, Norman C (2010) All-Source Information Acquisition and Analysis in the IAEA Department of Safeguards. In: Proc. International Safeguards Symposium - Preparing for Future Verification Challenges, Vienna
4. Additional Protocol. https://www.iaea.org/topics/additional-protocol. Accessed 28 Feb 2019
5. Rutkowski J, Keskinen A, Balter E, Steinmaus K, Rialhe A, Idinger J, Nussbaum S (2014) Digital Declarations: The Provision of Site Maps under INFCIRC/540 Article 2.a.(iii). In: Proc. Symposium on International Safeguards, Vienna
6. Pabian F (2015) Commercial Satellite Imagery as an Evolving Open-Source Verification Technology: Emerging Trends and Their Impact for Nuclear Nonproliferation Analysis. EUR27687, Publications Office of the European Union
7. IAEA (2007) IAEA Safeguards: Staying Ahead of the Game
8. Vienna Centre for Disarmament and Non-Proliferation (2016) Emerging Satellites for Non-Proliferation and Disarmament Verification. Vienna
9. Vilece K, Norman C, Baute J, Giaveri G, Kiryu M, Pellechi M (2012) Visualization of Environmental Sampling Results at Inspected Facilities. In: Proc. Institute of Nuclear Materials Management 53rd Annual Meeting, Orlando, USA
10. Wolfart E, Goncalves J, Ussorio A, Gutjahr KH, Niemeyer I, Loreaux P, Marpu P, Listner C (2009) Integrated Analysis of Satellite Imagery for Treaty Monitoring - The LIMES Experience. ESARDA Bulletin 43
11. Niemeyer I, Listner C, Canty MJ, Wolfart E, Lagrange JM (2014) Integrated Analysis of Satellite Imagery for Nuclear Monitoring - Results from G-SEXTANT. In. Proc. Institute of Nuclear Materials Management 55th Annual Meeting, Atlanta, USA

12. Wolfart E, Bril LV, Bellezza F, Contini S, Ussorio A, Mazza F (2006) A geo-portal for data management and integration in the context of the additional protocol.In: Proc. Symposium on international safeguards: Addressing verification challenges, Vienna, 2006.
13. Tsalas S, Ciccarello S, Goncalves J, Isoard O, Mazza F, Wolfart E (2008) GIS-based Information Management and Integration for EURATOM Safeguards. In: Proc. Institute of Nuclear Materials Management 49th Annual Meeting, Nashville, USA
14. Steinmaus K, Norman C, Ferguson M, Rialhe A, Baute J (2013) The Role of the Geospatial Exploration System in Integrating All-Source Analysis. In: Proc. Institute of Nuclear Materials Management 54th Annual Meeting, Palm Desert, California
15. Agboraw E, Johnson S, Creusot C, Poirier S, Saukkonen H, Chesnay B, Sequeira V (2006) IAEA experience using the 3-Dimensional Laser Range Finder. In: Proc. IAEA Safeguards Symposium: Addressing Verification Challenges, Vienna
16. Chare P, Lahogue Y, Schwalbach P, Smejkal A, Patel B (2011) Safeguards By Design – As applied to the Sellafield Product and Residue Store (SPRS). ESARDA Bulletin 46:72–78
17. Sánchez C, Taddei p, Ceriani S, Wolfart E, Sequeira V (2016) Localization and tracking in known large environments using portable real-time 3D sensors. Computer Vision and Image Understanding 149:197–208
18. Wolfart E, Ceriani S, Puig D, Sánches C, Sequeira V, Murtezi M, Turzak P, Zein A, Enkhjin L, Ingenieri M, Rochhi S, Yudin Y (2015) Mobile Laser Scanning for Nuclear Safeguards. ESARDA Bulletin 53:62–72
19. UN-CGIM (2015) Future trends in geospatial information management: the five to ten year vision - Second Edition
20. Mainetti L, Patrono P, Sergi I (2014) A Survey on Indoor Positioning Systems. IEEE SoftCOM, Split
21. Harle R (2013) A Survey of Indoor Inertial Positioning Systems for Pedestrians. IEEE Communications Surveys and Tutorials 15(3)
22. Finker D, Kocjan J, Rutkowski J, Cai R (2014) Evaluation of an Autonomous Navigation and Positioning System for IAEA Safeguards. In: Proc. Symposium on International Safeguards - Linking Strategy, Implementation and People, Vienna
23. IAEA (2016) New Indoor Positioning Technology to Improve Efficiency and Accuracy of Safeguards Inspections. https://www.iaea.org/newscenter/news/new-indoor-positioning-technology-to-improve-efficiency-and-accuracy-of-safeguards-inspections. Accessed 28 Feb 2019
24. Whitaker G, Becar JM, Ifyland N, Kirkgoeze R, Koevesd G, Szamosi L (2006) The IAEA Safeguards Information System Re-Engineering Project (IRP). In: Proc. International Safeguards Symposium on Addressing Verification Challenges, Vienna
25. IAEA (2015) Enhancing Effective Nuclear Verification: Upgrading IAEA Safeguards Capabilities. https://www.iaea.org/newscenter/news/enhancing-effective-nuclear-verificatio-upgrading-iaea-safeguards-capabilities. Accessed 28 Feb 2019

Remote Sensing Data Processing and Analysis Techniques for Nuclear Non-proliferation

Joshua Rutkowski and Irmgard Niemeyer

Abstract Remote sensing offers analysts the ability to observe buildings and structures within nuclear facilities without visiting the localities in person. This chapter highlights some of the modern methods and technologies related to remote sensing data as it relates to nuclear non-proliferation and arms control verification. Commercially available techniques are discussed and examples related to the nuclear fuel cycle are used to illustrate the application for nuclear non-proliferation and arms control practices used by analysts. This chapter explains the sources of remote sensing data, how the data is stored and how it can be processed. Specific processing techniques including supervised and unsupervised classification, pixel and object oriented classifications as well as change detection methods are examined through the lens of non-proliferation and arms control studies. Throughout the chapter, common use case scenarios are provided in order to give the reader a better idea of how the topics are applied to verification activities. These paragraphs provide examples where the section's topic is applicable in a non-proliferation or arms verification scenario.

1 Introduction

Since the launch of the IKONOS sensor in 1999, people outside national governments could obtain high resolution satellite imagery. The availability of such datasets meant that analysts working on non-proliferation and arms control issues could monitor activities where they did not have ground access. The IKONOS sensor was the first commercially available sensor which imaged the earth at a spatial resolution acceptable for discerning key features at the building level. The International Atomic Energy Agency's Board of Governors found, in 1995, that the use of commercial satellite imagery may improve the effectiveness of

J. Rutkowski (✉) · I. Niemeyer
Forschungszentrum Jülich GmbH, Jülich, Germany
e-mail: jerutko@sandia.gov; i.niemeyer@fz-juelich.de

I. Niemeyer et al. (eds.), *Nuclear Non-proliferation and Arms Control Verification*,
https://doi.org/10.1007/978-3-030-29537-0_23

Safeguards and therefore permitted its use by the Department of Safeguards to aid in carrying out their activities [1]. The Comprehensive Test Ban Treaty Organization documented the utility of remote sensing information when conducting their integrated field exercise in 2014 and continues to evaluate potential applications of earth observation data for their activities [2]. Research organizations like the Institute for Science and International Security[1] or 38 North[2] regularly cite satellite imagery when analyzing current events related to nuclear facilities. Remote sensing offers analysts the ability to observe buildings and structures within nuclear facilities without visiting the localities. The option to observe without physical access benefits many interested parties since some organizations will never have physical access, some may only have one-time access and some only limited access to the area of interest. Hinderstein highlights the need for imagery analysis to be an integral part of the verification process and calls for an 'International Satellite Verification Agency' (page xix) to fully exploit satellite imagery for verification purposes [3]. This chapter highlights some the modern methods and technologies related to remote sensing data as it relates to nuclear non-proliferation and arms control verification. The chapter explains the sources of remote sensing data, how the data is stored and how it can be processed. Specific processing techniques including supervised and unsupervised classification, pixel and object oriented classifications as well as change detection methods are examined through the lens of non-proliferation and arms control studies.

Remote sensing studies relate to many different aspects of our society. This chapter will illustrate the utility of remote sensing by using examples related to the nuclear fuel cycle. Throughout the chapter, common use case scenarios are provided in order to give the reader a better idea of how the topics are applied to verification activities. These paragraphs provide examples where the section's topic is applicable in a non-proliferation or arms verification scenario. The format of these examples will be to first explain a situation and follow it with an explanation of how remote sensing might offer insight into the situation.

2 Data Sources

2.1 Electro Optical Data

Kerekes explains how "the main function of electro-optical imaging sensors is to collect incident electromagnetic (EM) radiation and convert it to a stored representation useful for remote sensing analysis. These sensors operate in the optical region of the EM spectrum traditionally defined as radiation with wavelengths between 0.4 and 15 μm." [4]

[1]https://isis-online.org/satellite-imagery/. Accessed 28 Nov 2018.

[2]https://38north.org/. Accessed 28 Nov 2018.

Atmospheric transmission and absorption greatly affect the ability of any sensor to collect information about the earth. The specific wavelengths within the electromagnetic spectrum that are observable to satellite borne sensors are well understood and therefore the majority of earth observation satellites collect wavelengths in regions that have the highest potential for information to be collected by the sensor. These areas include visible wavelengths, near-infrared, thermal and radio wavelengths. The visible and near infrared (NIR) wavelengths are very common for image analysis since they are the easiest for humans to visually interpret as they closely match the human eye's own collection wavelengths. Commercially available sensors with high spatial resolution, like the Pleiades or Worldview constellations, collect information in these bands. Some of these sensors, such as the Worldview-3 sensor, also collect information in the shortwave infrared while others, like Kompsat 3A, collect in the thermal wavelengths. For a more in-depth review of such a sensor, please refer to chapter *Commercial Satellite Imagery: an Evolving Tool in the Non-proliferation Verification and Monitoring Toolkit* by Frank Pabian et al. in this book.

Situation *An organization wants to acquire remote sensing information over a large nuclear fuel cycle complex in a foreign country where the organization does not have ground access to the facility.*

Use *Analysts must carefully review their requirements before acquiring the remote sensing datasets since each dataset costs time and resources to process, and of course, money. While the analysts are not sure what they will observe beforehand, they must have a good idea about the facility operations in order to determine temporal and spatial resolutions of the dataset to satisfy the requirements of the investigation. To do so, the analysts will need to be in contact with various subject matter experts knowledgeable about the type of facility to determine the optimal spatial and spectral resolution for the facility as well as the best time of day to capture the datasets.*

2.2 Synthetic Aperture Radar (SAR) Data

Synthetic aperture radar (SAR) is a valuable active sensor type since it penetrates most cloud cover and offers a different set of information for interpretation compared to optical sensors. SAR data, as described later in this chapter, requires a different set of processing techniques and demands a different approach for processing and analysis compared to the other earth observation (EO) sensors mentioned above. Space borne radar sensors are routinely used for earth observation but have limited use for non-proliferation and arms control due to their complex data structure and interpretation. The frequency bands generally used for these activities are the X, C and L bands. Some of the commonly used SAR sensors include: TerraSAR-X, TanDEM-X, COSMOS-Skymed, Radarsat, and Sentinel-1 SAR. These sensors have well documented processing workflow, see e.g. [5–8]. SAR sensors offer the option to change resolution (both range and azimuth) which might offer benefits to an analyst since the area of interest is can be observed through more flexible means compared to electro-optical sensors since there are more parameters to adjust. SAR data, however, requires more sophisticated processing and training for interpretation.

3 Data Storage Formats

Data sent from the satellite borne sensors to the ground stations are initially converted to output formats by the satellite vendor or data provider in order to be used by satellite imagery analysts. When delivered to the customer, the data have traditionally been delivered in standard raster data file formats such as Geo-Tagged Image File Format (Geo-TIFF) or JPEG 2000. Such file formats are able to contain critical geographical information about the imagery and in some cases, valuable metadata information about the sensor during time of image acquisition. The large size of these files may pose problems for some users because of limited space when exchanging files among analysts or when storing files on a disk. Image compression techniques have been developed to reduce the file size with minimized loss of data quality. Oftentimes, image compression may come with tradeoffs for image quality in terms of pixel bit depth and quality degradation. Proprietary image compression techniques, such as the ERDAS Enhance Compression Wavelet (.ecw) are regularly employed in larger image processing organizations.

Delivering remote sensing data to desktop clients via web services has been gaining popularity in the past several years. With the launch of Keyhole, later to become Google Earth, the ability to stream tiles of data as a bundle of pixels across the internet meant that users no longer have to maintain large datasets locally if they wanted to use satellite imagery on their desktops. Many web services now exist and are typically provided in standardized formats agnostic to software packages or operating systems. The Open Geospatial Consortium (OGC) is the leader in defining clear protocols for geospatial data. The OGC works with governments, businesses, international organizations and non-profits to ensure that various IT systems, including image analysis software packages, in defining its mission "to advance the development and use of international standards and supporting services that promote geospatial interoperability. To accomplish this mission, OGC serves as the global forum for the collaboration of geospatial data/solution providers and users." [9] This means that server-side software is able to provide the satellite imagery to users without knowledge of client-side applications. Users and analysts consuming such data streams have a wide selection of options for controlling the data streams. OGC standards, such as the Web Coverage Services and Web Map Services, offer different benefits to the end user based on individual requirements. The OGC continually works to expand and standardize web services to improve performance and compatibility with emerging technologies. Some image analysis software packages now permit a wide selection of streaming datasets, including background imagery from Bing Maps,[3] Google Earth[4] and Open Street Maps.[5]

Situation *A large scale inspection team is deployed in the field to address specific concerns about the testing of a nuclear weapon in a remote area of a large country.*

Use *Streaming satellite imagery through web services permits the headquarters to transmit time sensitive data to field technicians and inspectors based on initial feedback from the inspection team after arriving on site. The upload and download data transmission rates are low due to limited communication. The data streams allow in-field users to restrict downloads to an optimized format of the specific areas of interest without compromising data quality. Through secure web connections, the in-field staff are able to directly view recently acquired and pre-processed satellite images from the inspection headquarters.*

4 Pre-processing Remote Sensing Data

Pre-processing remote sensing data describes the steps required during post acquisition but before the image is analyzed. The techniques prepare the data for general processing and analysis [10]. Before analysts use remote sensing datasets for their work, certain pre-processing techniques may be applied in order to increase quality and enhance data accuracy. Atmospheric and radiometric corrections are often performed by the data suppliers because they often have more information for pre-processing inputs, such as internal sensor calibration parameters and precise weather conditions at the time of acquisition. Many data providers offer the datasets at different processing levels based on the needs of the customer. Additional processing techniques may be applied by analysts in order to further enhance the datasets but require advanced software and knowledge.

Geometric distortions inherent with remote sensing data acquisition may also be corrected during pre-processing. For instance, orthorectification of imagery is critical when performing accurate ground measurements because the process eliminates distortions created from terrain relief and sensor acquisition perspective. Co-registration, another technique commonly employed by the provider before analysts receive imagery, geographically aligns multiple datasets such that observable features are spatially aligned among the data. Accurate temporal analysis relies heavily on such pre-processing workflows.

[3]https://www.bingmapsportal.com/isdk/ajaxv7#CreateMap1. Accessed 28 Nov 2018.

[4]https://www.google.com/earth/. Accessed 28 Nov 2018.

[5]http://wiki.openstreetmap.org/wiki/WMS. Accessed 28 Nov 2018.

Situation *An analyst is monitoring a facility by reviewing information provided by the facility, including engineering plots and facility maps. The analyst has access to several different high resolution satellite images over the area of interest.*

Use *By orthorectifying the satellite imagery, the analyst can confidently georeference the engineering plots and facility maps to a geographic coordinate system so the data may be visualized in layers within the analyst's exploitation software. The geographically co-registered datasets will save the analyst time and increase their utility for comparisons to other datasets.*

5 Processing Remote Sensing Data

5.1 Enhancements

Enhancing optical imagery for analysis oftentimes includes the sharpening of imagery whereby the panchromatic dataset's high spatial resolution is combined with each multi-spectral band's high spectral resolution to result in a so-called pan-sharpened image. The results vary according to the processing technique. An example study comparing the techniques details the results from processing a WorldView-2 scene using common imagery processing software [11]. Many proprietary algorithms like the Gram-Schmidt technique may be improved through changes in the parameters or through image preparation [12]. FuzeGo[6] and NNDiffuse[7] are newer techniques offering faster processing times and potentially better results than more traditional methods. Sensors with a unique set of bands, such as the Worldview-2 sensor, require specific pan-sharpening techniques [13]. While many common pan-sharpening techniques, like Intensity, Hue and Saturation (IHS) or Resolution Merge, are offered in various software processing packages, the calculation methods, resampling techniques and output options have a major impact on the resulting rasters. For this reason, analysts must take extra care when choosing the pan sharpening techniques for their own purpose. The ultimate decision for processing techniques lies in the desired result and may depend on the local land cover for the area of interest. For example, jungle canopy cover or bright desert sand may skew the algorithm and have a direct impact on other pixels in the image. Because each pan-sharpening algorithm computes the weighting between the spatial and spectral information differently, the outcome naturally differs too. On-the-fly pan-sharpening techniques are now available in some analyst software packages and reduce the time from data delivery to interpretation but come with major drawbacks

[6]http://www.fuzego.com/. Accessed 28 Nov 2018.

[7]http://www.exelisvis.com/docs/ENVINNDiffusePanSharpeningRaster.html. Accessed 28 Nov 2018.

such as limitations on the available sharpening methods and parameters. The on-the-fly techniques do offer advantages in time and disc space, however.

Enhancements of SAR imagery highly depends on the subsequent processing technique, e.g. radargrammetry, interferometry, polarimetry, etc., and may consist of speckle filtering and image segmentation [14].

Situation *An analyst is verifying the State declaration for a nuclear facility located in a tropical area with large canopy cover.*

Use *Panchromatically sharpening a Worldview-3 image will assist the analyst in discerning natural and human-created observable features that may be difficult to separate in higher resolution panchromatic data or lower resolution multi-spectral data.*

5.2 DEM Generation

A digital elevation model (DEM) is the general term used to describe earth surface or terrain models used within geographic information systems. They offer a very important source of data for remote sensing applications. They greatly increase accuracy when pre-processing geometric distortions in imagery. Additionally, such models permit for accurate 3D measurement and line-of-site analysis. The German Aerospace Institute (DLR) [15] demonstrated the ability to derive accurate elevation models from EO stereo pairs. Given the ability of some sensors to create triplet data acquisitions, the accuracy of the derived elevation models is even greater. Certain parts of treaty verification, especially design information verification and building-level descriptions, are possible using elevation models derived from satellite imagery. DLR demonstrated the utility of accurate volume measurements based on elevation models created from EO datasets [16]. Volume calculations assist in verification of mining activities and are valuable inputs to change detection analysis.

Radar techniques, including radargrammetry and interferometry, have been employed for over a decade to create accurate elevation models [17]. The derived elevation datasets offer analysts similar levels of detail compared to the optical counterparts [18]. The TanDEM-X mission released a worldwide elevation dataset at a higher resolution and accuracy than ever available to the scientific community.[8]

Situation *A large open-pit uranium mine reports their monthly rates of excavation on their website.*

Use *An analyst wants to calculate the monthly rate of excavation based on satellite imagery by generating high resolution digital elevation models at the beginning and end of the month. Afterward, the analyst performs a volume difference calculation between the resulting elevation models to verify the company's monthly report.*

[8]http://www.dlr.de/hr/desktopdefault.aspx/tabid-2317/3669_read-5488/.

5.3 Classification

Image classification is the designation of pixels within a remote sensing dataset where each pixel is identified as belonging to a specific pre-defined group. While visual interpretation is a common classification method used by analysts viewing satellite imagery, it does little to assist in further computations or geographic studies because the classification is made using annotations and therefore cannot be readily retrieved for further computation. Image classification, performed by digital processors, allows analysts to estimate land cover and proceed with further analysis on the resulting analysis. This section briefly explains the use of supervised and unsupervised classification methods as well the more advanced strategies of pixel- and object-based classification methods. Jensen [22] notes that ancillary data used for assistance in determining image classes, datasets such as socio-economic information or soil maps, are very helpful. For treaty verification, site plans of the facilities and neighboring environs are valuable.

Unsupervised vs. Supervised Classification methods based on a digital number for each pixel within a raster dataset typically fall into one of two initial categories—supervised or unsupervised. Supervised classification refers to the analyst as the 'supervisor' whereby the classification is initiated with some input from an analyst. The input guides the classification algorithm by supplying more information about the raster data than the digital numbers. Some examples of data input include the automatic or manual selection of representative training classes. Unsupervised classification is computed by using only the digital number values within the raster dataset and relying on mathematical or statistical calculations to derive a classification of the dataset. The analyst usually limits their input to the number of classification bins or groups. Supervised and unsupervised classification methods are regularly used, oftentimes as subroutines in a more elaborate analysis workflow. Both supervised and unsupervised classification methods are available in image processing software packages. For example, simple unsupervised classification methods are freely available for distinguishing agricultural and built-up areas within a large satellite image scene.

Situation *A new report alleges that a new nuclear facility has been constructed in an area of a country. The location of the facility is vague.*

Use *An analyst uses an unsupervised classification algorithm to determine newly built up areas in the region. Since the location of facility is unknown, the classification will assist the analyst by focusing their attention to areas where the land cover has recently changed.*

Pixel-Based vs. Object-Based With the improving spatial resolution and the amount of details visible on remote sensing data, new algorithms for analyzing very high resolution satellite imagery with regard to safeguards verification are also needed. Object-based approaches show promises against pixel-based one as they try to imitate the image understanding of an image analyst, see e.g. [19, 20].

An image analyst can easily identify image regions along with their colour, shape, texture and context, and categorize them into objects of interest, such as buildings, streets, forests etc. Computer-driven object-based image analysis is an approximation to human perception. Using specific object features (e.g. colour, shape, texture or context) of defined object classes through rule bases, each object can be assigned to an object class. The selection of the optimum features can either be performed automatically by applying statistics to a set of samples for each class or interactively by integrating knowledge on the object classes in the recognition process. Object-based image analysis approaches are far from being able to perform an image interpretation on its own, however, they offer some time and cost effective procedures for object recognition and feature extraction, especially when using very high resolution satellite imagery [21].

5.4 Change Detection

The changes in a specific geographic area may be detected by comparing two or more datasets covering the area of interest. The image acquisition time and date are usually provided for each dataset so detecting change between two or more scenes is possible. Visual interpretation of multiple datasets is reasonable for localized areas and has been proven to be an effective method if the area under study has been reviewed in the past. When several scenes are under consideration or if the area is very large, automated change detection offers faster methods for discovering the majority of changes. Because different sensors collect data at different spectral and spatial resolutions, comparing two or more scenes demands pre-processing preparation in order to normalize the data before analysis. One option for analysts is to first classify each image using the techniques mentioned in the previous section and then compare the resulting classification to determine change. The desired result of a change detection analysis is to determine the geographic areas which have undergone some type of change. The results are usually reported in a new raster dataset, a set of polygon vectors or an annotated image. Many image analysis software packages contain change detection tools which perform sophisticated, proprietary methods to detect change. For electro-optical datasets, each scene is unique because the sun's illumination affects the amount of energy received by the sensor. Such changes in illumination are difficult to normalize between scenes, especially when the geometric distortions from each scene cannot be fully adjusted.

Change detection oftentimes begins with two images for comparison. This bi-temporal change detection may be compared using visual analysis or using an algorithm to highlight spectral changes between the two images. Most image analysts now have access to many different sensors that produce large amounts of imagery. In the past few years, several companies started launching small satellite constellations capable of collecting several images over one area of interest per day.

Haddal et al. details a new public-private research partnership in the USA that will spend the next 3 years investigating this situation and potential benefits the small satellites may have on imagery analysis for nuclear non-proliferations. Integrating multiple datasets, either derived from satellite imagery, such as digital elevation models [22–24] or incorporating newsfeeds or other open sources [25] will prove to be a new challenge for the coming decade. Applying advanced change detection methods within the Google Earth platform offers a new way to use the increasing amount of freely accessible satellite imagery data for monitoring nuclear sites and activities [26].

The new deluge of datasets will also pose specific issues to organizations like the International Atomic Energy Agency as their mission is to monitor so many nuclear facilities worldwide. Warner et al. describe a possible path forward for such institutions to incorporate the new sensors and vastly larger amount of remote sensing data into their analysis lifecycles [27].

Situation *A complex nuclear facility has undergone some change since monitored in the past six months but the analyst has no supporting documentation about what parts of the facility were modified.*

Use *The analyst compares two images from the same sensor so that the spectral and spatial resolutions of the datasets are similar. By acquiring one image from an acquisition date before the changes occurred and one image from a recent acquisition, a change detection algorithm can alert the analyst to areas of possible change so they can investigate further.*

6 Conclusion

Since the first high spatial resolution datasets from IKONOS to the present hundreds of small satellite collecting imagery daily around the globe, remote sensing has proven to be valuable in assisting remote sensing analysts to investigate nuclear facilities worldwide. At the same time, also SAR imagery has gained in importance for This chapter briefly introduced the how remote sensing is for non-proliferation and disarmament analysis then discussed the data sources, how the data is stored and processed and finally some of the emerging techniques and workflows that are emerging in the field. The selection, acquisition, pre-processing, analysis and dissemination of information derived from remote sensing data will continue to evolve as new data sources become available and new techniques are created to process and interpret the data.

References

1. International Atomic Energy Agency (1995) Strengthening the Effectiveness and Improving the Efficiency of the Safeguards System. GOV/2807
2. Comprehensive Test Ban Treaty Organization (2014) Largest-ever CTBT on-site-inspection exercise concludes successfully. https://www.ctbto.org/press-centre/press-releases/2014/largest-ever-ctbt-on-site-inspection-exercise-concludes-successfully/. Accessed 28 Feb 2019
3. Hinderstein C (2010) Cultivating Confidence: Verification, Monitoring and Enforcement for a World Free of Nuclear Weapons. Hoover Press
4. Kerekes JP (2009) Optical Sensor Technology. In: Warner TA, Foody GM, Nellis MD (eds) The SAGE handbook of remote sensing. Sage Publications, pp 95–107
5. AIRBUS (2015) TerraSAR-X Application Guide. https://www.intelligence-airbusds.com/en/6583-terrasar-x-application-guide. Accessed 28 Feb 2019
6. Italian Space Agency (2016): COSMO-SkyMed. Mission and Products Description. https://www.asi.it/sites/default/files/attach/bandi/cosmo-skymed_mission_and_products_description_update.pdf Accessed 28 Feb 2019
7. Maxar Technologies (2018) RADARSAT-2 Product Description. https://mdacorporation.com/docs/default-source/technical-documents/geospatial-services/52-1238_rs2_product_description.pdf?sfvrsn=10. Accessed 28 Feb 2019
8. European Space Agency (2018) Sentinel-1 SAR Technical Guide. https://sentinel.esa.int/web/sentinel/technical-guides/sentinel-1-sar. Accessed 28 Feb 2019
9. Open Geospatial Consortium (2016) About OGC. Open Geospatial Consortium. http://www.opengeospatial.org/ogc. Accessed 28 Feb 2019
10. Van Der Meer F, de Jong SM (2001) Imaging Spectrometry: Basic Principles and Prospective Applications, vol 1., Springer Science & Business Media,
11. Li H, Jing L, Tang Y, Liu Q, Ding H, Sun Z, Chen Y (2015) Assessment of pan-sharpening methods applied to WorldView-2 image fusion. In: Proc. IEEE Geoscience and Remote Sensing Symposium (IGARSS) 2015, pp 3302–3305
12. Aiazzi B, Baronti S, Selva M, Alparone L (2006) Enhanced Gram-Schmidt spectral sharpening based on multivariate regression of MS and pan data. In: Proc. IEEE Geoscience and Remote Sensing Symposium (IGARSS) 2016, pp 3806–3809
13. Padwick C, Deskevich M, Pacifici F, Smallwood S (2010) WorldView-2 pan-sharpening. In: Proc. of the ASPRS 2010 Annual Conference, San Diego, CA, USA, 2010.
14. Richards JA (2009) Remote Sensing with Imaging Radar. Signals and Communication Technology. Springer, Berlin Heidelberg
15. d'Angelo P, Lehner M, Krauss T, Hoja D, Reinartz P (2008) Towards automated DEM generation from high resolution stereo satellite images. In: The International Archives of the Photogrammetry, Remote Sensing and Spatial Information Sciences, International Society for Photogrammetry and Remote Sensing, XXXVII (B4), pp 1137–1342.
16. d'Angelo P, Máttyus G, Reinartz P (2015) Skybox image and video product evaluation. International Journal of Image and Data Fusion:1–16
17. Crosetto M, Pérez Aragues F (2000) Radargrammetry and SAR Interferometry for DEM Generation: Validation and Data Fusion. In: SAR workshop: CEOS Committee on Earth Observation Satellites, 2000. p 367
18. Minet C, Eineder M, Niemeyer I (2011) High Resolution Radar Satellite Imagery Analysis for Safeguards Applications. ESARDA Bulletin 46:57–64
19. Nussbaum S, Menz G (2008) Object-Based Image Analysis and Treaty Verification: New Approaches in Remote Sensing - Applied to Nuclear Facilities in Iran. Springer, Berlin
20. Listner C, Niemeyer (2011) Object-based Change Detection. Photogrammetrie, Fernerkundung, Geoinformation 4:233–245
21. Nussbaum S, Tueshaus J, Niemeyer I (2013) Images Objects vs. Pixels: A comparison of new methods from both domains. ESARDA Bulletin 49:66–74

22. Tian J, Chaabouni-Chouayakh H, Reinartz P, Krauß T, d'Angelo P (2010) Automatic 3D change detection based on optical satellite stereo imagery. In: Proc. ISPRS TC VII Symposium 2010, Vienna, July 5–7, 2010, XXXVIII (7B), pp 586–591
23. d'Angelo P, Rossi C, Minet C, Eineder, M, Flory M, Niemeyer I (2014) High Resolution 3D Earth Observation Data Analysis for Safeguards Activities. In: Proc. Symposium on International Safeguards: Linking Strategy, Implementation and People, Vienna, Austria, November 2014
24. d'Angelo P, Rossi C, Eineder, M, Rutkowski J (2017) High Resolution 3D Earth Observation Data Analysis for Safeguards Activities. In: Proc. ESARDA Symposium 2017, Düsseldorf.
25. Haddal R, Ohlinger J, Smart H, Lee D, Puccioni A (2018) Detection via Persistence: Leveraging Commercial Imagery from Small Satellites. In: Proc. Symposium on International Safeguards: Building Future Safeguards Capabilities, Vienna, Austria, November 2018
26. Rutkowski J, Canty MJ, Nielsen AA (2018) Site monitoring with Sentinel-1 dual polarization SAR imagery using Google Earth Engine. Journal of Nuclear Materials Management XLVI(3):48–59
27. Warner T, Rutkowski J, Keskinen A, Duckworth S (2018) Exploitation of high frequency acquisition of imagery from satellite constellations within a semi-automated change detection framework for IAEA safeguards purposes. In: Proc. Symposium on International Safeguards: Building Future Safeguards Capabilities, Vienna, Austria, November 2018

Commercial Satellite Imagery: An Evolving Tool in the Non-proliferation Verification and Monitoring Toolkit

Frank V. Pabian, Guido Renda, Rainer Jungwirth, Lance K. Kim, Erik Wolfart, Giacomo G. M. Cojazzi, and Willem A. Janssens

Abstract Although commercial satellite imagery has already been proven to be an effective and accepted means for nuclear monitoring, verification, and mission planning for IAEA safeguards purposes, it is continuing to advance as a result of significant new improvements in terms of temporal, spatial, and spectral resolutions from increasingly diverse and rapidly growing international satellite constellations. It remains a critical verification technology that provides an optimal, non-intrusive, capability to both follow-up on geospatial cueing information from other open-sources and to remotely "peer over the fence" to obtain new and unique information from otherwise inaccessible areas, anywhere on earth, on a consistently repetitive basis. The improving ("faster, better, cheaper") means of access to this multi-resolution imagery diversity is providing increased opportunities for open-source information augmentation and unexpected data-fusion synergies. The open-source geospatial tools (e.g., Google Earth) continue to keep pace as efficient and cost-effective means to contextually visualize that imagery in 3D as well as promote greater global transparency. This ongoing commercial satellite imagery (r)evolution continues to add to the expanding and transforming open-source tool-kit to derive and assess new nuclear non-proliferation relevant information critical for enhanced global nuclear security.

1 Introduction

One evolving means of verification of a State's compliance with its obligations involves the application of publicly available commercial satellite imagery. The International Atomic Energy Agency (IAEA) has publicly acknowledged commer-

F. V. Pabian (✉) · G. Renda · R. Jungwirth · L. K. Kim · E. Wolfart · G. G. M. Cojazzi · W. A. Janssens
EC Directorate General Joint Research Centre, Directorate G - Nuclear Safety and Security, Ispra, Italy
e-mail: guido.renda@ec.europa.eu; rainer.jungwirth@ec.europa.eu; erik.wolfart@ec.europa.eu; giacomo.cojazzi@ec.europa.eu; willem.janssens@ec.europa.eu

I. Niemeyer et al. (eds.), *Nuclear Non-proliferation and Arms Control Verification*, https://doi.org/10.1007/978-3-030-29537-0_24

cial satellite imagery as "*a particularly valuable open source of information*" [1]. The IAEA first established an in-house Satellite Imagery Analysis Unit (SIAU) in 2001 to provide an independent capability within the Department of Safeguards for "*the exploitation of satellite imagery which involves imagery analysis, including correlation/fusion with other sources (open source, geospatial, and third party")* [2, 3]. Over the past 17 years, the quantity and quality of commercial satellite imagery has improved significantly. This chapter: (1) updates earlier work on the topic by these authors, [4, 5] and, (2) explains how commercial satellite imagery can play an ever increasingly important role for nuclear non-proliferation monitoring and verification in the future.

2 The Number of Satellites Is Rapidly Increasing, and Capabilities Are Improving

The number and variety of commercial imaging satellites that provide high resolution imagery sufficient for monitoring and verification applications continues to grow. In early 2000, there was only one commercial imaging satellite, Ikonos, which was capable of providing electro-optical[1] (EO) imagery at a resolution of less than 2 m, see Fig. 1 [7].

There are now over 50 earth orbiting commercial imaging satellites/systems with electro-optical spatial resolutions of **2 m** or finer with the sharpest imagery currently available, via WorldView-3, at ~**31 centimeters (cm)**.

While there has been a substantial consolidation of the US commercial satellite imagery industry in the US over the past few years (DigitalGlobe merged with GeoEye, which itself had earlier merged with Orbimage) there has also been a period of greater market segmentation with new imagery capabilities beginning to fill distinct market niches that may serendipitously potentially benefit open source non-proliferation analysts. New constellations of "refrigerator-sized" "small sats" and "shoebox-sized" "nano-sats" are now providing commercial satellite imagery with complimentary capabilities of those provided by the larger vendors such as DigitalGlobe and Airbus Industries [8]. One US company, Terra Bella (formerly Sky-Box recently acquired and renamed by Google [9, 10] operating two "small sats" known as SkySats 1 and 2 [11, 12], which it is acquiring **90 cm** resolution color and near infrared imagery, providing new opportunities to augment data from existing higher (finer) resolution imaging constellations) [13]. Terra Bella intends to acquire multiple images by multiple satellites of any point on Earth at several different times per day (or night) when tasked, eventually having as many as 24 satellites in its constellation [14]. Previous acquisition time windows had

[1]For additional information on other satellite-based imaging sensors in addition to electro-optical (e.g., synthetic aperture radar, hyperspectral, etc.) see [6].

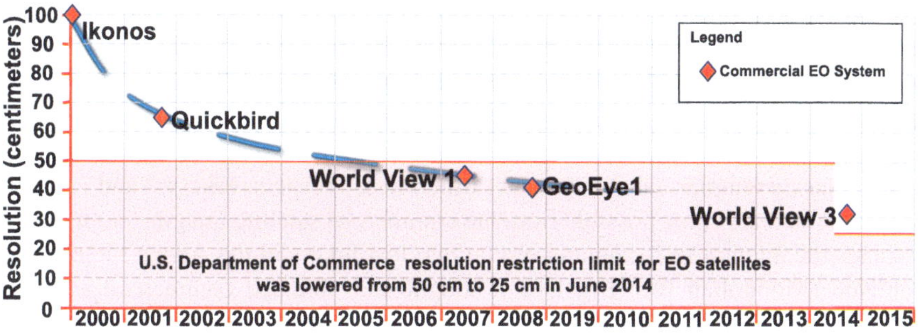

Fig. 1 Benchmarks for increasingly fine resolution of commercial satellite imaging systems over time. Image credit: EC JRC

generally been limited to around 10:30–11:30 a.m. local time (although Digital Globes WorldView-3 acquires imagery around 1:30 local time) [15].

Planet has already placed 101 "nano-sats" into orbit, which although limited to **3 m** resolution, will be able to image the entire planet in 1 day, every day [16]. The constellation will effectively provide, as the company claims, "*a line scanner for the planet*" [17]. Urthecast is also in the process of creating its own free-flying satellite constellation of 16 satellites that is expected to provide frequent EO and EO video at ~**50 cm**, with near coincident radar imagery [18].

Competition is increasing between various commercial satellite imagery vendors from the United States, France, India, Russia, China, Israel, Japan, South Korea, etc. Companies like the British Surrey Satellite Technology (SST) LTD, provide 1 m imaging capabilities for purchase on a turnkey basis. Three SST DMC-3 "mini-satellites" were successfully launched in 2015 [19]. This competition could put downward pressure on prices, making commercial satellite imagery more affordable. The diversity will also provide greater access to what might otherwise be denied areas with some vendors, and also ensure the integrity and validity of the data obtained for the historical record.

3 Temporal Resolution Improvements: Observing Activity

With such a large number of commercial imaging satellites and sensors orbiting the earth at one time (which one senior US Government official termed an "*explosion*" of geospatial information due in large part to the proliferation of small satel-lites [20]), the previous concerns, regarding temporal resolution or the timeliness of revisit between image acquisitions, will be less of an impediment with regard to monitoring and verification applications.

Commercial satellite coverage of the Chernobyl and Fukushima disasters shows how much has already changed in the past quarter century. For Chernobyl, commercial satellite images of **10 to 30 m** spatial resolution **were taken days apart**

[21]. In monitoring Fukushima, DigitalGlobe not only acquired two ~**50 cm** spatial resolution images of the Fukushima reactor site on the same day (March 14, 2011, using two different satellites in its constellation), but those images were acquired **one minute before**, and **only three minutes after**, the building housing Reactor Unit 3 exploded [22].

Another exemplar showing the timeliness, responsiveness, and global coordination that is now possible from the commercial satellite imaging community involves the attempt to locate and rescue two lost hikers in the Andes Mountains. TomNod (Mongolian for "Big Eye"), sought and obtained commercial satellite imagery from DigitalGlobe, and within hours engaged nearly 800 globally-linked volunteers in the search. The imagery was not only made available within 2 days, but was of sufficient resolution (~**50 cm**) to see the lost and doomed hikers' footprint tracks in the snow [23].

The introduction of video capabilities offers new advantages over still images in that it can allow more recognizable observation of plant operation signatures (rising cooling tower plumes) and other activity (vehicular and construction equipment movement) at sites of monitoring and verification interest. The Terra Bella SkySat satellites can also acquire High Definition (HD) videos [24], with durations of up to 90 s, utilizing ~**1.1 m** spatial resolution sensors capable of night time imagery (in one example, automobile headlights can be observed moving down the streets of Las Vegas, Nevad [25]). The International Space Station (ISS) now also includes a 1 m resolution capable camera (operated by Urthecast), which can also acquire 60 s long Ultra-High Definition (UHD) videos, over any location that the space station orbits.

This shortening time gap capability is bringing us ever closer to the asymptote of "persistent surveillance" on a global scale. Though varying in resolution from coarse to fine, *"the types of spacecraft being developed by providers such as Skybox, UrtheCast, and Planet Labs are intended to "darken" the skies with sensors. Their advantage is the ability to revisit a target multiple times a day, offering more intelligence on the patterns of life and activities taking place there."* [26] The more frequently any point on the globe is imaged (or videoed), the more difficult it becomes to conceal illicit operations. The resulting high repetitive revisit rate (from the sum of all the existing and planned EO systems) will also make it much easier to detect changes associated with the construction of larger features like roads and major buildings of potential relevance to monitoring and verification for non-proliferation applications. It should also become easier to detect such changes in an automated way using feature extraction tools (e.g., advanced machine learning algorithms) that are also currently under development [27]. In late 2017, Planet had a sufficient number of Dove satellites in orbit (3 m resolution) that it could acquire an image of every point on earth every day [28].

One other aspect of this new era of observation satellites that should not be overlooked is that they are increasingly agile, providing another way to reduce the time gap (hence improve the spatial resolution) between multiple image captures. High -Resolution Optical Imaging Constellation of CNES [29] shows three pictures on the Mecca "tower clock" acquired by Pléiades 1B every 90 s in a single pass. In the images it is possible to notice the movement of minutes needle. Airbus

Industries' Pléiades 1B, is able to supply images at **70 cm** resolution, which can be resampled at **50 cm**. While each image is still just a snapshot in time, the gaps between each snapshot can be reduced, potentially capturing on-the-ground activity not otherwise possible.

4 Spatial Resolution Improvements: Seeing Greater Detail

Among the significant developments to have occurred since the beginning of the twenty-first Century included the public availability of 1 m resolution commercial satellite imagery (the first 1 m resolution images were provided by US commercial companies, allowed under a 1992 US federal law [30]). By 2008, **50 cm** resolution imagery had become available via the GeoEye-1 satellite, and in 2014, **31 cm** imagery first became available via the WorldView-3 "superspectral" satellite (a June 2014 change in US Federal law permitted US companies to sell satellite imagery with resolution as fine as **25 cm** [31]). Nine pixels at **30 cm** resolution cover the equivalent footprint of one pixel at **90 cm**, such that the resulting resolution actually represents a 9:1 improvement in image detail.

Figure 2 helps one to more fully appreciate the significance of such improved resolution. What is particularly noteworthy is that it is now possible to distinguish between automobile types (e.g., sedans versus station wagons). This capability will lead to improved monitoring and verification of nuclear facilities (e.g., potentially allowing differentiation of different types of UF_6 cylinders in open storage (see lower section Fig. 2), the identification of critical fuel cycle equipment either in transit or in open storage, or operational details associated with electrical power conditioning, and heating, ventilation, and cooling (HVAC) related equipment and infrastructure).

5 Spectral Resolution Improvements: Seeing Beyond the Visible

To date, commercial satellite imagery being utilized for open-source non-proliferation and potential treaty verification applications has primarily involved electro-optical (EO) multi-spectral bands in the visible and near-infrared combined to create panchromatic sharpened naturally appearing color imagery. While it is not possible here to address all of the implications of applying other commercial satellite-based sensor suites to promote additional synergies, it is important to at least be aware of their complementary strengths that other sensors can add and that they are also evolving with improving resolutions.

Synthetic aperture radar (SAR) imagery, with resolutions now available down to ~**25 cm** [35], is not only useful for all hours, day or night, monitoring of activity, but it is particularly helpful in detecting security perimeters that might otherwise be

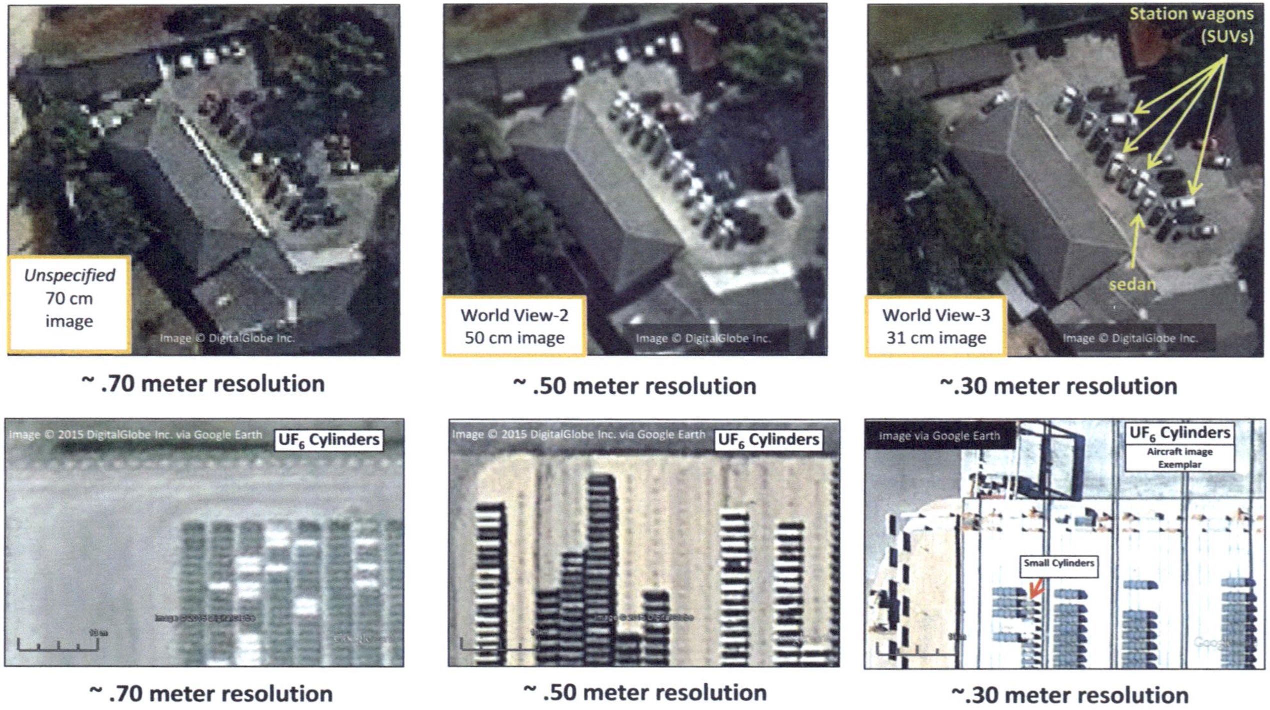

Fig. 2 Interpretability comparison: the top images are exclusively from commercial imaging satellites over Sao Paolo, Brazil; but each was acquired on a different day. Slideshare [32] The bottom images are of UF_6 cylinders. Lower right is aircraft derived, but is illustrative of what is possible at ~**30 cm** resolution. Such imagery is capable of revealing stiffening rings on large UF_6 cylinders that are critical for definitive identification (and eight small cylinders). This graphic was also published in [33] and [34]. Image Credit: top row: Image ©DigitalGlobe Inc. Bottom row, first and second from the left: Image ©2015 DigitalGlobe Inc. via Google Earth. Third from the left: image via Google Earth

obscured by vegetation. Hyperspectral imagery (as is currently available from the US Hyperion satellite [36], the European Space Agency's (ESA) Proba-1 satellite [37], and soon to be on the German EnMap satellite [38]) is derived from data covering up to hundreds of bands (at resolutions of between **17 and 34 m**) which can provide the capability to discriminate between materials and/or minerals, e.g., chalcopyrite (copper-iron-sulfide) versus hematite (iron oxide). Such a capability could play an important role in geologic mapping and where such minerals are reported as occurring, or are suspected of occurring, in association with uranium bearing minerals when assessing a suspect uranium mine or ore processing facility.[2] DigitalGlobe's newest WorldView-3 satellite, advertised as "super-spectral" with 29-bands (including panchromatic **31 cm** at Nadir and multispectral **1.24 m** at Nadir), may provide similar unique insights. The additional new **3.7 m** resolution shortwave infrared (SWIR) capability makes it possible to see through thick haze and smoke. DigitalGlobe [40] Such a capability might be requisite for detecting some critical equipment movement of verification relevance not otherwise apparent by alternative remote-sensing means.[3]

Thermal infrared (TIR) satellite imagers, even with resolutions as coarse as **100 m**, can potentially provide unique operational information when combined with other geospatial information. One new development in the field of satellite-based thermal imaging is that the South Korean operated satellite, KOMPSAT-3A, was successfully launched in March 2015 and is now operational. The KOMPSAT-3A has a **5.5 m** resolution mid-wavelength infrared (MWIR) thermal imaging sensor, operating in the 3.3–5.2 μm range, with a coincident capacity for acquiring a **55 cm** EO imagery for cross-correlation. Most significantly, KOMPSAT-3A now provides significantly finer spatial resolution thermal imagery than both Landsat and ASTER, and that imagery can be also acquired at night. To what degree KARI's TIR imagery will be applicable for non-proliferation monitoring and verification has yet to be proven, but it is expected that it should provide the capability to readily detect warm water effluents and more precisely locate "hotspots" arising from operations at a variety of nuclear-related facilities.

6 Improvements in Accessibility and Pricing

No one previously foresaw the huge volumes of commercial satellite imagery being provided at little or no direct cost as found on the various virtual globes, which are also viewable on smart phones today via the internet (albeit not necessarily with all the functionality available on a computer). Virtual Globes (AKA Digital Earths), including *Google Earth, Here, Bing Maps,* Esri's *ArcGIS*, and the new maps app

[2]The ECOSTRESS spectral library provides a compilation of over 2400 spectra of natural and anthropogenic materials [39].

[3]For more examples of the advantages provided by SWIR, please see [41].

for Windows 10, etc., provide the capability to synoptically view multi-resolution, three-dimensional, virtual representations of the earth and have created a new venue to "navigate through space and time" with very high resolution commercial satellite imagery (augmented in many areas by even higher resolution aircraft imagery) [42, 43]. Global transparency via overhead observation has become the new norm for a growing global cadre consisting of everyone and anyone having an interest in nuclear non-proliferation and arms control verification.

Interestingly, when Google Earth first came on the scene in 2005, one common complaint was that the imagery was often too old and out of date (or more recently, that did not represent "a perfect planetary mirror" [44]), with the implication that separate imagery purchases would always be necessary for temporal currency. However, in some cases, particularly in those areas of high media interest, including nuclear facilities in Iran and the Democratic People's Republic of Korea (DPRK), the imagery can be quite current. For example, at the DPRK's Punggye-ri nuclear test site there are eight different high-resolution commercial satellite images archived in the Google Earth historical layer that were acquired between January 1, 2013 and February 12, 2013, the date of the third identified underground nuclear test. Google Earth's historical layer view permits high-resolution visual inspection of each of those images in sequence.

The price of both acquisition and processing of imagery, which had been another major concern in the 1990s and early 2000s (in that individual frames of archived imagery originally cost about $3000–4000), has, in some but not all cases, dropped substantially such that archived imagery from DigitalGlobe (for example) can be purchased for as little as $350 for a 25 km^2 area, or around $500 for a similar area special order request. One interesting development is a smart phone app (for both Android and iPhone) that makes possible the purchase of a 1 km^2 area of some archived commercial satellite imagery (including SAR) from up to 36 satellites for as little as $10 [45] (see lower left inset in Fig. 4). In April 2015, the ESA began providing SPOT and Pléiades data for free for research and application development through project proposals submitted via ESA's Earth Online Portal [46].

Searching the multitude of available imaging archives of the increasingly diverse vendor options would have been a daunting task if not for the creation of web-based applications such as Apollo Mapping's *Image Hunter* [47] or GeoCento's *Earth Images* [48], which provide quick access to multiple image archives of different vendors from different countries. However, as of this writing, there is still not a complete "one-stop-shop" for all vendors of commercial satellite imagery, but may emerge if present trends continue. (And even thumbnail versions of archived images can supply new useful information, as is exemplified in the upper right inset of Fig. 4.)

With respect to hardware and software expenses, those costs have also dropped dramatically, and the large file sizes associated with such imagery are also much easier to transfer, process, and store than ever before. In-the-cloud storage and processing of big data in a distributed environment across clusters of computers using simple programming models via open open-source frameworks like *Apache Hadoop*, now empowers anyone, anywhere with unimagined, effectively cost-free computing power [49].

Additionally, in 2015, Google Earth Pro (previously priced at $399) became freely downloadable. It offers the additional key benefits (not previously available with the original free version) of allowing area measurements for facility size determinations; video movie-making of fly-throughs for pre-inspection familiarization, etc.; and the capability to create "view-sheds" to highlight all areas within a given line of sight to assess ground level building and terrain masking, etc., that might also be useful for safeguards pre-inspection planning [50].

7 Multi-Satellite/Multi-Sensor Data Fusion: Creating Innovation in Verification Applications

This section highlights the utility of combining information from multiple electro-optical satellites (e.g., each having different resolutions) but also applies to multiple satellites having different sensors suites (e.g., high resolution EO imagery with synthetic aperture radar imagery or thermal infrared imagery). With each combination, sometimes referred to as "multi-sensor data fusion,"[4] it is possible to derive a unique synergy, such that the whole of the information that can be gleaned from the combination is really greater than the simple sum of the information from the multiple images if each were to be analyzed in the absence of the others [51]. Two examples are provided below.

The first example was derived when observing activity at the Punggye-ri nuclear test site in the DPRK using two different satellites from two different vendors with differing resolutions (~**50 cm** and ~**90 cm**) acquired only 59 min apart (See Fig. 3). The first ~50 cm image from DigitalGlobe (as found on Google Earth, Fig. 3, top left) shows a multi-vehicle entourage involving two black sedans or SUVs (and two possible escort vehicles) near the South Portal, but not at the West Portal (Fig. 3, top right), which is separated from the South Portal by about 1 km. Although the subsequent **90 cm** SkyBox imagery (Fig. 3, bottom left) showed that the previous four vehicle entourage was no longer present at the South Portal, that imagery did reveal that two dark-toned vehicles (albeit fuzzier due to the lower resolution) had arrived near the West Portal (Fig. 3, bottom right).

Taking into account the timing and cross-correlation of the two images together, it is possible to hypothesize that the two dark-toned vehicles imaged by the SkyBox satellite near the West Portal were likely the same two dark-toned sedans or SUVs observed 59 min earlier by the DigitalGlobe satellite at the South Portal. If we assume that the vehicles were indeed the same, the evident synergy provides us with at least the suggestion that some kind of an official visit/inspection had initially involved the South Portal and then moved to the West Portal, and, moreover, that those visit/inspections occurred less than 1 h apart.

---

[4] *"Data fusion is defined as the combination of data from multiple sensors such that the resulting information is better than would be possible when the sensors are used individually"* [51].

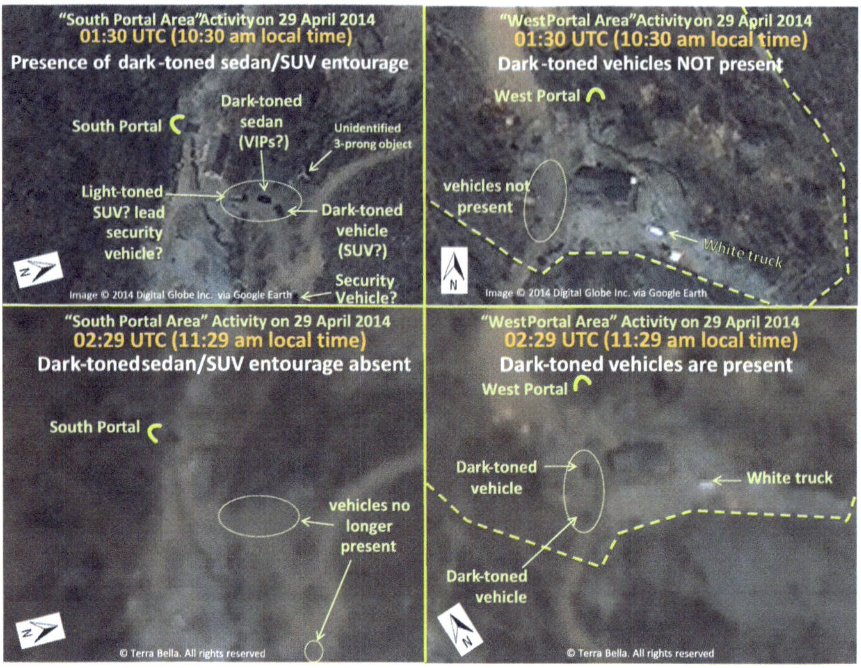

Fig. 3 Multi-satellite coverage can provide synergy by allowing us to derive dark-toned vehicle movement between two nuclear test portal areas within 1 h using different satellites from different vendors (top row left ~50 cm resolution images from DigitalGlobe, bottom row right 90 cm resolution from Skybox -now Terra Bella- imaging) that is suggestive of an official visit/inspection to each of the two active test tunnels within an hour of each other. This graphic was also published in [34]. Images credit: top row: Images ©2014 DigitalGlobe Inc. via Google Earth. Bottom row: Images ©Terra Bella. All rights reserved

The second exemplar (see Fig. 4) shows how, through the use of multiple EO images from multiple commercial satellites provided free by Google Earth (along with a separately purchased image at low-cost) it was possible to newly locate, identify, and characterize a possible nuclear waste site by simply following-up on a single report from Iranian news [52].

In late 2014, Iranian news reported that the, *"Chief of the Atomic Energy Organization of Iran (AEOI) Ali Akbar Salehi paid a visit to a long-term nuclear waste storage facility in the central province of Isfahan. Salehi visited the city of Anarak... to get update on the latest developments regarding the construction of the nuclear waste stabilization and storage facility"* [52]. A simple search of the area around the town of Anarak, Iran, using the most recent imagery available from Google Earth (July 16, 2013, Fig. 4, large bottom image), made possible the location of a candidate site.

A subsequent review of DigitalGlobe imagery archives (Fig. 4, top left) revealed multiple acquisitions centered on that same site, indicating that this site first began

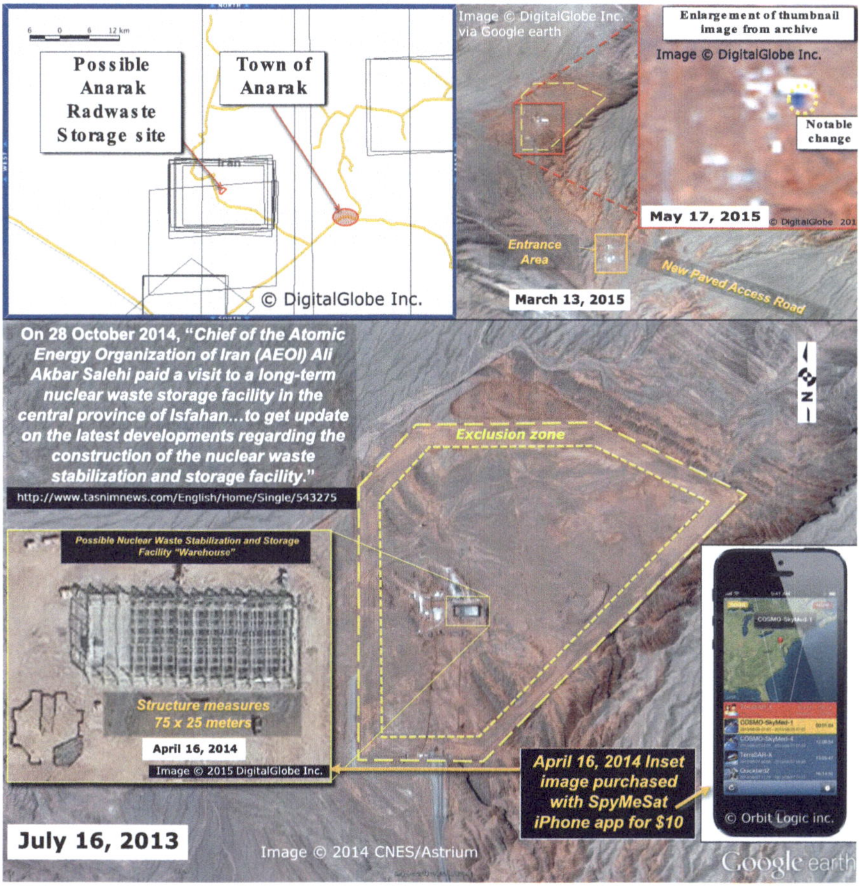

Fig. 4 In following-up to a single news report (see quote in inset), with commercial satellite imagery as is available on Google Earth it was possible to identify a new "Nuclear Waste Stabilization and Storage Facility" nearing completion near Anarak, Iran (33.364208 N, 053.467296 E). Significantly, the April 16, 2014 inset photo (bottom left), showing substantial concrete and steel construction, was purchased via SpyMeSat for $10 and is more recent than that currently found on Google Earth. This graphic was also published in [33] and [34]. As this chapter was undergoing final edit, Google Earth has more recently posted very high resolution Airbus imagery from December 10, 2015. Moreover confirmation of the identification has also become available that includes videos of the interior of the main vault-structure building and aerial drone videos of the site [53, 54]. IPhone picture: Image ©2000–2014 Orbit Logic Inc. All Rights Reserved. IPhone picture source: http://www.orbitlogic.com/products/SpyMeSat.php

attracting continuing interest by others (NFI) in early 2014 (although Google Earth shows that construction at the site was first underway by October 2011). The two most recent images (Fig. 4, top right, from March 13, 2015 and May 17, 2015) are viewable for free directly from those archives (albeit at reduced resolution). Even at the reduced resolutions, they are sufficient to show continued construction and changes on-site, e.g., like the addition of a new blue roof to a small building in close proximity to the largest building on-site. The overall site, which is served by a newly paved access road and an outer entrance facility (Fig. 4, top right), exhibits the requisite features for a secure storage vault-type radwaste structure situated in dry, stable, geology that is also not susceptible to flash flooding. As this chapter was undergoing final edit, these authors obtained additional open source information sufficient to confirm the identification of the radwaste site and the vault-type structure which is now officially open and identified in two YouTube videos as the Pasmangoor Nuclear Waste Storage and Stabilization Facility in Anarak.[5]

For a discussion of how multi-satellite, multi-sensor commercial satellite imagery analysis could blend with other open source information, please see the accompanying chapter on *Open Source Analysis* by G. Renda et al. in this book.

8 Imagery Analysis Is Critical for Effective Use of Commercial Satellite Imagery

Imagery analysis is the extraction of meaningful information from images. In previous studies, it was stated that commercial satellite imagery, as it is available for purchase today in its raw form, is nothing more than "a pile of pixels". Pabian et al. [5] As one practitioner has pointed out, *"Pixels are becoming a commodity, and the real value is in extracting that data, making it understandable, and making it useful to customers,"* and *"Albert Einstein observed that information, however, is not knowledge. Raw data means nothing without interpretation"* [55].

Imagery analysis is the means by which a body of pixels is holistically interpreted, and creatively combined, in light of other collateral open-source information (e.g., safeguards pertinent data) to synergistically derive new, "value-added," information from the raw un-annotated imagery. That information can then be added to the overall existing body of knowledge with respect to a particular facility, activity, or program. Imagery analysis is critical to verification and monitoring, particularly as it applies to identifying possible clandestine "undeclared facilities and activities." Commercial satellite imagery must be competently interpreted by well-trained and experienced imagery analysts who are fully cognizant of the nuclear fuel cycle and its infrastructure, equipment, and operations, and who are able to distinguish them from those associated with other industrial processes.[6] Imagery analysts should also be acutely aware of the possibilities for deception and signature suppression. Not all observable facilities and equipment are going to exhibit the near perfect signature correlations as found in the infrastructure training exemplar shown in Fig. 5.

The imagery analysis process is also heavily dependent on the eye of the beholder. John Jensen (2007) described factors that distinguish a superior imagery

[5]Please see [53, 54].

[6]One useful reference regarding imagery analysis and the observable signatures associated with various industrial processes is available at [56].

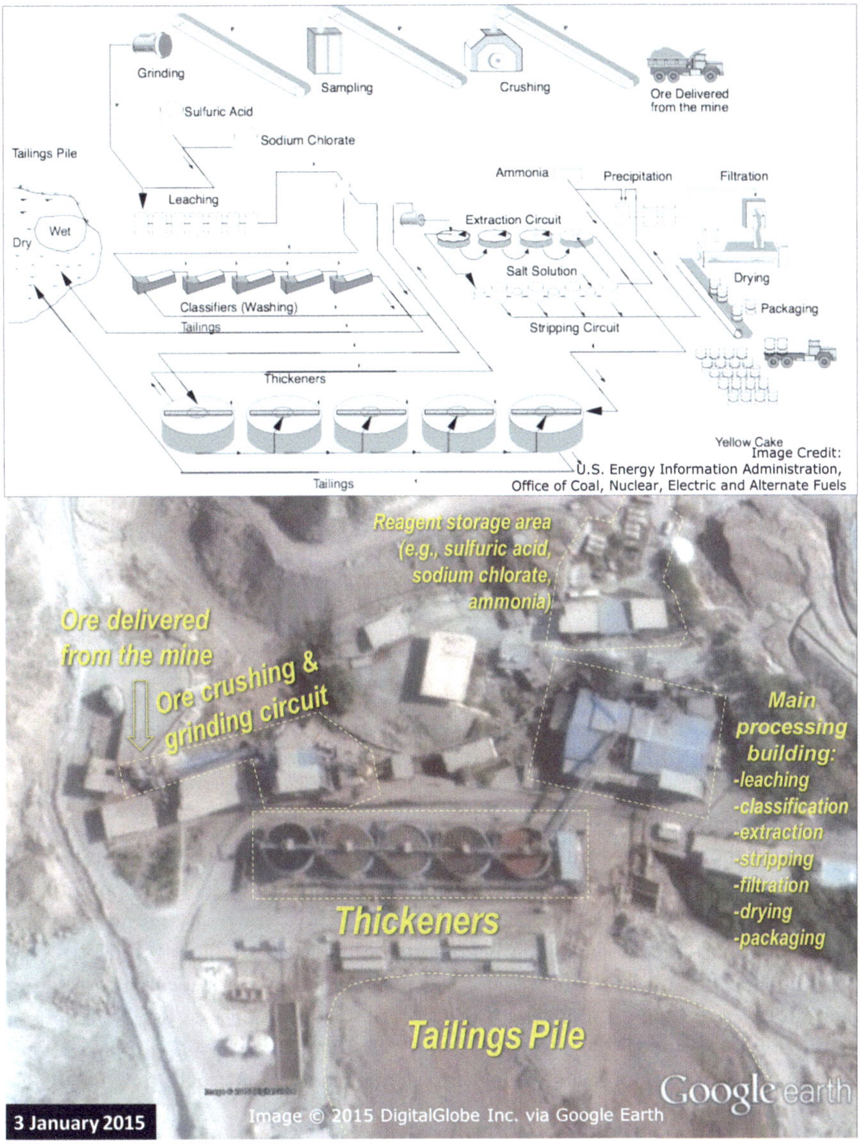

Fig. 5 "Signatures" can play a critical role in nuclear fuel cycle infrastructure analysis. The top view provides a flow diagram of the key components in the operation of a uranium ore concentration plant. The lower image is an operational uranium ore concentration plant. However, such near perfect correlation with signatures is unusual and unlikely to be found in most other cases. This graphic was also published in [34]. Diagram credit: Energy Information Administration, Office of Coal, Nuclear, Electric and Alternate Fuels. Accessed via [57]

Fig. 6 Left side shows the Arak, IR-40 (40 MWth), radioisotope production reactor under construction in 2005; the right side shows a large concrete structure under construction in 2009 near Shiraz, Iran, having somewhat comparable attributes. This graphic was also published in [34]

analyst. He said, *"It is a fact that some image analysts are superior to other image analysts because they: (1) understand the scientific principles better, (2) are more widely traveled and have seen many landscape objects and geographic areas, and/or (3) they can synthesize scientific principles and real-world knowledge to reach logical and correct conclusions. Thus, remote sensing image interpretation is both an art and a science"* [58].

The possibility of misinterpretation, or discovering "false positives," is always possible in the course of any imagery analysis for any purpose. With respect to nuclear non-proliferation, innocuous facilities can sometimes have the potential to be mistaken for undeclared nuclear facilities of interest for treaty monitoring and verification, particularly during early construction phases (adapted from [59]). One example (see Fig. 6) shows how the concrete foundation work located in a large, deep circular excavation that was observable in 2009 in an area north of Shiraz, Iran, exhibited some of the physical features that were somewhat similar to what had previously been observed during the early phase of construction of the IR-40 radioisotope production reactor located near Arak, Iran.

However, in this particular case, an imagery analyst would only have had to conduct some simple background research with open sources to quickly dispel any possible nuclear connection (by incorporating social media user content, e.g., *Panoramio* posted ground photos and virtual globe labelling as provided by the Google Earth Community and *Wikimapia*, along with associated links to appropriate descriptive construction websites). Drawing from such information, it could readily be shown that the circular concrete foundation was merely that for a then projected 30-story cylindrically shaped hotel adjacent to what is now one of the largest shopping malls in the Middle East [60]. See Fig. 7. Although originally

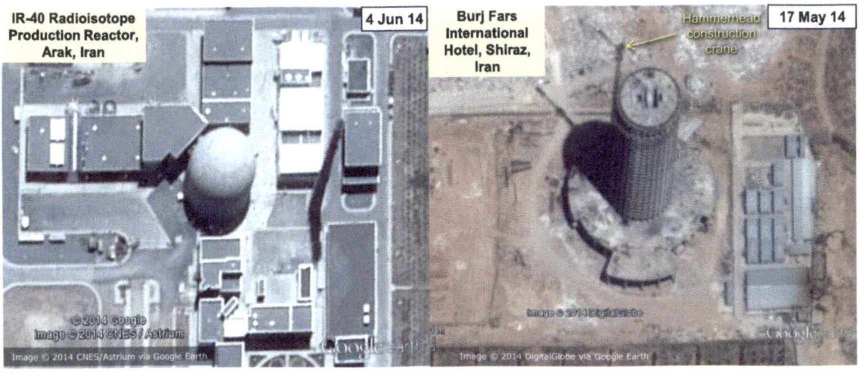

Fig. 7 Left side shows the Arak IR-40 (40 MWth) radioisotope production reactor that was externally complete in 2014; the right side shows a 30-story cylindrical hotel nearing completion in 2014. At this point, there is little possibility of mistaking one for the other. This graphic was also published in [34]

projected for completion in 2012, the hotel was still incomplete as of December 28, 2014, based on the most recent commercial satellite imagery available on Google Earth. Moreover, the Google Earth historical imagery layer provides the additional information that, as of 2005, the site was the location of an amusement park, which was easily accessible to the local populace in the northern Shiraz suburbs.

Finally, some relevant tenets: (1) never assume that which can be discovered is already known, and (2) never assume that which can be discovered is not already known. One must do the requisite background research, using all available open sources, to verify each discovery and report and caveat at the most appropriate level of objective certainty.

To conclude with a quote: *"Without adequate experience and training, it can be relatively easy to see proof of sinister intent in a benign image, or to miss details that would be obvious to a knowledgeable photo interpreter"* [61].

9 Advances in 3D Modeling

Previous studies described the utility of using commercial satellite imagery to create 3D models (as one of several potential products which may be derived from satellite imagery) of nuclear relevant facilities, infrastructure, and equipment for better assessments including walk-arounds and fly-arounds (e.g., for IAEA inspection planning, etc.) [62]. Such a capability was then limited to only manual creation with digital modeling software like *SketchUp* combined with geospatial context visualization provided by virtual globes. Now, virtual globes like Google Earth are in the process of mapping the whole planet in 3D using digital stereo-photogrammetry to automatically create *"3D cityscapes, complete with buildings,*

Fig. 8 The Pickering Nuclear Power Station, Canada, as viewed on Google Earth with the 3D Building layer turned ON and OFF (inset). The 3D rendering was derived from multiple aircraft overflight images. The 3D modelling (including buildings, trees, bushes, and even vehicles) was created entirely via an automated photogrammetric process operated by Google. This graphic was also published in [34]

terrain and even landscaping, from 45-degree aerial imagery" [63] at higher rates. The 3D rendering shown in Fig. 8 was derived from multiple aircraft overflight images acquired on April 10, 2013. The 3D modelling (including buildings, trees, bushes, and even vehicles) was created entirely via an automated process run by Google. To date, however, such cost-free modeling has been limited only to areas accessible by aircraft, and many governments restrict aerial over-flights of their nuclear facilities, if not all of their territories. With the advent of highly agile, highly accurate, capable commercial satellite imaging satellites (e.g., DigitalGlobe's WorldView-3 ∼30 cm resolution and Airbus's Pléiades 50 cm resolution), such automated 3D modelling is now also possible for any place on Earth with nearly the same fidelity. Notably, in May 2015, a new joint venture company was announced between DigitalGlobe and Saab, called Vricon, which has begun offering, on a "for-fee" basis, such automated 3D modeling, including that derived solely from commercial satellite imagery. Vricon [64] One initial exemplar, provided in a promotional video, shows an automated 3D model of the Natanz Uranium Enrichment Facility in Iran (See Fig. 9). Vricon [65] Similarly, Airbus Defence and Space has proven its ability to successfully and accurately produce a 3D model

Fig. 9 Among the very first 3D models to ever be created of a nuclear-related facility via an entirely automated photogrammetric process using only high resolution commercial satellite imagery (from DigitalGlobe) was displayed by Vricon in a promotional video [65]. This graphic was also published in [34]

using thirteen Pléiades images, forming nine stereo pairs, combined to produce the 3D model, also viewable in a promotional video [35]. For the foreseeable future, however, manual 3D modeling will continue to be required for building interiors. One example of such manual 3D modeling for nuclear facilities can be found in the modeling of the interior of the uranium enrichment facility at Yongbyon in the DPRK, which was based on a description provided by Dr. Siegfried Hecker who had briefly visited the site in 2010 [66].

10 Conclusions

Commercial satellite imagery has already been proven to be a timely and accurate openly accessible data source to support, supplement, and/or enhance nuclear and other non-proliferation related treaty monitoring and verification activities. Not only has it been supporting onsite inspection planning and monitoring and verification of declared activities, [67, 68] but, because of the accompanying global transparency, it also *"increases the possibility of detecting proscribed nuclear activities"* [1]. With technological advancement, a new era has begun with expanded new capabilities in earth observation that include large constellations of more agile and capable satellites having improved spatial, spectral, and temporal resolutions (that includes high definition video). Those capabilities are synergistic in that the sum of derivable information is greater, in aggregation, than the total of the information would otherwise be were the images viewed in isolation. Commercial satellite imagery provides global coverage, and much of that imagery is freely available via digital virtual globes (but not yet including **31 cm** WorldView-3 imagery [69], which could eventually provide auto-generated 3D viewing of all nuclear infrastructures worldwide in proper terrain context). Moreover, such freely available imagery is increasingly easy to supplement with focused additional imagery purchases from multiple platforms, with multiple sensors, from multiple companies, from multiple nations, and at least some can be purchased with only a smart phone for as little as ten US dollars. Such imagery will also continue to be an enabler of independent action, in that it democratizes the availability and use of the medium by anyone in the world with digital information technology access (including hand held devices). However, we must also be fully aware that such democratization also increases the potential for misinformation via deception, signature suppression, or simple human error in interpretation (and with the internet, misinformation can easily "go viral"). As commercial satellite imagery capabilities and usage continue to evolve, the consequences for future monitoring and verification efforts will only grow in significance.

Acknowledgements The reflections here presented stem from the activities performed within the project, Information Analysis and methodologies for Nuclear Nonproliferation and security, work package Open source information for non-proliferation and nuclear security, funded within the European Commission (EC) Euratom Horizon 2020 Research and Training Programme. The views expressed in this chapter are those of the authors and do not necessarily represent the official views of the European Commission. This chapter is partially based and adapted from the EUR report [34].

References

1. IAEA (2007) IAEA Safeguards: Staying Ahead of the Game, p. 21. https://www.iaea.org/sites/default/files/safeguards0707.pdf. Accessed 28 Feb 2019
2. Brown PJ (2009) IAEA's not-so-secret satellite game, Asia Times. http://www.atimes.com/atimes/Middle_East/KJ10Ak02.html. Accessed 28 Feb 2019
3. Steinmaus K (2009) Satellite Imagery as a Verification Tool for International Safeguards, Technical Seminar for Diplomats, IAEA, Vienna, February. http://www-pub.iaea.org/mtcd/Meetings/PDFplus/2009/36489/p36489/Top6.2K.Steinmaus.pdf. Accessed 28 Feb 2019
4. Pabian FV (2008) Commercial Satellite Imagery: Another Tool in the Non-proliferation Verification and Monitoring Tool-Kit. In: Doyle J (ed) Nuclear Safeguards, Security and Nonproliferation. Butterworth-Heinemann, Burlington, pp 221–249
5. Pabian FV, Renda G, Jungwirth R, Kim LK, Wolfart E, Cojazzi GGM (2015) Recent Developments Promoting Open-Source Geospatial Synergy: Emerging Trends and Their Impact for Nuclear Non-proliferation Analysis. Proceedings of the INMM 56th Annual Meeting, Indian Wells, California USA
6. Niemeyer I (2015) Space-Borne Verification: Satellite Imagery within nuclear non-proliferation and Arms Control Verification Regimes. Proceedings of the INMM 56th Annual Meeting, Indian Wells, California USA

7. Baker JC, O'Connell KM, Williamson R (2001) A pioneering and milestone reference study marking the advent of 1 m resolution commercial satellite imagery heralded by Ikonos in Commercial Observation Satellites: At the Leading Edge of Global Transparency, Santa Monica, CA: RAND Corporation. http://www.rand.org. Accessed 28 Feb 2019.
8. Patton T (2014) Workshop Report: New Technologies for Information Analysis to Support Non-Proliferation and Disarmament Verification, Vienna, Austria. http://vcdnp.org/150210_new_technologies_verification.htm. Accessed 28 Feb 2019
9. Huet E (2014) Google Buys Skybox Imaging—Not Just For Its Satellites, Forbes. http://www.forbes.com. Accessed 28 Feb 2019
10. Protalinski E (2016) Google rebrands Skybox as Terra Bella, will launch more than a dozen satellites over the next few years, VentureBeat. http://venturebeat.com. Accessed 28 Feb 2019
11. Skybox. http://www.firstimagery.skybox.com. Accessed 28 Feb 2019
12. Safyan M (2017) Planet Doubles Sub-1 Meter Imaging Capacity With Successful Launch Of 6 SkySats. https://www.planet.com/pulse/planet-doubles-sub-1-meter-imaging-capacity-with-successful-launch-of-6-skysats/. Accessed 28 Feb 2019
13. SkySkat (2015) SkySat Imaging Program of Skybox Imaging. https://directory.eoportal.org. Accessed 28 Feb 2019
14. Caleb H (2014) Skybox Imaging Exec Discusses Constellation Plans, Via Satellite. http://www.satellitetoday.com. Accessed 28 Feb 2019
15. Satimagingcorp. http://www.satimagingcorp.com. Accessed 28 Feb 2019
16. Haldeman B (2015) 101st Dove Satellite Reaches Space. https://www.planet.com. Accessed 28 Feb 2019
17. Marshall W (2014) Tiny satellites show us the Earth as it changes in near-real-time, TED Talk. http://www.ted.com. Accessed 28 Feb 2019
18. Amos J (2015) Urthecast video cameras to circle globe, BBC News. http://www.bbc.com. Accessed 28 Feb 2019
19. SSTL (2015) SSTL's DMC3 Constellation demonstrates 1-metre capability. http://www.sstl.co.uk. Accessed 28 Feb 2019
20. Butler (2015) National Geospatial-Intelligence Agency Eyes Role for Small Satellites, Aviation Week and Space Technology. http://aviationweek.com. Accessed 28 Feb 2019
21. Sadowski F, Covington S (1987) Processing and Analysis of Commercial Satellite Data of the Nuclear Accident near Chernobyl, U.S.S.R., U.S. Geological Survey Bulletin 1785. http://pubs.usgs.gov/bul/1785/report.pdf. Accessed 28 Feb 2019
22. Earth Imaging Journal (2011) Disaster Response in Japan. http://eijournal.com. Accessed 28 Feb 2019
23. Wilkens J (2012) Company uses the crowd to solve mysteries, U-T News. http://www.utsandiego.com. Accessed 28 Feb 2019
24. Youtube (2015) https://www.youtube.com/watch?v=fCrB1t8MncY. Accessed 28 Feb 2019
25. Milonopoulos N (2014) Nighttime HD Satellite Video 2014. http://www.skyboximaging.com. Accessed 8 Oct 2015
26. Butler (2015) Quoting from National Geospatial Intelligence Agency Director, Robert Cardillo. http://aviationweek.com. Accessed 28 Feb 2019
27. Descartes Labs. http://www.descarteslabs.com. Accessed 28 Feb 2019
28. Quartz (2017) The company photographing every spot of land on earth, every single day, https://qz.com/1126301/the-company-photographing-every-spot-of-land-on-earth-every-single-day/. Accessed 20 June 2018
29. Pleiades-HR (High-Resolution Optical Imaging Constellation of CNES). https://directory.eoportal.org, http://seso.com, https://pleiades.cnes.fr. Accessed 28 Feb 2019
30. Weber RA, O'Connell KM (2011) Alternative Futures: United States Commercial Satellite Imagery in 2020, p 16–17. https://nsarchive2.gwu.edu//NSAEBB/NSAEBB404/docs/37.pdf. Accessed 28 Feb 2019
31. Avery G (2014) DigitalGlobe gets OK to sell more detailed satellite images, Denver Business Journal UPDATED. http://www.bizjournals.com. Accessed 28 Feb 2019 and Loyd C (2014), Mapbox. https://www.mapbox.com. Accessed 28 Feb 2019; See also: DigitalGlobe's Satellite

Pics Are So Good They're Almost Illegal, NBC News, http://www.nbcnews.com. Accessed 28 Feb 2019
32. Slideshare (2015) 30 cm Satellite Imagery Comparison, DigitalGlobe, February 25, 2015. http://www.slideshare.net. Accessed 28 Feb 2019
33. Pabian FV, Renda G, Jungwirth R, Kim L K, Wolfart E, Cojazzi G G M (2015) Recent Developments Promoting Open-Source Geospatial Synergy: Emerging Trends and Their Impact for Nuclear Non-proliferation Analysis. Proceedings of the INMM 56th Annual Meeting, Indian Wells, California USA
34. Pabian FV (2015) Commercial Satellite Imagery as an Evolving Open-Source Verification Technology: Emerging Trends and Their Impact for Nuclear Nonproliferation Analysis; EC JRC, EUR27687
35. Airbus Defense and Space. http://www.geo-airbusds.com. Accessed 28 Feb 2019
36. NASA (2015) Hyperion. http://earthobservatory.nasa.gov. Accessed 28 Feb 2019
37. ESA (2015) CHRIS. https://earth.esa.int. Accessed 28 Feb 2019
38. ENMAP (2015) Welcome to EnMAP, The German Spaceborne Imaging Spectrometer Mission. http://www.enmap.org. Accessed 28 Feb 2019
39. NASA jet Propulsion Laboratory (2018) ECOSTRESS spectral library. https://speclib.jpl.nasa.gov/. Accessed 28 Feb 2019
40. DigitalGlobe (2014) Revealing the Hidden World with Shortwave Infrared (SWIR) Imagery. http://www.digitalglobeblog.com. Accessed 28 Feb 2019
41. Dupont P (2013) White Paper: Short Wave Infrared (SWIR) for Surveillance Applications in Defense. https://alliedscientificpro.com/blog/welcome-to-our-blog-1/post/white-paper-short-wave-infrared-swir-for-surveillance-applications-in-defense-25/. Accessed 28 Feb 2019
42. Blaschke T, Donert K, Gossette F, Kienberger S, Marani M, Qureshi S, and Tiede D (2012) Serving Science and Society. Information 3(3):372–390
43. Craglia M, de Bie K, Jackson D, Pesaresi M, Remetey-Fülöpp G, Wang C, Annoni A, Bian L, Campbell F, Ehlers M, van Genderen J, Goodchild M, Guo H, Lewis A, Simpson R, Skidmore A, Woodgate P (2012) Digital Earth 202: Towards the vision for the next decade. International Journal of Digital Earth 5(1):4–21. doi:10.1080/17538947.2011.638500
44. Fallows J (2013) Google's Michael Jones on how Maps Became Personal. http://www.theatlantic.com. Accessed 28 Feb 2019
45. Spymesat. http://www.spymesat.com. Accessed 28 Feb 2019
46. ESA Earth Online. https://earth.esa.int/web/guest/pi-community/apply-for-data/3rd-party. Accessed 28 Feb 2019
47. Apollo Mapping. https://imagehunter.apollomapping.com. Accessed 28 Feb 2019
48. Geocento. http://www.geocento.com. Accessed 28 Feb 2019
49. SAS Hadoop. http://www.sas.com/en_us/insights/big-data/hadoop.html. Accessed 28 Feb 2019
50. Google Earth Blog. http://www.gearthblog.com/blog/archives/2015/02/google-earth-pro-viewshed-tool.html. Accessed 28 Feb 2019
51. Jung H-S, Park S-W (2014) Multi-Sensor Fusion of Landsat 8 Thermal Infrared (TIR) and Panchromatic (PAN) Images. Sensors (Basel, Switzerland). http://www.mdpi.com. Accessed 28 Feb 2019
52. Tasnim News Agency (2014) AEOI Chief Visits Iran's Nuclear Waste Repository. http://www.tasnimnews.com. Accessed 28 Feb 2019
53. Youtube (2016). https://www.youtube.com/watch?v=YbzzVeXlVOA. Accessed 28 Feb 2019
54. Youtube (2016). https://www.youtube.com/watch?v=_FIsYR7IaiM. Accessed 28 Feb 2019
55. Rice J (2014) All Source Analysis:Longmont's Global Intelligence Hub, Longmont Compass. http://longmontcompass.com. Accessed 8 Oct 2015
56. Defense Mapping Agency (1996) Photo Interpretation Student Handbook. http://sites.miis.edu/geospatialtools2013/files/2012/07/Photo-Interpretation-Student-Handbook.-Photo-Interpretation-Principles.pdf. Accessed 28 Feb 2019
57. New Mexico Bureau of Geology & Mineral Resources. https://geoinfo.nmt.edu/resources/uranium/mining.html. Accessed 28 Feb 2019

58. Jensen JR (2007) Geography 551: Principles of Remote Sensing, Slide 18. http://www.slideserve.com. Accessed 28 Feb 2019
59. Pabian FV, Renda G, Jungwirth R, Kim LK, Wolfart E, Cojazzi GGM (2014) Open Source Analysis in Support to Non-proliferation Monitoring and Verification Activities: Using the New Media to Derive Unknown New Information. Proceedings Symposium on International Safeguards: Linking Strategy, Implementation and People, vol. IAEA-CN-220, paper #312. http://www.iaea.org/safeguards/symposium/2014/home/eproceedings/sg2014_eproceedings_online.pdf. Accessed 28 Feb 2019
60. Persian Gulf Complex (20018) https://en.wikipedia.org/wiki/Persian_Gulf_Complex Accessed 28 Feb 2019
61. Schmidt LJ (2001) New Tools for Diplomacy, Remote Sensing Use In International Law, NASA Earth Observatory, January 12, 2001. https://earthobservatory.nasa.gov/Features/Diplomacy/ Accessed 28 Feb 2019
62. Pabian FV (2013) The New Geospatial Tools: Global Transparency Enhancing IAEA Safeguards. Proceedings of the INMM 54th Annual Meeting, Atlanta, Georgia USA
63. Google (2012) The next dimension of Google Maps. https://www.youtube.com/watch?v=HMBJ2Hu0NLw. Accessed 28 Feb 2019
64. Vricon (2015) Saab and DigitalGlobe Announce Vricon Joint Venture to Create the Globe in 3D. https://www.vricon.com/news/saab-digitalglobe-announce-vricon-joint-venture-create-globe-3d/. Accessed 28 Feb 2019
65. Vricon. http://www.vricon.com/capabilities (see 3:32 in the five minute long video) Accessed October 8, 2015
66. Patton T (2015) Approximate 3D Visualization of Yongbyon U-Enrichment Facility. https://prezi.com. Accessed 28 Feb 2019
67. Rutkowski J (2012) Satellite Imagery Analysis at the International Atomic Energy Agency IAEA. http://ungiwg.org. Accessed 28 Feb 2019
68. Jasani B, Niemeyer I, Nussbaum S, Richter B, Stein G (eds) (2009) International Safeguards and Satellite Imagery: Key Features of the Nuclear Fuel Cycle and Computer-based Analysis. Springer
69. de Selding PB (2015) Why You Won't Find DigitalGlobe's Best Imagery on Google Maps. Space News. http://spacenews.com/why-you-wont-find-digitalglobes-best-imagery-on-google-maps. Accessed 28 Feb 2019

Strategic Trade Analysis for Non-proliferation

Cristina Versino and Giacomo G. M. Cojazzi

Abstract Strategic trade analysis is about national and global trade flows of items with strategic military value. Strategic items include goods with exclusive military use, but also dual-use items that have both military and civil uses. Because of their strategic nature, dual-use goods are controlled as a measure of non-proliferation of Weapons of Mass Destruction. Trade analysis in dual-use items, in particular the nuclear related ones listed for export controls by the Nuclear Suppliers Group, is the focus of this contribution. Despite the global dimension of strategic trade and controls, no specific data are available on global strategic trade. We thus suggest using open source data on global trade to support strategic trade analysis. The data are not specific to strategic trade, but related to it, providing a portrait of strategic trade in the absence of more specific data. It can be expected the effectiveness of controls be enhanced in several ways by the analysis of trade data related to strategic goods. By understanding the national and global trade in strategic-related commodities, officials may better focus controls, select companies for in-reach and audit, target transactions for analysis, verification, and inspection, and assess potential economic impacts of trade control regulations. The chapter covers data sources, a methodology for data selection, underpinning tools and data use cases. The context for strategic trade analysis is briefly recalled by introducing lists of items subject to export controls, with a focus on the nuclear-related ones.

1 Introduction

Strategic trade analysis is about gaining or providing insights into national or global trade flows of items with strategic military value. Strategic items include items with exclusive military use, but also dual-use items that have both military and civil uses.

C. Versino (✉) · G. G. M. Cojazzi
EC Directorate General Joint Research Centre, Directorate G - Nuclear Safety and Security, Ispra, Italy
e-mail: cristina.versino@ec.europa.eu; giacomo.cojazzi@ec.europa.eu

I. Niemeyer et al. (eds.), *Nuclear Non-proliferation and Arms Control Verification*,
https://doi.org/10.1007/978-3-030-29537-0_25

Trade analysis in dual-use items, in particular the nuclear related ones, is the focus of this contribution.

Because of their strategic nature, dual-use items are export-controlled as a measure of non-proliferation of Weapons of Mass Destruction. Controls involve a range of actors and processes taking place at national and world level. At national level, controls of strategic items are first defined in legislation. In-reach actions to exporters promote awareness of control measures and related obligations, such as the requirement to apply for authorisation prior to exporting. National licensing authorities are responsible to decide upon export applications. Their judgment needs to address proliferation concerns while not disrupting legitimate trade. In the European Union (EU), a comprehensive export control policy sets a common context for controls to be effectively and consistently applied by the Union's members. At world level, consistent and high-standard controls by all exporters of strategic goods can reduce the risk that dual-use goods are procured from countries applying weak controls. To this goal, outreach programs promote best practices for effective controls worldwide. Finally national and coordinated customs controls constitute a further barrier to illegal exports of strategic items.

Despite the inherent global dimension of strategic trade and controls, *no specific data are available for the analysis of global strategic trade*. National export license statistics, for instance, are not published by most exporting countries.

It can be expected the effectiveness of controls be enhanced in several ways by the analysis of trade data related to strategic goods. By understanding the national and global trade in strategic commodities, officials may better focus controls, select companies for in-reach and audit, target transactions for analysis, verification, and inspection, and assess potential economic impacts of trade control policies.

This contribution suggests using open source data on global trade to support strategic trade analysis. The data are *not specific* to strategic trade, but *related to it*, providing a portrait of strategic trade in the absence of more specific data. The chapter covers data sources, a methodology for data selection, underpinning tools and data use cases. The context for strategic trade analysis is first briefly recalled by introducing lists of items subject to export controls, with a focus on the nuclear-related ones.

2 Reference Lists of Strategic Items

Exports of strategic items are subject to authorisation in States adhering to internationally agreed controls, including the Wassenaar Arrangement on military capabilities [1], the Missile Technology Control Regime [2], the Nuclear Suppliers Group (NSG) [3, 4], the Australia Group on chemical and biological weapons [5], and Chemical Weapons Convention [6]. To a large extent, controls relate to items enumerated and defined in lists issued by the regimes. Control lists are revised over time in order to reflect the evolution of items and technologies.

In the EU, Annex I to Regulation 428/2009 on dual-use controls[1] [7] merges the regimes' lists into a single, comprehensive list of items. Since its origin[2] in 1994, Annex I has emerged as a de facto international reference for defining the bulk of dual-use items subject to controls. Annex I is structured in 10 Categories of items, listing over 1800 items[3] [9] (Fig. 1, top chart).

Nuclear items included in Annex I to the dual-use Regulation reflect the NSG two sets of Guidelines on nuclear exports (Part 1, [10]) and on nuclear-related exports (Part 2, [11]). Part 1 is the Guidelines for nuclear transfers [10], identifying about 130 especially designed or prepared (EDP) items for nuclear use[4] (Fig. 1, middle chart). This list corresponds to Category 0 in Annex I on 'Nuclear materials, facilities and equipment'. Part 2 of the NSG controls is the Guidelines for transfers of nuclear-related dual-use[5] equipment, materials, software and related technology [11], identifying about 170 items (Fig. 1, bottom chart) *'that can have a major contribution to an unsafeguarded nuclear fuel cycle or nuclear explosive activity, but have non-nuclear uses as well, for example in industry'* [4]. In Annex I to the dual-use Regulation, these items are distributed in four Categories: Category 1 on 'Special materials and related equipment', Category 2 on 'Materials processing', Category 3 on 'Electronics', and Category 6 on 'Sensors and lasers'.

The NSG lists are also related to the Model Additional Protocol [13] (AP) by the International Atomic Energy Agency (IAEA). States signing a AP are required, *inter alia*, to provide the IAEA with information about certain activities and imports/exports of nuclear relevance. Specifically, information about activities referred to in article 2.a.(iv) of the AP and listed in its Annex I (e.g., the manufacture of centrifuge rotor tubes or the assembly of gas centrifuges) is related to items listed in the NSG Part 2 Guidelines (e.g., item 1.B.1. Flow-forming machines, spin-forming machines). Import/export information referred to in article 2.a.(ix) of the AP concerns specified equipment and non-nuclear material listed under Annex II of the AP, this list being referrable to the NSG Part 1 Guidelines.[6]

---

[1]Regulation 428/2009 is referred to in this chapter also as 'the dual-use Regulation' or simply 'the Regulation'.

[2]See Davis [8], pp. 57–68, for an historical account on the development of the EU Regulation on dual-use exports.

[3]Number of leaf nodes counted in Annex I tree of items.

[4]Equipment or material especially designed or prepared for the processing, use or production of special fissionable material, as defined in article III.2 of the Treaty on the Non-Proliferation of Nuclear Weapons [12]. At its origin, the NSG Part 1 list incorporated the 1974 Trigger List by the Zangger Committee on items whose transfer would 'trigger' a requirement for safeguards (see [4], p. 3).

[5]The term 'dual-use' as used in the NSG context indicates items that can have both nuclear and non-nuclear uses. The same term 'dual-use' is used differently in the broader context of strategic trade controls, where it refers to items that can have both military and civil uses, as seen in the Introduction.

[6]Annex II of the Additional Protocol is based on Annex B of the NSG Guidelines Part 1 published in October 1995.

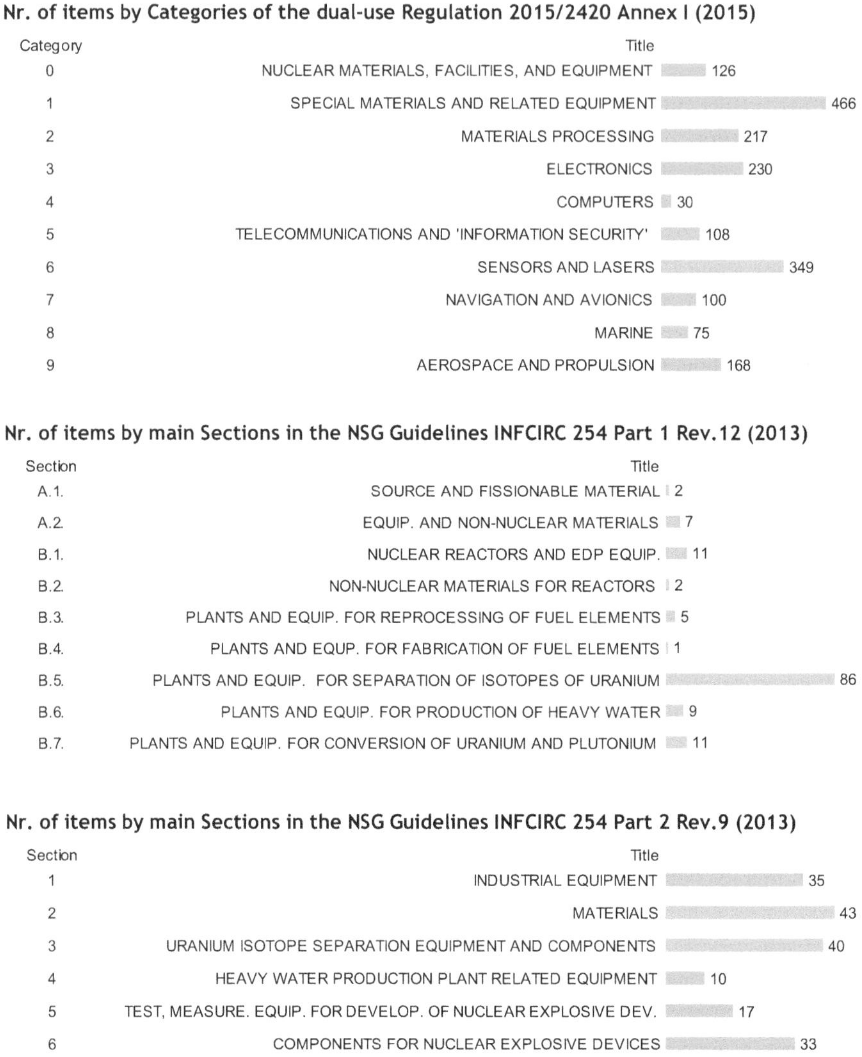

Fig. 1 Main headings and number of items listed in: Annex I of the EU dual-use Regulation for export controls (top chart); the NSG Part 1 Guidelines on nuclear exports (middle chart); the NSG Part 2 Guidelines on nuclear-related exports (bottom chart)

3 Data Sources for Trade Analysis

The world merchandise trade statistics is available from open source databases on international trade. A catalogue on main trade data sources resulting from a survey is provided in [14]. The data originate from declarations of goods to customs, with trade flows reported independently by the importer and the exporter. Customs data

Table 1 Some trade data sources presented by key features

Data source name	Data granularity	Temporal granularity	Geographical coverage	Access cost
UN COMTRADE	Statistical	Year/month	Global	Free
BACI	Statistical	Year	Global	Free
Global Trade Atlas	Statistical	Year/month	Multi national	Pay
COMEXT	Statistical	Year/month	EU Member States	Free
CTI	Transactional	Day	National (China)	Pay
INFODRIVE	Transactional	Day	Multi national	Pay

are collected and processed by national statistical offices, and shared[7] in global databases making the world trade measurable and amenable to analysis.

Key features describing trade sources are [14]: (i) the trade granularity (transactional or statistical), (ii) the temporal granularity (day, month, year), (iii) the geographical coverage (national, multi-national, global), and (iv) the cost of access (free, pay). Trade granularity refers to the level of aggregation of the data. Transactional data are based on import-export declarations to customs or shipment data (e.g., bills of lading). Statistical data result from the aggregation of transactional data records along several dimensions, including the commodity traded and time, and the countries reporting imports and exports. Examples of trade databases are UN COMTRADE [15], BACI [16], Global Trade Atlas [17], COMEXT [18], CTI [19], INFODRIVE [20]. Table 1 describes these sources by key features.

The choice of the data source to use depends on the scope of the analysis and level of trade detail sought. For example, to inform policies for export controls or to shape outreach programs, statistical trade sources are the appropriate choice. For enforcement, audit or specific verification tasks, the use of transactional sources may be necessary.

From the data access perspective, merchandise trade data can be retrieved either by queries on the data provider web site, or by bulk downloads of data records by whole files or using a dedicated Application Protocol Interface, when available. The latter two modalities are preferable when large trade data sets are needed for the analysis. For example, analyses in support to policies for export controls are based on measuring trade flows between countries on several hundreds of commodities, requiring the download and management of large data sets (see Sect. 5.2).

Typical statistical trade data records include standard dimensions and measures such as:

- the data *reporting country*;
- the *country partner* in trade;
- the *commodity traded*, e.g. described in the Harmonized System [21] product nomenclature, or in the reporter's national subdivision of the HS (read below);

[7] Respecting national provisions on data confidentiality.

Table 2 Example of trade data record

Reporter	Partner	HS code	HS description	Year	Flow	Value M USD	Quantity Kg
USA	Canada	2845.10	Heavy water "deuterium oxide"	2013	Import	11.041	23,024

Data source: BACI [16]

Table 3 Examples of HS codes and their description

HS code	HS description
Section V	Mineral products
26	Ores, slag and ash
2612	Uranium or thorium ores and concentrates
2612.10	Uranium ores and concentrates

- the trade *flow*, as import or export, re-import or re-export;
- the *time period* of when the trade has taken place;
- measures of *value* (e.g, in US Dollars) and *quantity* (e.g., in kg) of trade aggregated along the above dimensions.

An example of statistical data record on the trade of heavy water is shown in Table 2.

To be noted, traded commodities are reported in import-export databases according to the Harmonized System [21] (HS), the commodity classification system designed and maintained by the World Customs Organization (WCO).

The HS is based on about 5000 commodity groups organized within 22 Sections in a hierarchy made up of Chapters, Headings, Subheadings. Each level in the hierarchy is identified by an HS code and a description. Codes are 2-digit for Chapters, 4-digit for Headings and 6-digit for Subheadings. Table 3 shows a few HS codes with the corresponding textual descriptions.

The HS is amended every 5 years. New products are added and deprecated products removed to ensure the nomenclature reflects the evolution of the world's trade. The HS is adopted by States adhering to the WCO, but also by trade associations and statistical offices in the majority of countries. Import-export declarations to customs are also based on the HS. This is why HS descriptors are used in trade databases with customs origin.

At country level, national subdivisions of the HS at 8 or more digits exist that describe goods more precisely in relation to customs declarations. For instance, EU Member States use a 8-digit subdivision of the HS, the Combined Nomenclature [22] (CN), which includes about 15,000 commodity groups. As an example, Table 4 shows the CN codes subdividing HS code 2612.10 'Uranium ores and concentrates'. Some trade databases report data also in national nomenclatures (e.g., the 'Tariff lines' by UN COMTRADE [15], or GTA [17]). Because national nomenclatures generally differ between countries, world trade analysis can only focus on the HS level. However analysis on the trade by specific countries can

Table 4 Examples of CN codes and their description

CN code	CN description
2612.10.10	Uranium ores and pitchblende, with a uranium content of >5% by weight [Euratom]
2612.10.90	Uranium ores and concentrates (excl. uranium ores and pitchblende, with a uranium content of >5% by weight)

'drill-down' to greater detail making use of data reported according to national nomenclatures whenever available.

4 Mapping Strategic Items to Trade Descriptors

Because trade is reported in import-export databases by the Harmonized System, in order to retrieve relevant data for strategic trade analysis it is first required to map strategic items of interest to HS descriptors.

As a starting point, use can be made of existing tables relating export control lists with the HS. One such a table is the Correlation Table[8] [23] (CT), developed and maintained by the European Commission (EC), Directorate General on Taxation and Customs (DG TAXUD). The CT maps about 600 major items[9] listed in Annex I to the EU dual-use Regulation with about 1200 CN descriptors.[10] The table is revised annually to reflect the yearly amendment of Annex I (if any) and of the CN nomenclature.

Since CN codes are valid only in the EU, to make the Correlation Table usable at the world scale, it can be converted to HS codes by truncating the CN codes at the 6-digit level. By doing so, a reduced table is obtained of about 700 distinct HS

[8] Why was the Correlation Table developed? The CT is an instrument to inform EU exporters and customs officers on the restrictions that apply to the export of items defined in Annex I to the dual-use Regulation. As such, the CT is part of the 'Integrated tariff of the European Communities' or TARIC [24], a system which incorporates all Community legislation on trade concerning tariff suspensions, quotas, import/export prohibitions, surveillance, restrictions. TARIC identifies goods by 10-digit codes, a subdivision of the CN. Querying the TARIC database on such codes returns all legislative measure that may apply to the trade of the corresponding goods. For example, by querying TARIC on 2612101000 (code for 'Uranium ores and pitchblende', see Table 4) traders are made aware that an export authorization is required if the good to be exported is item 0C001 'Natural uranium ...' listed under the dual-use Regulation. It is to the exporter the final determination of whether the specific good to be exported fully corresponds to the descriptive text of the related controlled entry listed in Annex I.

[9] By major items it is meant items identified by a code of length 5, such as 0A001.

[10] BAFA, the German Authority for export controls, also developed and maintains a correspondence table (Umschlüsselungsverzeichnis) between items in Annex I of the dual-use Regulation and CN commodities. The table is published with explanatory notes in German under the site http://www.ausfuhrkontrolle.info/.

codes related to the 600 major items listed in Annex I to the dual-use Regulation. Since Annex I is a complete reference for strategic dual-use items (see Sect. 2), this HS-reduced CT offers a comprehensive list of trade descriptors for use in strategic trade analyses on the basis of import-export data sources with global geographical coverage.

If a focus on the nuclear-relevant part is of sought, entries for dual-use items related to the NSG lists can be selected in the CT. For example, all entries for Category 0 items of Annex I map to HS descriptors relevant to the NSG Part 1 list. In the same way, selected items from Categories 1, 2, 3, and 6 that map to the NSG Part 2 list, provide other HS descriptors of nuclear relevance. This approach for generating a Nuclear Correlation Table out of the original CT was first proposed and used in [25].

For strategic trade analysis, a further pre-compiled set of relevant HS descriptors is available from the Strategic Trade Control Enforcement (STCE) project [26] by the WCO. The scope of the STCE list is not as large as Annex I of the dual-use Regulation. The WCO identified a total of 94 'example' HS Headings and Subheadings related to strategic commodities. The purpose of the STCE list is awareness raising and training of customs and border agencies active in the prevention and detection of the trafficking of strategic goods.

Irrespective of which basket of HS/CN commodities is used for strategic trade analysis, one needs to be aware of the semantic gap existing between strategic items, as described in control lists such as the dual-use Regulation, and the HS/CN closest descriptors. In the general case, the correspondence between control lists and the HS/CN is a many-to-many relationship: a controlled item may correspond to one or more HS/CN codes, and vice versa. This happens primarily because the HS/CN languages are less expressive than the very technical language used in control lists.

For example, machine tools listed in Annex I to the Regulation (e.g., item 2B201) meet certain technical specifications that make them export-controlled. Technical specifications concern the number of axes of the machine, its positioning accuracy, and the possibility for the machine to be numerically controlled. These machine tools are described in HS/CN categories for numerically controlled machines which do not meet all other technical specifications captured in the Regulation. As a consequence, trade data records on those HS/CN categories will encompass export-controlled machine tools, but not exclusively.

There are favorable cases for which very precise HS/CN descriptors exist. For example, in the nuclear domain, precise descriptors include: uranium ores (2612.10), thorium ores (2612.20), natural uranium (2844.10), enriched uranium, plutonium (2844.20) depleted uranium, thorium (2844.30), spent fuel (2844.50), heavy water (2845.10), nuclear reactors (8410.10), machinery for isotopic separation (8410.20), fuel elements (8401.30), nuclear reactor parts (8410.40).

Because HS/CN descriptors have limited precision, the use of import-export databases for strategic trade analysis will provide, in most cases, upper bound figures of the 'truly' strategic trade flows between countries.

A recent study [27] estimated the amplification factor of truly strategic trade flows when measured on import-export databases using correlation tables. The study made use of COMEXT [18] data on EU trade, CN descriptors included in the EU CT, and aggregated figures of EU export licenses published by the EC [28]. The study found that the value of EU licenses appears amplified about 6.5 times, when measured on COMEXT data using the CT. This factor gives an overall quantitative indication of the level of data non-specificity to be expected when analyzing trade records with a view to strategic trade. Yet, to date, these trade data sources are the only openly accessible sources that can be related to strategic trade with a world coverage.

A final practical consideration on the use of correlation tables for trade analysis is that, given the size of control lists and the high number of trade descriptors of interest, part of the analytical work can be made easier by tools that relate control lists with the language of trade. Developed by the EC Joint Research Centre (JRC), The Big Table (TBT) [29–32] is such a tool. TBT is a search tool working on a collection of reference documents on lists of strategic items. The collection includes regulatory documents,[11] technical handbooks,[12] and trade nomenclatures.[13] An extended document collection [33] set-up for IAEA-only use includes also IAEA's Physical Model [36]. TBT stores documents both in PDF format and in database tables. It allows searching the document collection by text, but also by a large set of correspondence tables that relate items listed in different documents of the collection. For example, items in the NSG lists are related to Harmonized System descriptors via the dual-use Regulation and the EU Correlation Table. TBT is distributed to EU licensing and customs authorities who request it. Developed initially as a desktop application, TBT is currently being re-programmed for web-based use.

5 Use Cases for Strategic Trade Analysis

Use cases in strategic trade analysis can be presented by the granularity and amount of trade data required to address them, leading to micro or macro views of strategic trade.

Micro views find uses in verification tasks against detailed information, where few traded strategic commodities are involved, import or export concern specific entities, in reference to given points in time [29–31, 37, 38]. Typical applications of micro views include the validation of the correctness and completeness of State-declared information to the IAEA e.g. under the Additional Protocol (see Sect. 2), or investigations by enforcement agencies in cases of alleged illicit trade.

Macro views operate on regional and world scales. They involve hundreds of strategic commodities (e.g., related to whole control lists), in time series of up to a

[11] E.g., the EU dual-use Regulation, the NSG Guidelines, the Model Additional Protocol.

[12] E.g., the MTCR [34] and the AG handbooks [35].

[13] The Harmonized System and the Combined Nomenclature.

decade of years [27, 39–41]. Macro views are of use in the definition and monitoring of policies for export controls. They can help evaluating the industrial capability of countries in strategic goods by means of measuring a country's comparative advantage in the export of these goods. This information can be of value also for tailoring outreach programs and training activities in areas of technical relevance for the participating countries. Macro views can also find application in the planning of coordinated enforcement operations by customs, whereby authorities are offered awareness about global strategic-related trade flows in addition to national views only [41].

Micro and macro views are attained by different build-up and analytical processes, hereafter described and illustrated by means of examples.

5.1 Micro Views

Micro views are instrumental to the conduct of strategic trade case studies. A case may start by a piece of information or a hypothesis about trade activities on strategic items subject to export controls, between given entities, in a reference time frame. The goal of a study is to confirm or deny the information/hypothesis by 'an independent source', in this case, trade data from import-export databases.

Before retrieving any data, a first step is to identify clearly the items of interest to the case at hand. Regulatory documents and technical handbooks on materials and equipment subject to controls are of use here. Tools such as The Big Table are also a support to the process.

Identified items are then mapped to HS descriptors (e.g. with the help of correlation tables) in order to retrieve relevant trade data, notwithstanding two limitations: (i) HS descriptors generally only approximate targeted items; (ii) exporters and importers may (intentionally or unintentionally) have declared trade to customs in HS categories aside from the 'correct' ones.

With HS codes selected, a plan of queries on trade databases is designed to reflect the questions addressed by the case study. For example, querying for statistical data allows to specify as query parameters the country reporting trade, the HS commodities, the trade flow, the country partner in trade, the time period (see Sect. 3).

Results of queries return value and quantity of trade (if any) for assessment by the case analyst. The criteria adopted for the analysis are, in general, specific to the commodities at hand. However a generic useful criterion is the 'trade unit value', defined as the ratio between value and quantity traded for a given HS, in a temporal point (e.g. month or year). For limited traded quantities, the trade unit value approximates the price at which commodities have been traded. Knowing the market price for items targeted in a case study, one can compare it with the trade unit values of returned data records, thus reducing the data ambiguity introduced by the use of the HS. Data points compatible with the market price of targeted items are

Table 5 Country A exports to Country B on machine tools

HS code	HS description	Unit	Measure
8457.10	Machining centres for working metal	Value (USD)	1,076,148
		Quantity (no)	2
		Unit value	538,074
8459.61	Milling machines for metals, numer. controlled	Value (USD)	2,815,177
		Quantity (no)	1
		Unit value	2,815,177

highlighted as points of interest. These may confirm alleged information or provide new insights in the case study.

Table 5 illustrates a micro view for a case study of an alleged illicit export of machine tools between two countries. The table features trade records compatible with the case dimensions: HS codes relevant for machine tools, month of trade, and a high-unit value, expected for high-precision machines subject to controls. These records provide positive indication that an export of controlled machine tools may have taken place between the countries indicated in the case study.

5.2 Macro Views

Macro views are for describing, exploring, estimating global trade flows on large sets of strategic-related commodities. The commodity space can be as large as the whole Annex I to the dual-use Regulation [7], or as large as full control lists published by a specific regime (e.g., the NSG Guidelines [10, 11]), or as large as the WCO selection of commodities for strategic trade control enforcement [26] (see Sect. 4). Of interest is to profile the world trade in these groupings of goods.

Existing correlation tables provide an useful starting point to determine HS descriptors in scope for the chosen reference control list(s).

Data access on hundreds of HS commodities for trade reported by close to 200 countries over several years entails significant data downloads from import-export databases, and an infrastructure to store the data in a permanent, repository to be refreshed over time.

On the bulk kernel of raw data records for the standard dimensions and measures of trade (i.e., reporter, partner, flow, HS commodity, period, value, quantity), meta-data can be added to enable, at analysis time, the presentation, organisation, aggregation/disaggregation, exploration, and filtering of the raw kernel data by meaningful levels of abstraction, to 'tame' the large amount of data at hand, and facilitate data understanding and insights. Examples of meta-data to enrich raw trade records are:

- The HS hierarchy (Chapters, Headings) above the Subheadings in scope. This allows aggregating/disaggregating the data on HS compact or detailed descriptors to provide analysts with data overviews or details on demand.
- A geographical hierarchy by world's regions and sub-regions for the data reporters and/or partners. This allows reading the trade data by geographical areas, when of use.
- The reference framework of control lists. This allows referring the HS trade records to control entries using correlation tables, notwithstanding the inherent precision limit of correlation between nomenclatures.

These data preparation steps are functional to the setting-up of strategic trade atlases, examples of macro views representation of strategic-related trade organised by country, covering the whole world. As an illustration of atlases, Fig. 2 shows a template for a nuclear-related country profile. The profile is built on a basket of close to 200 HS commodities related to the NSG Guidelines Parts 1 & 2 by a Nuclear Correlation Table derived from the EU Correlation Table on the dual-use Regulation (see Sect. 4). The profile features a country's main import and export trade flows on HS commodities, as well as the country's main partners in trade.

Specifically, the profile includes the following elements (Fig. 2):

a. *Meta-data (Text)*—The source of trade data, date of data download, revision of the NSG lists, revision of the EU dual-use Regulation, revision of the Correlation Table used to relate NSG items to HS commodities.
b. *Country name (Text)*—Name of selected country (reporter of trade data).
c. *Select years (Menu)*—Selection of years to be included in the view.
d. *Records in view (Text)*—Number of database records included in the view.
e. *Trade balance (Heat map)*—The reporter's import-export balance by value.
f. *Trade by regions (Heat map)*—The reporter's trade value by world sub-regions [42].
g. *Top imp to, Top exp to (List)*—The reporter's top partners in trade by value, in import and export.
h. *(Maps with bubbles)*—The reporter's import and export value by partner.
i. *Trade by NSG headings (Heat map)*—The reporter's trade by value on headings of the NSG Guidelines Part 1 [10] (Fig. 1, middle chart) and Part 2 [11] (Fig. 1, bottom chart). The data view is generated by relating HS commodities to NSG items using the Nuclear Correlation Table.
j. *Top HS2 commodities (Heat map)*—The reporter's top HS Chapters (HS2) by value in import and export.
k. *Top HS6 commodities (Bar chart)*—The reporter's top HS Subheadings (HS6) by value in import and export. Bars are sliced by colors reflecting a partner origin in world sub-regions. Main partners are labelled by ISO2 3166 country codes [43].
l. *Footnotes (Table)*—The table relates HS commodities to NSG items. Codes and descriptive texts in the respective nomenclatures are provided.

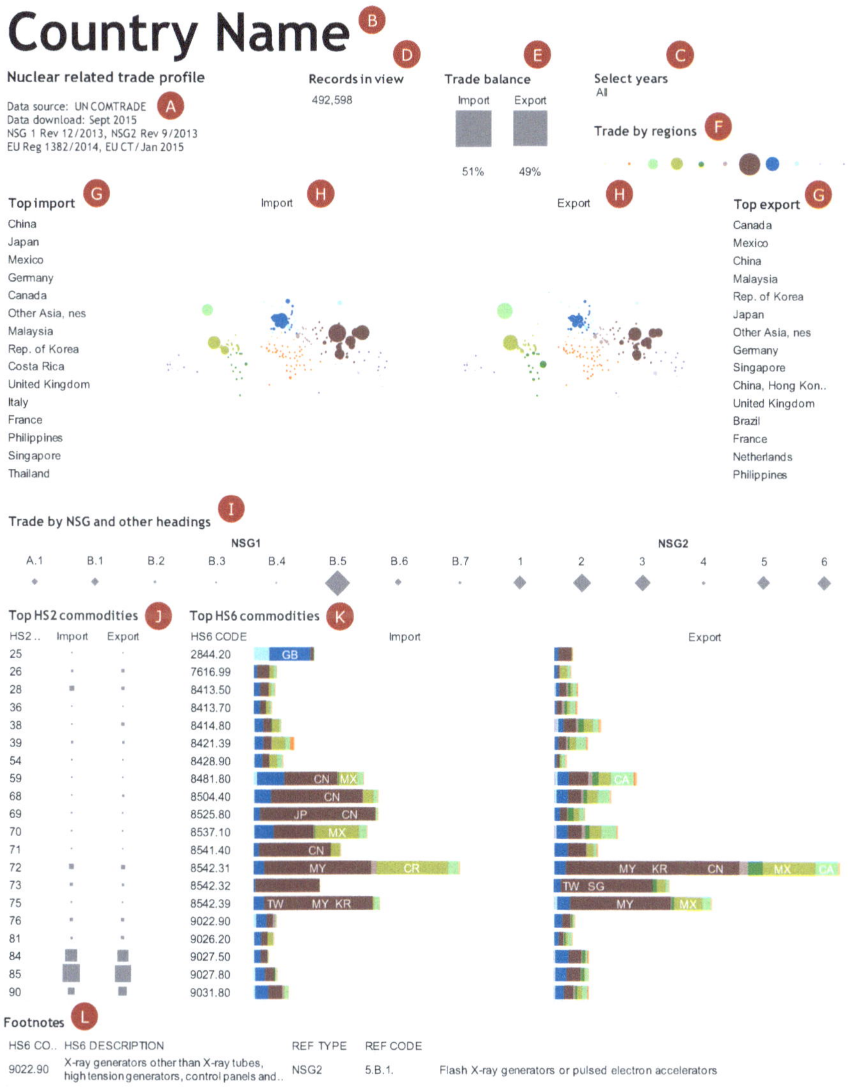

Fig. 2 Nuclear related country profile

The profile, built with [44], is designed to respond to interaction by the analyst. Each graphical element in the profile implements filters or highlight actions that specialise what is being displayed in the view. An example of the result of a filtering action is provided in Fig. 3. By selecting the NSG Part 1 heading B.2 on ‘Non-nuclear materials for reactors’, all data in the profile is filtered to keep in scope only the trade on HS commodities related to this NSG heading. This includes heavy water (related to HS 2845.10) and nuclear-grade graphite (related to HS 3801.10, 3801.90,

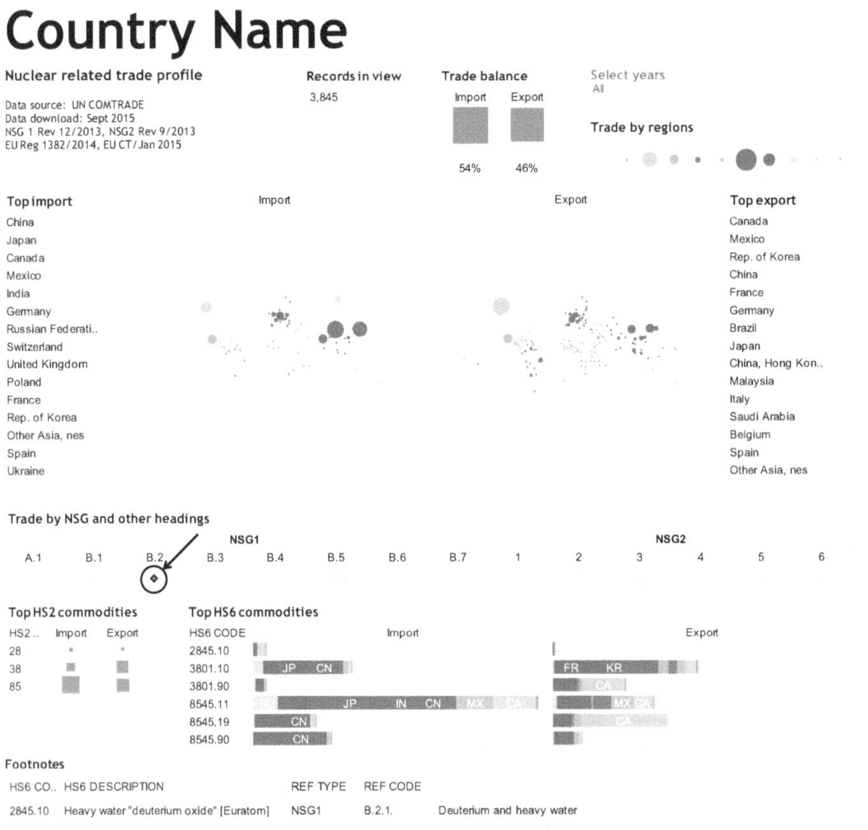

Fig. 3 How the profile reacts to interaction by the analyst

8545.11, 8545.19, 8545.90). Note that all elements in the view (nr. of records, trade balance, trade by regions, top partners in import and export) are now referred only to this limited set of goods. Used in this way, the profile becomes an effective tool to browse very large data sets in relation to analytical goals.

A related example of strategic trade atlas was produced for print in a public report on the STCE list of commodities [26] set-up by the WCO [40, 41]. Distributed to Member States of the WCO, the atlas represents an effort to promote and facilitate the use of trade data to enhance the effectiveness and efficiency of strategic trade controls.

6 Conclusions

The book on verifying treaty compliance [45] published in 2006 featured an article by JRC on open source information collection, processing and applications [46]. The article included a Section[14] on export control of goods and technologies, referring to the NSG lists, the EU dual-use Regulation, and the *'growing need to identify multiple and independent sources of information that would allow cross-checks and the verification of the implementation of this Regulation'*. The use of databases on international trade as a data source for monitoring the application of such agreements and regulations was first suggested there. The article contained in seed what developed, over a 10 year period, as a series of JRC activities and collaborations outlined here.

The first JRC activity in the field of strategic trade analysis started in 2007 as an EC Support Task to the IAEA Department of Safeguards. The Task, still active, explores uses of import-export databases in support to safeguards [47–49]. JRC work focuses on methodological aspects, with outcomes: (i) the catalogue of web data services on global trade [14] (Sect. 3); (ii) the development of The Big Table tool for relating reference lists of strategic items with trade nomenclatures [29–33] (Sect. 4); (iii) the development of a time-efficient process for the analysis of large data sets [39] based on trade data warehousing and the use of a commercial tool [44] for data visualisation and analysis, as illustrated in Sect. 5 by the example on the nuclear-related trade atlas. On the use of these data sources for safeguards, the IAEA reported that trade data of customs origin can be used by the IAEA for the State Evaluation Process to [50]: *(a) identify possible nuclear related cooperation between States by monitoring relevant trade flows over time; (b) identify indicators of safeguards relevant activities (link between AP Annex I activities and dual-use items); (c) identify possible indicators of incomplete declarations of exports of AP Annex II items; (d) contribute to identify safeguards relevant capabilities or missing capabilities of States.*

As of 2011, a second line of work on strategic trade is the analytical, substantive work performed in relation to EU dual-use Regulation. The EC Directorate General for Trade (DG TRADE) chairs the Dual-Use Coordination Group (DUCG) participated by EU Member States (MS). The DUCG examines primarily questions related to the implementation of dual-use controls in the EU. In this context, JRC supports the DUCG by collecting quantitative data about licenses. EU aggregated figures about licenses are published annually in an EC report to the Council and the European Parliament (e.g., [28]). License data are complemented by statistical figures obtained using the COMEXT database [18] on EU trade and the Correlation Table [23]. As noted in Sect. 4, the trade area identified by the Correlation Table is larger than the strict dual-use trade because of the incomplete descriptive power of the HS/CN languages. JRC measured this large trade area to value about 20% of the

[14] Written by one of the authors of the present chapter.

intra- and extra-EU total export [28]), while the dual-use part of it is estimated to value about 3% of the intra- and extra-EU total export [27]. These figures, together with the underlying estimation method [27], help in assessing (i) the impact of strategic trade regulations, (ii) the economic value of the dual-use sector in the EU, and (iii) policy options in the on-going review of the EU export control system [51]. Recently, trade analyses have been extended to cover non-EU countries in support to the definition of the EU Outreach Programme on export controls [52] funded by the EC Directorate General for International Cooperation and Development (DG DEVCO). Of interest is to tailor the programme by the technical competence areas in strategic trade of participating countries. This can be indicated by trade flow analysis performed in the scope of the dual-use Regulation reference list applied to the world scale. To this goal, a dedicated atlas is hence being developed on the world dual-use related trade.

A third line of work on research and development for strategic trade analysis is carried out jointly in a collaborative project with Argonne National Laboratory [40, 41]. Activities include: (i) establishing new approaches for profiling strategic trade by countries or by commodities; (ii) designing and developing interactive tools for independent use by trade analysts; and (iii) raising further awareness on uses of import-export databases for strategic trade analysis. The atlas [53, 54] produced jointly on the Strategic Trade Control Enforcement list of commodities set-up by the WCO is a tangible outcome of this collaborative project.

Acknowledgements The work presented in this chapter stem from activities performed in the project Information Analysis and methodologies for Nuclear Non-Proliferation and security, work package on Strategic Trade Analysis for Non-Proliferation (STANP), funded by (i) the European Commission (EC), Euratom Horizon 2020 Research and Training Programme, (ii) Administrative Arrangement JRC.ITU.AA.32608-2011 between DG TRADE and JRC, (iii) Administrative Arrangement IFS/2012/301-675 (as amended in 2014) between DG DEVCO and JRC. STANP contributes to the EC Support Programme to the IAEA under Task EC D 1662, and to the Collaboration Agreement between the United States Department of Energy (DOE) and the European Atomic Energy Community (EURATOM) under Action Sheet 57.

References

1. Wassenaar Arrangement. http://www.wassenaar.org. Accessed 28 Feb 2019
2. Missile Technology Control Regime. http://www.mtcr.info. Accessed 28 Feb 2019
3. Nuclear Suppliers Group. http://www.nuclearsuppliersgroup.org. Accessed 28 Feb 2019
4. Nuclear Suppliers Group (2015) The Nuclear Suppliers Group: Its Origins, Role and Activities. INFCIRC/539/Rev.6
5. Australia Group. http://www.australiagroup.net. Accessed 28 Feb 2019
6. Chemical Weapons Convention. http://www.australiagroup.net. Accessed 28 Feb 2019
7. Council Regulation (EC) No 428/2009 setting up a Community regime for the control of exports, transfer, brokering and transit of dual-use items (Recast)
8. Davis I (2002) The Regulation of Arms and Dual-Use Exports: Germany, Sweden and the UK. Oxford University Press, ISBN 0-19-925219-X

9. Commission Delegated Regulation (EU) No 2015/2420 of 12 October 2015 amending Council Regulation (EC) No 428/2009 setting up a Community regime for the control of exports, transfer, brokering and transit of dual-use items
10. Nuclear Suppliers Group (2013) Guidelines for Nuclear Transfers, INFCIRC/254/Rev.12/Part 1
11. Nuclear Suppliers Group (2013) Guidelines for Transfers of Nuclear-Related Dual-Use Equipment, Materials, Software and Related Technology, INFCIRC/254/Rev.9/Part 2
12. Treaty on the Non-Proliferation of Nuclear Weapons (NPT). http://www.un.org/disarmament/WMD/Nuclear/NPTtext.shtml. Accessed 28 Feb 2019
13. IAEA (1997) INFCIRC/540 (Corrected). Model Protocol Additional to the Agreement(S) Between State(S) and the International Atomic Energy Agency for the Application of Safeguards
14. Versino C, Tsukanova M, Cojazzi GGM (2010) Catalogue of WEB Data Services on Global Trade. EUR 24657 EN, ISBN 9789279189197, JRC62363
15. UN COMTRADE. http://comtrade.un.org. Accessed 28 Feb 2019
16. BACI. http://www.cepii.fr/CEPII/en/bdd_modele/presentation.asp?id=1. Accessed 28 Feb 2019
17. Global Trade Atlas. https://www.gtis.com/gta/. Accessed 28 Feb 2019
18. COMEXT. http://epp.eurostat.ec.europa.eu/newxtweb/. Accessed 28 Feb 2019
19. CTI China Trade Information. http://info.hktdc.com/chinastat/gcb/index2.htm. Accessed 28 Feb 2019
20. INFODRIVE India. http://www.infodriveindia.com. Accessed 28 Feb 2019
21. Harmonized System by the World Customs Organization. http://www.wcoomd.org/en/topics/nomenclature/overview/what-is-the-harmonized-system.aspx. Accessed 28 Feb 2019
22. Combined Nomenclature site by DG TAXUD. http://ec.europa.eu/taxation_customs/customs/customs_duties/tariff_aspects/combined_nomenclature. Accessed 28 Feb 2019
23. EU Correlation Table document by DG TAXUD, updated January 2015. http://trade.ec.europa.eu/doclib/docs/2015/january/tradoc_153050.xlsx. Accessed 28 Feb 2019
24. TARIC consultation site. http://ec.europa.eu/taxation_customs/dds2/taric. Accessed 28 Feb 2019
25. Versino C (2007) Export Control of Dual-Use Items and Technology: The Nuclear Correlation Table. JRC40413
26. World Customs Organization (2014) Strategic Trade Control Enforcement (STCE) Implementation Guide. http://www.wcoomd.org/en/media/newsroom/2014/june/~/link.aspx?$_$id=A3AD81E1BFC742F2806DCAC55C54B743&$_$z=z. Accessed 16 May 2018
27. Versino C (2015) Dual-use Trade Figures and How They Combine. JRC97664, EUR 27514 EN, ISBN 978-92-79-52715-9, https://doi.org/10.2789/439924
28. Report from the Commission to the European Parliament and the Council on the implementation of Regulation (EC) No 428/2009 setting up a Community regime for the control of exports, transfer, brokering and transit of dual-use items. COM(2015) 331 final, 10.7.2015.
29. Versino C, Contini F, Cojazzi GGM (2011) The Big Table. An Information Tool on Items Listed for Export Controls. In: Proc. 33rd ESARDA Annual Meeting, Budapest
30. Versino C, Contini F, Cojazzi GGM (2011) An Information Tool on Items Listed for Export Controls. In: Proc. 52nd INMM Annual Meeting, Palm Desert, CA USA
31. Versino C, Contini F, Chatelus R, Cojazzi GGM (2011) The Big Table. A Software Tool for Nuclear Trade Analysis. EUR 24823 EN, ISBN 978-92-79-20297-1, JRC62804
32. Cojazzi GGM, Versino C, Wolfart E, Renda G, Janssens W (2014) Tools for Trade Analysis and Open Source Information Monitoring for Nonproliferation. In: Proc. Symposium on International Safeguards: Linking Strategy, Implementation and People. IAEA, Vienna
33. Versino C, El Gebaly A, Marinova E, Pasterczyk C, Cojazzi GGM (2013) Integrating IAEA's Physical Model with JRC's The Big Table document search tool. EUR 26215 EN, ISBN 978-92-79-33675-1, JRC77541
34. Missile Technology Control Regime (2017) Annex Handbook. http://mtcr.info/mtcr-annex/. Accessed 28 Feb 2019

35. Australia Group Common Control Lists Handbooks. http://www.australiagroup.net/en/controllisthandbooks.html. Accessed 28 Feb 2019
36. Liu Z, Morsy S (2007) Development of the Physical Model. IAEA-SM-367/13/07, IAEA, Vienna
37. Versino C, Cojazzi GGM, Contini F (2010) Understanding Nuclear Trade: Data Sources and Tools. In: Proc. Symposium on International Safeguards: Preparing for Future Verification Challenges, Vienna
38. Cojazzi GGM, Versino C, Wolfart E, Renda G, Janssens W (2014) Trade Analysis and Open Source Information Monitoring for Non Proliferation. Information Analysis Technologies, Techniques, and Methods for Safeguards, Nonproliferation and Arms Control Verification Workshop. INMM, Portland, Oregon USA
39. Versino C (2014) Pattern Recognition by Humans and Machines. In: Proc. Symposium on International Safeguards: Linking Strategy, Implementation and People, Vienna
40. Heine P, Versino C (2015) A Strategic Trade Atlas For Non-proliferation and Export Control. In: Proc. 56th) INMM Annual Meeting, Indian Wells, CA USA
41. Versino C, Heine P (2016) Strategic Trade Atlas, Country-Based views. EUR 27739 EN, ISBN 978-92-79-55003-4, JRC100392
42. Geonomenclature site by Eurostat. http://ec.europa.eu/eurostat/ramon/other_documents/geonom/. Accessed 28 Feb 2019
43. ISO 3166 Country codes. The International Standard for country codes and codes for their subdivisions. http://www.iso.org/iso/home/standards/country_codes.htm. Accessed 28 Feb 2019
44. Tableau software. http://www.tableausoftware.com. Accessed 28 Feb 2019
45. Avenhaus R, Kyriakopoulos N, Richard M, Stein G (Eds) (2006) Verifying Treaty Compliance. Limiting Weapons of Mass Destruction and Monitoring Kyoto Protocol Provisions. Springer-Verlag Berlin Heidelberg. ISBN 978-3-540-33854-3
46. Bril L-V, Goncalves JGM (2006) Open Source Information Collection, Processing and Applications. In: Avenhaus R, Kyriakopoulos N, Richard M, Stein G (Eds) Verifying Treaty Compliance. Limiting Weapons of Mass Destruction and Monitoring Kyoto Protocol Provisions. Springer-Verlag Berlin Heidelberg. ISBN 978-3-540-33854-3
47. Versino C, Chatelus R, Cojazzi GGM, Contini F, Tarvainen M, Tsois A (2009) Exploring the Use of Trade Data to Support Safeguards Verification Activities. In: Proc. 31st ESARDA Annual Meeting, Vilnius
48. Versino C, Chatelus R, Cojazzi GGM, Contini F, Tarvainen M, Tsois A (2010) World Trade Data: Can They Help Safeguards Verification Activities? In: Proc. 51st INMM Annual Meeting, Baltimore, MD USA
49. Versino C, Chateluy R, Cojazzi GGM, Contini F, Tarvainen M, Tsois A (2010) Global Trade Data to Support IAEA Safeguards. ESARDA Bulletin 45:14–22
50. Marinova E, Ardhammar M, El Gebaly A, Shot P, Ferguson M (2014) Analysis of Nuclear Relevant Information on International Procurement and Industrial Activities for Safeguards Purposes. In: Proc. Symposium on International Safeguards: Linking Strategy, Implementation and People. IAEA, Vienna
51. Communication from the Commission to the Council and the European Parliament. The Review of export control policy: ensuring security and competitiveness in a changing world. COM(2014) 244 final, 24.4.2014
52. EU P2P Export Control Programme. https://export-control.jrc.ec.europa.eu/Home/Dual-use-trade-control. Accessed 28 Feb 2019
53. Versino C, Heine P, Carrera J. Strategic Trade Atlas. Country-Based Views. (2018) ISBN 978-92-79-84030-2, 2018. http://publications.jrc.ec.europa.eu/repository/handle/JRC111470. Accessed 28 Feb 2019
54. Versino C, Heine P, Carrera J. Strategic Trade Atlas. Commodity-Based Views. (2018) ISBN 978-92-79-84031-9, 2018. http://publications.jrc.ec.europa.eu/repository/handle/JRC111471. Accessed 28 Feb 2019

Part VI
Applications

Is it practical to drive the development of systems concepts for verification of compliance on a state level given the complexity of the arms control and nonproliferation issues? Different groups of experts and possible customers around the world, from politics, military and research, engaged in discussions, workshops and projects to explore new ideas and innovation. Some of the concepts were tested on scenarios to evaluate the feasibility of applying them to particular problems and identify gaps for further research. It is hoped that these results will provide guidance for future initiatives to implement innovation and a system concept for successful nuclear nonproliferation and arms control.

A Systems Approach for Evaluating a Warhead Monitoring System

Cliff Chen, Sharon DeLand, and Thomas Edmunds

Abstract Future agreements that limit and reduce total warhead stockpiles may require monitoring nuclear warheads, potentially necessitating new monitoring approaches, technologies, and procedures. A systems approach for evaluating a warhead monitoring system can help ensure that it is evaluated against all requirements, and that the evaluation is objective, standardized, transparent, and reproducible (not analyst dependent). All engineered systems are designed to meet a set of requirements based on inputs and assumptions about the environment and processes. For arms control monitoring, identifying requirements is challenging because treaty objectives are often strategic or political, and must be translated into technical monitoring objectives that are agreed upon by multiple stakeholders. The monitoring objectives are the foundation of the evaluation framework, which also includes evaluation scenarios that reflect strategic concerns, performance metrics based on detection goals, and a functional architecture of how objectives are achieved by system components. This framework embodies the system requirements against which the monitoring system can be evaluated. Computational simulations, when validated by experimental activities and field exercises, can help characterize monitoring system performance. One such class of simulations is the discrete-event simulation, which can cohesively model weapons enterprise processes and monitoring and inspection activities. In conjunction with analysis algorithms that correlate inspection outcomes with declarations and other information streams, these simulation results can be used to quantify the confidence with which a monitoring system can detect and differentiate scenarios. The capability to evaluate the effectiveness of monitoring options and explore tradeoffs is essential in supporting

C. Chen (✉) · T. Edmunds
Lawrence Livermore National Laboratory, Livermore, CA, USA
e-mail: chen41@llnl.gov; edmunds2@llnl.gov

S. DeLand
Sandia National Laboratories, Albuquerque, NM, USA
e-mail: smdelan@sandia.gov

I. Niemeyer et al. (eds.), *Nuclear Non-proliferation and Arms Control Verification*, https://doi.org/10.1007/978-3-030-29537-0_26

technical design activities, guiding future R&D investments and supporting future treaty negotiations.

1 Introduction

Past and existing agreements that limit the numbers of nuclear weapons focused on nuclear weapon delivery systems and the warheads deployed on these systems. Any future agreements to limit or reduce total numbers of weapons—deployed and non-deployed—may require monitoring nuclear warheads in other parts of the weapons enterprise. This may pose a new set of verification challenges requiring new monitoring approaches, technologies, and procedures. The monitoring system for such agreements may be more intrusive, particularly as additional elements of the warhead lifecycle are included, resulting in significant operational, safety, and security challenges.

Nuclear warheads are small, difficult to authenticate while also protecting sensitive information, and subject to regular movements and lifecycle processes (e.g., deployment, storage, maintenance, transportation, and dismantlement) Changes in policies and technologies could impact enterprise procedures and further compound the dynamic nature of the enterprise. The ability to reduce these uncertainties may be restricted due to security considerations, necessitating alternate approaches.

The definition of treaty objectives helps establish verification objectives, which in turn informs the requirements for the monitoring system. The monitoring system's performance in providing confidence that declarations are correct and complete must be balanced against additional acceptability factors, such as the risk of information disclosure and the impact on facility operations. To assist technical design activities, guide future R&D investments, and support future treaty negotiations, the capability to evaluate the effectiveness of monitoring options and explore tradeoffs is essential.

This chapter will summarize an evaluation methodology [1], describe how it might be adapted to support the evaluation of a warhead monitoring system, and illustrate its use through an example analysis. This chapter does not analyze what will be needed for future warhead monitoring, but rather is intended to present an approach that could be adapted to support such an analysis. The focus is on the cooperative monitoring[1] of declared facilities that operates in conjunction with other existing monitoring activities, such as national technical means, to support a comprehensive compliance assessment. The detection of undeclared facilities, an important aspect of a monitoring regime, is not considered in this chapter.

[1]"Monitoring" is used in this chapter as shorthand for "cooperative monitoring", which refers to procedures and activities agreed upon by all parties. In other contexts, it refers to the totality of data-collection activities related to a compliance judgment.

2 Evaluation Methodology

A standard systems engineering design process includes identifying system requirements, designing a system that meets those requirements, evaluating performance at the system, sub-system, and component levels, and using assessments in a spiral development process to improve system design [2].

Figure 1 illustrates the evaluation process. System requirements are translated into an evaluation framework against which monitoring options are evaluated. This framework helps ensure that an evaluation is objective and standardized, transparent, and reproducible (not analyst dependent). The evaluation assesses the monitoring system's performance against the evaluation framework through a combination of analytic and mathematical analysis, simulations, component testing, and integrated field testing as the system matures. The evaluation results are then factored into the continued development of the approach until an optimal or consensus system is achieved. Since the terms of future agreements are subject to negotiation and thus unpredictable, a number of monitoring options that span the design space may need to be developed, tested, and matured.

2.1 *Evaluation Framework*

All engineered systems are designed to meet a set of requirements based on inputs and assumptions about the environment and processes. These requirements identify

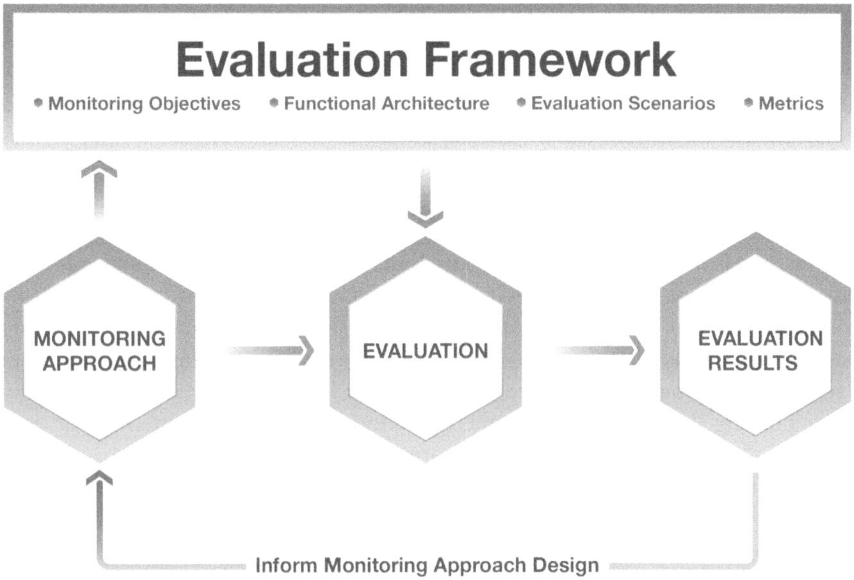

Fig. 1 Evaluation methodology overview

what the system needs to do, the conditions under which the system should perform, and the level of performance the system must achieve. A system's performance is then evaluated against these requirements. The evaluation framework presented here focuses on performance against monitoring objectives and is comprised of those objectives, a functional architecture, evaluation scenarios, and performance metrics. This approach described here has analogues to the Department of Defense's Capabilities-Based Assessment process [3].

A state's evaluation framework is often very sensitive as it may reflect its strategic concerns and risk perceptions. However, an evaluation framework that is open and transparent can help frame options for domestic stakeholders and facilitate understanding and confidence, to both domestic and international parties, that a monitoring approach has been carefully considered.

Monitoring objectives are derived from a treaty's central limits, and usually reflect where states desire greater confidence in a treaty partner's compliance. States enter arms control treaty negotiations having carefully considered the strategic and political context, along with the costs and benefits of reciprocal monitoring. This analysis supports states' positions on a treaty's central limits and the verification regime. For a complex verification regime, a systematic analysis of pathways that could be employed to circumvent the treaty might be helpful in identifying where monitoring might best detect or deter cheating.

For example, one possible monitoring objective for a treaty limiting total warhead inventories is to confirm the declared inventory of treaty-accountable items (TAIs) at all locations within the declared enterprise. The term TAI is used because treaty definitions are carefully negotiated, and accountable items might be defined in ways that do not solely include real warheads. The declared enterprise would include all sites at which TAIs may be present.

The functional architecture describes the relationship between monitoring objectives and potential monitoring functions and sub-functions (what the system does and how it accomplishes it). The functional architecture maps the monitoring objectives onto a hierarchy of monitoring functions and sub-functions, and provides the structure for the development of derivative performance metrics.

A notional functional architecture is shown in Fig. 2. The lowest level sub-functions represent technology implementation options for a given function. The architecture is not specific to an approach, but rather a superset of what needs to be done and different ways of achieving it. A given monitoring approach implements some combination of the functions and sub-functions. The hierarchical structure creates traceability between the functions performed by the monitoring system and the monitoring objectives they support. This helps organize the diverse array of monitoring functions, supports the definition of performance metrics for each function, and facilitates a modular comparison of implementation options.

Evaluation scenarios describe the possible conditions under which the monitoring system is expected to perform. Evaluation scenarios are derived from verification concerns. They serve to specify the types and patterns of discrepancies that may be encountered that either need to be resolved when assessing compliance or are indicators of non-compliance. It is important to develop a comprehensive set

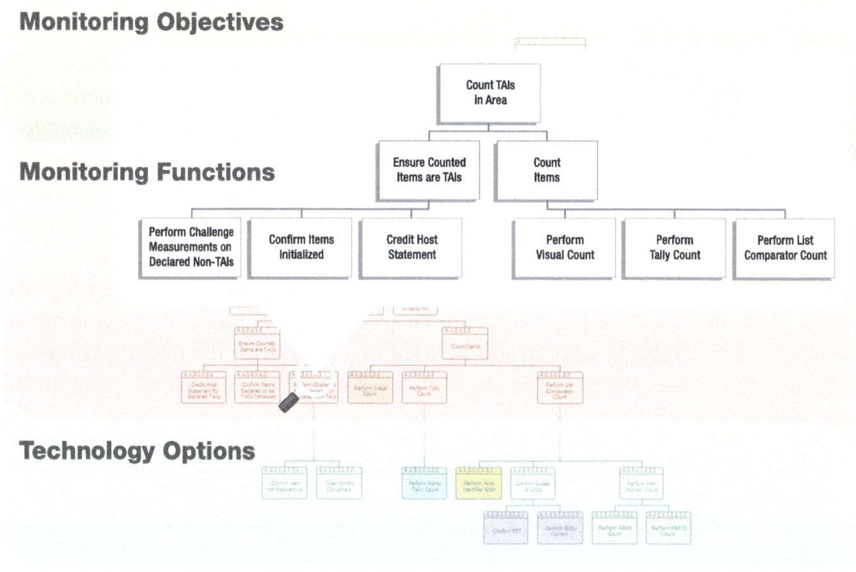

Fig. 2 Functional architecture

of scenarios that reflect verification concerns and potential cheating strategies. These scenarios can then be used to quantify system performance. Evaluation scenarios should also take into account honest mistakes that may occur, such as clerical errors in declarations or logistical and operational errors. Understanding the types of potential mistakes and their likelihoods may help identify activities an inspector can be authorized to undertake in order to resolve certain low-level discrepancies during an inspection and minimize the uncertainty introduced. This is important because these factors degrade system performance by introducing uncertainty in the information collected and can result in incorrect conclusions.

Identifying the correct metrics is a fundamental challenge in any evaluation activity. In this context, the objective of the monitoring system is to provide confidence that activities and inventories are consistent with declarations. The evaluation scenarios identify the types of manifestations that the monitoring system should be able to detect. The monitoring system should also be able to differentiate between expected and unexpected results (cheating scenarios vs. compliance), and to identify if anomalous or unanticipated conditions are present.

The inverse problem of linking a set of observations to a quantitative estimate and confidence level in the true state of a treaty partner's nuclear enterprise is challenging. Multiple evaluation scenarios, along with honest mistakes and technology errors may generate similar sets of events within the monitoring system and thus it can be difficult to distinguish among them. In other contexts, such as assessing indicators for nuclear terrorism scenarios, situational awareness frameworks have been used [4] and may be adaptable to treaty monitoring.

As a first step in metrics development, the monitoring system can be framed as a detection system. In this construct, the goal of a monitoring system is the timely detection of discrepancies. For example, the international safeguards system implemented by the International Atomic Energy Agency has specific definitions for timeliness goals for detecting the diversion of significant quantities of nuclear materials. These metrics are derived from a technical analysis of the risks of material diversion and are applied to monitor the multinational Nuclear Nonproliferation Treaty. In the arms control context, the measure of timeliness (or timely detection of noncompliance) and threshold for significant quantity may be state-specific and is related to avoiding instability in a bilateral or regional strategic relationship [5]. While requisite detection goals may not be specified until close to the conclusion of an agreement, these goals can still be used to characterize the potential range of system performance. For example, the high-level metrics for a detection system can be formulated as [6]:

- Time to first detection of a discrepancy as a function of the number of discrepancies (X), denoted as $X \geq 0$
- Time to achieve a given level of confidence in the magnitude of X throughout the enterprise

The performance of a monitoring approach can be characterized by these metrics for each evaluation scenario. Accounting for realistic uncertainties, the performance metrics are slightly modified such that the monitoring system provides detection confidence above an expected level of false positives or honest mistakes. Additional metrics at lower levels of the functional architecture may also provide insight into specific questions of interest or inform design elements.

2.2 Monitoring Approach Specification

Defining a monitoring approach involves specifying a broad range of information, including:

- the content and frequency of declarations
- monitoring functions and technologies
- inspection and/or unattended monitoring processes, procedures, and frequencies
- anticipated inspector decision logic
- analysis algorithms, and
- access and time constraints

Declarations include data exchanges, where treaty partners routinely exchange inventories and other associated information, notification of events or activities, and site declarations in support of an on-site inspection. The monitoring approach must define the monitoring functions and technologies available to the regime, as well as the processes and procedures on when and how they can be used. This includes, for example, inspection activities, measurement technologies, the collection of

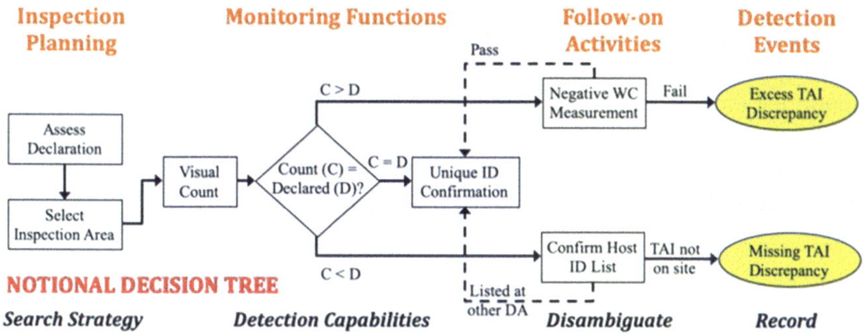

Fig. 3 Decision trees define the anticipated inspector logic process in response to monitoring activity observations and connect the range of potential observations to scenarios within the evaluation framework

unattended monitoring data, and when and by what process data can be accessed or transferred.

Algorithms for assessing technical monitoring data during inspections and state-level analyst assessments that correlate inspection outcomes with prior inspection reports, declarations, and other information sources are examples of analysis algorithms. They may vary in sophistication, from simple statistical techniques [7] to game theoretic approaches [6]. Performance metrics for classifying complex scenarios with similar indicators using multi-dimensional monitoring data may be aided by the use of machine learning techniques [8].

A monitoring approach's performance is only as good as the ability to make use of the monitoring data and observations being generated. Performance metrics may improve if better algorithms are implemented, even if monitoring functions, technologies, and procedures remain the same. Analysts must be careful to consider whether more, better, additional data is needed, or whether existing data can be better utilized.

The anticipated inspector logic process in response to monitoring activity observations can be mapped using decision trees. These decision trees connect the range of potential observations to scenario indicators within the evaluation framework. A notional decision tree is shown in Fig. 3.

The decision tree documents the steps in the inspection process, following defined procedures. This includes inspection planning, carrying out monitoring functions, potential follow-on activities, and the recording of detected discrepancies.[2] For example, in the inspection-planning phase, an inspector assesses the site declaration in relation to previous declarations and current monitoring data. The inspector then strategically selects designated areas to visit based on a combination

[2]In this context detected discrepancies may also be due to honest mistakes or technology false positives.

of factors such as data discrepancies, prior inspection activities, randomness, etc. The outcomes of monitoring functions guide the follow-on activities needed to disambiguate discrepancies and identify detection events. These decision trees support the identification of monitoring gaps and are part of the analysis logic for quantifying performance metrics. A decision tree needs to be developed for each proposed monitoring approach. While aspects may be modularized and portable between approaches, the consistency of the logic must be carefully considered.

2.3 Evaluation: Modeling and Simulations

The functional architecture and analysis of the inspection logic through the decision trees can help confirm that a monitoring approach contains functions that address each requirement. This analysis can be supported by semi-quantitative assessments from subject-matter experts, such as through an analytic hierarchy process [9].

Prior to or concurrent with experimental or operational field-testing, modeling and simulations may be helpful in quantifying performance metrics. Computational models can be used to generate simulated data for preliminary assessments and to explore a broader range of conditions than feasible through experiments.

Analytic methods require restrictive assumptions about processes and distributions. The system will monitor a dynamic nuclear enterprise, with frequencies and schedules that may change over time. Cheating scenarios may involve patterns of discrepancies that do not fit within standard probability models, or involve host and inspector strategies that adapt to previous observations. Simulations can accommodate these types of behaviors to empirically characterize and experiment with enterprise and monitoring activities.

There are multiple classes of simulations that can be used to model monitoring system performance. Physics modeling, such as using the Monte Carlo N-Particle (MCNP) transport code [10], is useful for assessing the performance of radiation measurement technologies against a range of objects and environmental conditions. Physical and event-based models such as those used in physical security simulations [11] can help assess the performance of chain-of-custody measures and unattended monitoring technologies. Process models are commonly used in operations research for assessing the process efficiency of systems with probabilistic behaviors. One type of process model is the discrete-event simulation (DES). The DES can serve as a forward model to generate simulated observations by modeling enterprise processes and frequencies, inspections, and the stochastic performance of technologies and monitoring functions under the evaluation scenarios of interest. This chapter describes the use of a DES for quantitative assessments.

Figure 4 gives an example of the architecture for a DES model for items accountable under a hypothetical treaty [12, 13]. The TAI Module models nuclear enterprise activities and processes such as deployment exchange, maintenance and refurbishment, inter-site transport, site movements, and dismantlement of the items. The Declaration Module generates data exchanges and site declarations, which may

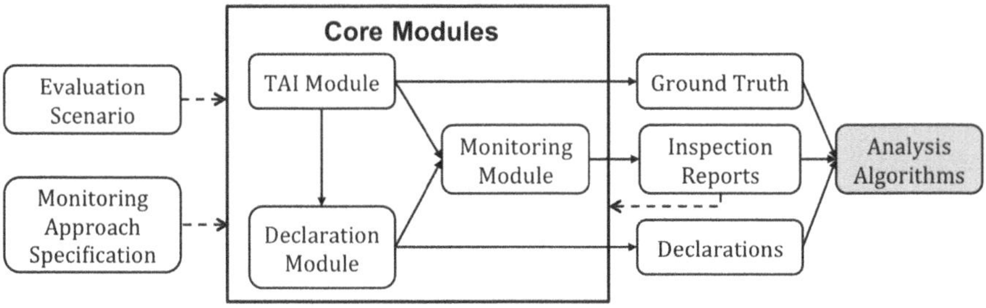

Fig. 4 Example software architecture of a Discrete-Event Simulation. The DES models enterprise processes, declarations, and monitoring activities to generate simulated declarations and inspection reports

depend on the evaluation scenario. Monitoring processes and decision trees are incorporated into the Monitoring Module, which generates monitoring data from unattended monitoring activities and inspections.

The simulation can be instantiated with an evaluation scenario. For example, in an excess undeclared TAI scenario, the TAI module creates excess TAIs, the Declaration module does not declare them, and the monitoring module may detect them during inspections. The outputs of the simulation are the ground truth of what actually happened, a set of declarations, including site declarations and data exchanges, and inspection reports. The simulation can be run for a large number of realizations of various stochastic elements to generate probability distributions of possible results for analysis. Analysis algorithms can use these results to quantify performance metrics. These results must, of course, eventually be verified and validated against realistic experiments and exercises.

3 Warhead Monitoring Example

The evaluation methodology will be illustrated with a specific example of monitoring nuclear warheads in a state's nuclear weapons enterprise. This example includes many assumptions and does not constitute a comprehensive analysis nor reflect an assessment of priorities for warhead monitoring—it is intended for illustrative purposes only.

3.1 Notional Treaty Provisions and Monitoring Approach

The notional treaty requires all treaty partners to limit and reduce total nuclear warheads. Treaty-Accountable Items (TAIs) include all nuclear warheads, except those that have been dismantled into constituent parts (fissile material, high-explosives, and non-nuclear components). Another regime that governs dismantled nuclear components is assumed to be in place. Both deployed and non-deployed nuclear warheads are subject to monitoring. The treaty incorporates a New START-like monitoring regime for deployed warheads on missile systems [14]. Since non-deployed warheads are monitored, warhead-counting rules for bombers, employed in New START, are superseded by this agreement. For simplicity, all warheads are assumed to be of a single type.

A notional monitoring approach is extrapolated from current regimes that involve unique identifiers on all TAIs and permits random inspections of sites where TAI are declared to be present (with certain access restrictions or controls). To demonstrate the quantitative evaluation of a monitoring approach, the details below are provided to illustrate the types of information that should be specified.

States assign a unique identifier to each TAI. Upon entry-into-force, each state declares the sites where TAIs are allowed to be present (e.g., reserve storage, regional storage, forward-deployment, and maintenance/ refurbishment/ dismantlement facilities) and provides an inventory for each site. This data is subsequently exchanged every 6 months, with a 14-day delay for operational security. Exhibitions of TAIs are performed and template measurements may be made. During baseline inspections, inspectors initialize each TAI by observing the host apply a unique identifier, such as an irreproducible tag, to each TAI container, along with container seals and tamper indicators. Non-portable measurement technologies may also be distributed to sites, tested, and placed under dual control during baseline inspections.

Twelve random inspections are permitted per year. At each inspection, the host provides the inspection team an up-to-date site declaration for that site. The site declaration includes the location and identity of each TAI, inter-site transport and deployment exchange logs, and whether any TAIs require re-initialization (routine maintenance or deployment exchanges may result in breaking of container seals).

The inspectors select a subset of monitored areas for inspection, where they can perform a visual count, confirm a sample of unique identifiers against the site-declaration, and check seal/container integrity. Inspection constraints also need to be specified, such as a maximum time, number of locations at the site that may be visited, or number of TAIs that may be sampled. At each site inspection, the inspectors may also request a specified number of TAIs to undergo warhead confirmation measurements. At deployment sites inspectors also confirm the number of warheads on missile systems according to New START procedures. Under New START, inspectors have the ability to count the number of shrouded reentry vehicles on a delivery vehicle, which count toward treaty limits. If the host state declares that certain shrouded reentry vehicles are non-nuclear, the inspector can use radiation detection equipment to confirm the item is not radioactive. The inspector does not confirm the items declared to be warheads are in fact real warheads.

3.2 Considerations When Applying the Evaluation Framework

For the notional treaty outlined above, one verification objective could be to "Confirm the number of declared TAIs". Potential cheating pathways may include withholding TAIs from baseline declarations, diversion and substitution of declared TAIs with spoofs of varying sophistication, diverting TAIs declared to be dismantled, or undeclared production of TAIs. Some of these pathways can be addressed through warhead monitoring. For example, potential monitoring objectives may include the following:

- Confirm the TAI inventory at all declared sites
- Confirm declared TAIs are warheads
- Confirm declared facilities are not being used to produce or refurbish undeclared TAIs
- Confirm that a TAI to be dismantled is a warhead, and that the dismantlement of that warhead was carried out

Detection goals for each verification objective are translated through the path analysis to establish performance requirements for each specified monitoring objective. For example, a goal of detecting the existence of a population of undeclared TAIs greater than 10% of the declared inventory within 1 year sets quantitative requirements for the accuracy of inventory confirmation and production and dismantlement monitoring. There may also be multiple detection goals. If a strategic objective is to greatly reduce nuclear inventories in the long-term, an additional long-term goal may be to detect the existence of a population of undeclared TAIs greater than 1% of the declared inventory within 20 years.

Note that verification and monitoring objectives must be carefully crafted, as slight modifications may impact monitoring requirements. For example, a TAI is defined as an item subject to treaty limits that a treaty partner submits for accounting purposes. In New START, the shrouded reentry vehicles are TAIs unless declared and demonstrated otherwise. While these items reflect the deployed warheads limited by the treaty, at no point are they confirmed to be warheads.

Confirming that declared TAIs are warheads may be an important function for confirming the declared number of TAIs. However, if the objective were to confirm the declared number of warheads, confirming TAIs are warheads becomes a core requirement rather than a supporting function, potentially constraining the set of alternate solutions. As with any problem statement, one must be careful to distinguish true objectives from convenient but artificial surrogates.

Potential scenarios for the monitoring objective "Confirm declared TAIs are warheads" may include the following:

- All TAIs are warheads (honest scenario)
- Some declared TAIs are not warheads at treaty initialization
- TAIs are diverted or substituted with a non-warhead over time while in storage
- TAIs are substituted with non-warheads during declared maintenance, refurbishment, or deployment exchange

Confirming declared TAIs are warheads involves multiple monitoring functions such as performing warhead confirmation measurements and maintaining chain-of-custody on TAIs. Maintaining chain-of-custody may involve perimeter monitoring or individual TAI monitoring through unique identifiers and tamper-indicators.

One key lower-level metric is the performance of warhead confirmation technologies. If warhead confirmation technologies are part of a future treaty regime, the treaty will need to specify a warhead definition, whether as a set of attributes or against a set of reference templates. This negotiated definition will be informed by assessments of what is needed in a definition, what technologies are capable of measuring, and what measurements are allowed subject to an acceptable risk of information disclosure. The set of characteristics that constitute a sufficient warhead definition has historically been discussed at length and research continues in this area.

For technical analysis of the monitoring system, performance can be characterized by the ability of a measurement to distinguish between selected reference (warhead) and evaluation (other) objects. These evaluation objects may represent other objects in an enterprise, alternate warhead definitions, or potential spoofs. The specific performance metric can be defined as a set of receiver operating characteristic (ROC) curves for the comparison of each test object to the reference. In this classification approach, an adjustable discrimination threshold impacts the true positive (TP), and false positive (FP) rates, which reflect the accuracy of correctly accepting the reference object and incorrectly accepting the test object, respectively. Their dependence can be mapped out on the ROC curve. For simplicity, points on the ROC curve can be shown as a table of FP rates for a specified TP rate. For example, Table 1 depicts the notional classification performance of a technology as false positive rates for three evaluation objects against a 99% true positive identification of the reference object. This table depicts how stochastic performance improves with longer measurement times, resulting in lower FP rates and therefore better discrimination.

In practice, it will not be possible to experimentally measure all objects of interest, particularly for a treaty partner's TAIs. Estimates based on prior knowledge, simulations, extrapolation, or informed judgment may be necessary to characterize performance. Objects that are may be confused with TAIs and are prevalent in the enterprise should generally be included in the evaluation.

Systematic errors arising from nuisance parameters are likely to degrade technology performance. Nuisance parameters represent the operational, environmental, and detector specific conditions that result in systematic uncertainty. For example,

Table 1 Notional technology performance for distinguishing reference and evaluation objects

For true positive = 99%	Object A	Object B	Object C
False positive [60 min]	30% ± 6%	80% ± 3%	8% ± 2%
False positive [120 min]	15% ± 3%	35% ± 2%	2% ± 1%
False positive [180 min]	5% ± 2%	7% ± 1%	0% ± 0.5%

some technologies may be much more sensitive to alignment or the position of a TAI within its container. It will be important to assess these sensitivities, evaluate their impact on classification performance, and potentially identify procedures that mitigate the performance degradation.

The objectives, scenarios, metrics, and functional architecture comprise the evaluation framework for assessing the capabilities of the monitoring system. Examples provided here were illustrative; realistic evaluations should establish an evaluation framework that encompasses the perspectives of all stakeholders.

3.3 Quantitative Evaluation

The following illustrative analysis examines the case where discrepancies (non-warheads) are only introduced into the population of declared TAIs prior to initialization at the start of the treaty.[3] It assumes that discrepancies are randomly distributed across the enterprise population (a greatly simplifying assumption for illustration). It further assumes that the notional enterprise is comprised of 2000 TAIs, with 1000 deployed and 1000 non-deployed. The TAIs are allocated across 8 defense sites and one site with refurbishment/dismantlement facilities. Non-deployed TAIs are stored in storage bunkers with a maximum capacity of 20 TAIs each at approximately 50% capacity.

As previously discussed it is assumed that there are 12 random inspections per year, where 5% of the TAIs are measured with a notional warhead confirmation (WC) technology that has a performance of TP = 99%, FP = 10% against all substitutions of concern.

With assumptions on enterprise process frequencies, the DES can be used to generate a 3-dimensional probability distribution (cumulative discrepancies detected, discrepancies introduced, time). A slice of the probability tensor at 36 months is depicted in Fig. 5. Each horizontal line represents a set of DES runs to calculate the probability distribution of cumulative discrepancies detected after each inspection for a given actual number of discrepancies. Multiple sets of runs initialized with different numbers of actual discrepancies are used to build the probability tensor. Note that even with zero discrepancies, the FN rate results in a finite number of detection events.

With proper normalization, vertical slices across the probability matrix for a given time provide a set of likelihood functions for estimating the number of enterprise discrepancies for a specific sequence of detection events within a single realization. In Fig. 5, if 16 discrepancies were detected after 36 months the vertical

---

[3]TAIs may be diverted and substituted at initialization to maintain consistency with other information. For example, the treaty may stipulate the exchange of production records, or a state may suspect the treaty partner has access to this information through other means.

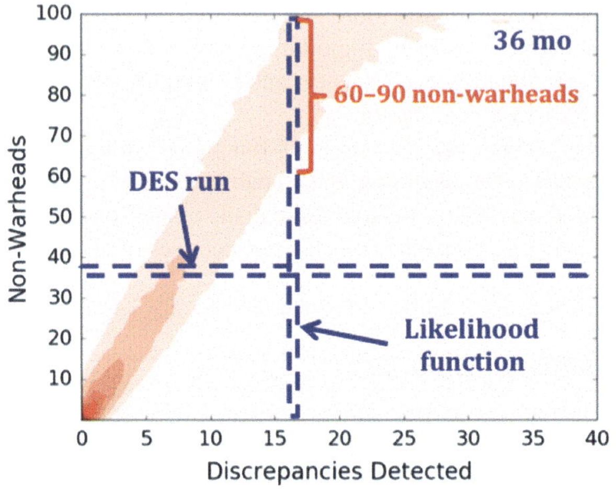

Fig. 5 Probability matrix for expected cumulative detection events after 36 months as a function of the number of discrepancies present

slice shows that there are likely between 60 and 100 non-warheads within the overall enterprise.

Figure 6 depicts a specific realization where the simulation was initialized with 32 non-warheads (shown by the yellow dashed line). The red curve shows the cumulative detections over time and the black bars represent the 90% confidence interval on the number of inferred discrepancies through the enterprise. The black 'x' within the confidence band is the most probable value. As the number of measured TAIs during inspections accumulate over time, the confidence bands narrow in on the true number of discrepancies. For simplicity we assumed that chain-of-custody for the TAI was maintained and verified such that TAIs did not need to be re-measured. Relaxing this assumption would change the numerical values shown.

This analysis may, for example, enable the assessment of the impact of varying warhead confirmation technology performance levels. Figure 7 shows potential high-level metrics for two cases, one of ideal warhead confirmation technology performance (TP = 100%, FP = 0%), and a notional performance (TP = 99%, FP = 10%). Figure 7a shows the time to 90% confidence a discrepancy exists. For ideal performance, this is equivalent to the time a discrepancy is first detected. The statistics are dominated by stratified sampling of the site to be inspected and the fraction of TAIs sampled at the site. Realistic performance would result in a finite number of false positive detection events. If technology performance is known to be better than a prescribed value, confidence that a discrepancy exists requires detection of events beyond what is statistically expected. Figure 7a represents a constraint between the number of discrepancies, the time to detection, and the confidence level. If two of these values are specified, the third value is constrained. For example, the monitoring system is required to detect discrepancies with 90% confidence within

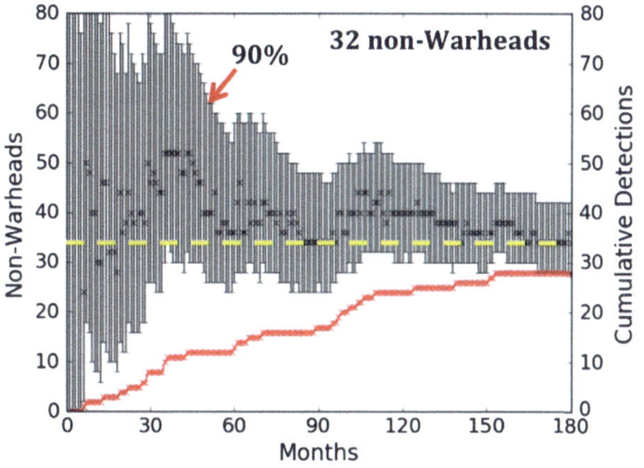

Fig. 6 For a given simulated sequence of detections, the likelihood functions are used to calculate the 90% confidence intervals on the estimated number of non-warheads across the enterprise. The estimates are compared to the yellow line representing actual discrepancy numbers, not the red line representing detection events to date

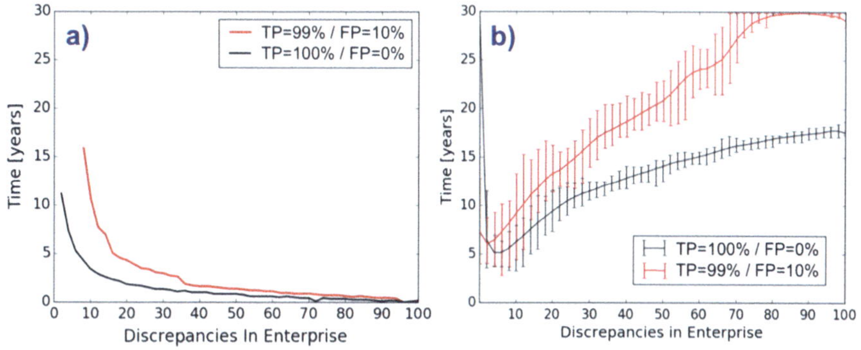

Fig. 7 (**a**) Time to 90% confidence a discrepancy exists for ideal and notional warhead confirmation technology performance. (**b**) Time required for 90% confidence in the estimated number of discrepancies

3 years, then Fig. 7a provides the detection threshold for the monitoring system for this scenario.

Figure 7b illustrates the time required for 90% confidence that the discrepancy magnitude is within 10 TAIs of the most probable value, using the analysis illustrated in Fig. 6. For low sampling rates, even with ideal technology performance the time to confidence is on the order of 5–15 years. Should a discrepancy be determined to exist, a mechanism to increase sampling rates will more quickly achieve confidence in the magnitude of the discrepancy.

Note that this example analysis is simplified and assumes a single scenario, discrepancies are randomly distributed at treaty initialization, process frequencies

Table 2 Notional assessment against monitoring objectives and scenarios

	Monitoring approach 1		
Scenario	Detection threshold	Key function	Key parameter
1	2%	Visual count	# of storage areas sampled
2	5%	Unique identifier confirmation	Fraction of unique identifiers sampled
3	18%	Warhead confirmation measurement	Fraction of TAIs measured; technology performance

are known, and that statistical approaches are used for performance quantification. In a more realistic application there could be discrepancies corresponding to multiple scenarios, such as inaccurate TAI inventories concurrent with substituted TAIs. Furthermore, the distributions may not be random and may be introduced over time, enterprise process frequencies may not be well known, and a dishonest host may take actions to intentionally confuse detection events. More advanced analysis algorithms, including machine.learning approaches that can analyze a high-dimensional space of indicators, may be beneficial in improving confidence estimates.

More broadly, the evaluation methodology can assess whether a given monitoring option addresses all key monitoring objectives and evaluation scenarios with an acceptable level of confidence. Table 2 notionally illustrates what such an assessment may look like for the assumed monitoring approach and confidence/detection time requirement. For example, Scenario 1 represents undeclared TAIs randomly distributed across declared facilities. If the requirement is to detect these undeclared TAIs with 90% confidence within 1 year, visual counting of 30% of designated areas will be able to detect undeclared TAIs in excess of 2% of the declared inventory. This threshold would depend on how many storage areas were sampled during each inspection.

The evaluation methodology enables the assessment of lower level metrics and tradeoffs to support design choices. As previously shown, specifying detection requirements (and a warhead definition) helps define performance requirements for warhead confirmation technologies. Other studies may include, for example, the impact of tag and seal robustness on overall chain-of-custody to help define robustness requirements. If portal and perimeter monitoring were to be employed, the methodology could help define requisite detection thresholds and the accuracy of directionality sensors.

It can also support the assessment failure modes. For example, the functional decomposition and analysis of scenario indicators can help find single points of failure in the monitoring system (e.g., data system failure) and quantitatively assess the implications of these failures. This information could help designers enhance the robustness of monitoring options.

4 Conclusions

A systems approach for assessing the performance of a warhead monitoring system may provide a useful framework to cohesively integrate various system design, technology research, and algorithm development efforts. Such a systems approach enables the quantification and structuring of performance metrics against high-level policy objectives, as specified by the monitoring objectives, scenarios, and performance requirements. It also supports the analysis of specific design tradeoffs across the monitoring system, so that each component is not optimizing within its own "sandbox."

Acceptance criteria for a monitoring system, such as the risk of information disclosure, the impact on facility operations, and financial costs, are as important, if not more important than performance. These factors have not been specifically addressed in this chapter, but would need to be integrated into a cohesive framework in order to effectively assess tradeoffs. Some of the components in the evaluation framework can support such an assessment. For example, simulations of declarations and monitoring data can help determine what aggregation of knowledge may disclosure protected information over time, either through sophisticated analysis and/or in conjunction with prior or external knowledge.

This evaluation methodology exists within a system structure that includes path analysis of the enterprise and strategic assessments of treaty objectives. Those assessments help define the requirements and constraints for the monitoring system. The political environment, strategic landscape, and ongoing activities of the treaty parties' nuclear enterprises will strongly influence the context in which the monitoring system is expected to perform.

The views and opinions of authors expressed herein do not necessarily state or reflect those of the United States government of Lawrence Livermore National Security, LLC. The work was performed under the auspices of the U.S. Department of Energy by Lawrence Livermore National Laboratory under Contact DE-AC52-07NA27244. LLNL-BOOK-708658. Sandia National Laboratories is a multimission laboratory managed and operated by National Technology and Engineering Solutions of Sandia, LLC, a wholly owned subsidiary of Honeywell International, Inc., for the U.S. Department of Energy's National Nuclear Security Administration under contract DE-NA0003525.

Acknowledgements The authors acknowledge the contributions of Crystal Dale for her contribution to the functional decomposition and general conceptual discussions. The authors also acknowledge Douglas Keating, Robert Brigantic, Angela Waterworth, Casey Perkins, and Matthew Oster for their work on the development of the discrete-event simulation.

References

1. Chen CD, Dale CB, DeLand SM, Waterworth A, Edmunds TA, Keating D, Oster M (2016) Developing a System Evaluation Methodology for a Warhead Monitoring System. In: Proceedings of the 57th Annual Meeting of the Institute for Nuclear Material Management, Atlanta
2. Defense Acquisition University Press (2001) Systems Engineering Fundamentals, Fort Belvoir, VA
3. Joint Chiefs of Staff, J-8 (2016) (2009) Capabilities-Based Assessment (CBA) User's Guide, Force Structure, Resources, and Assessments Directorate,Version 3. http://acqnotes.com/wp-content/uploads/2014/09/Capabilities-Based-Assessment-CBA-Users-Guide-version-3.pdf. Accessed 28 Feb 2019
4. Ni K, Faissol D, Edmunds TA, Wheeler R (2013) Exploitation of Ambiguous Cues to Infer Terrorist Activity. Decision Analysis 10(1):42–62
5. Nitze, PH (1985) On the Road to a More Stable Peace. Bureau of Public Affairs, Department of State, Current Policy No. 657
6. Avenhaus R, Canty MJ, Calogero F (2005) Compliance Quantified: An Introduction to Data Verification. Cambridge University Press, New York
7. Edmunds TA, Chen CD (2016) Statistical Sampling Methods for Treaty Verification in Dynamic Environments. In: Proceedings of the 57th Annual Meeting of the Institute for Nuclear Material Management, Atlanta, 2016
8. Sentz K, Hemez F (2014) The Future of Intelligent Systems for Safeguards, Nonproliferation, and Arms Control Verification. In: Proceedings of the Information Analysis Technologies, Techniques, and Methods for Safeguards, Nonproliferation and Arms Control Verification Workshop, Portland
9. Saaty TL (1990) How to make a decision: The Analytic Hierarchy Process. European Journal of Operational Research 48:9–26
10. X-5 Monte Carlo Team (2008) MCNP – A General N-Particle Transport Code, Version 5, Los Alamos Report LA-UR-03-1987, Los Alamos, NM
11. Gottlieb E, Harrigan R, McDonald M, Oppel F, Xavier P (2001) The Umbra Simulation Framework, Sandia Report SAND2001-1533, Albuquerque, NM
12. Perkins C, Brigantic RB, Keating D, Liles K, Meyer N, Oster M, Waterworth A (2015) Using simulation to evaluate warhead monitoring system effectiveness. In: Proceedings of the 56th Annual Meeting of the Institute for Nuclear Materials Management, Indian Wells
13. Oster M, Waterworth A, Keating D, Dale CB, DeLand SM (2016) Using Simulation to Support Monitoring System Design and Evaluation. In: Proceedings of the 57th Annual Meeting of the Institute for Nuclear Material Management, Atlanta
14. The Treaty Between the United States of America and the Russian Federation on Measures for the Further Reduction and Limitation of Strategic Offensive Arms, Signed 8 Apr 2010

Physical Model and Acquisition Path Analysis: Applicability to Nuclear Disarmament Verification

Irmgard Niemeyer, Joshua Rutkowski, and Gotthard Stein

Abstract This chapter describes how a systems based approach to physical model and acquisition path analysis may be used for nuclear disarmament studies by expanding the physical model to include both civilian and military nuclear domains and extending the model to the final stage of weapon deployment. The results show how such an approach may contribute to future treaty design and implementation. The approach shows the principal applicability of this concept to evaluate effectiveness and efficiency of verification activities in countries with commercial and military fuel cycles and activities. This chapter also demonstrates how other verification regimes, specifically biological weapons, may also be evaluated using a systems based approach to acquisition path analysis.

1 Introduction

1.1 The Physical Model in IAEA Safeguards

The International Atomic Energy Agency (IAEA) developed the physical model during so-called Programme 93+2 as an attempt to identify, describe and characterize various components of the nuclear fuel cycle, in order to provide a technical tool to aid enhanced information analysis [1]. The physical model of a nuclear fuel cycle, as defined in the IAEA Glossary [2], is *"a detailed overview of the nuclear fuel cycle* [...], *identifying, describing and characterizing every known technical process for converting nuclear source material to weapon usable material, and identifying each process in terms of the equipment, nuclear material and nonnuclear*

I. Niemeyer (✉) · J. Rutkowski
Forschungszentrum Jülich GmbH, Jülich, Germany
e-mail: i.niemeyer@fz-juelich.de; jerutko@sandia.gov

G. Stein
Consultant, Bonn, Germany

I. Niemeyer et al. (eds.), *Nuclear Non-proliferation and Arms Control Verification*,
https://doi.org/10.1007/978-3-030-29537-0_27

material involved. The physical model is used by the IAEA, inter alia, for acquisition path analysis [...] *and for safeguards State evaluations* [...]." The nuclear fuel cycle is defined as *"a system of nuclear installations and activities interconnected by streams of nuclear material."*

The IAEA Department of Safeguards has developed the physical model in 12 volumes: (1) Mining/Milling, (2) Conversion 1, (3) Uranium Enrichment, (4) Conversion 2, (5) Fuel Fabrication, (6) Nuclear Reactors, (7) Heavy Water Production, (8) Reprocessing of Irradiated Fuel, (9) Spent Fuel Management, (10) Radioactive Waste Management, (11) R&D Activities with Hot Cells, (12) Development of Nuclear Explosive Devices [1]. Currently, the IAEA is running an updating process for Volumes 1–11 with support Member States. The physical model has become an integral part of the State evaluation process in the Department of Safeguards, including the development and update of State-level safeguards approaches. It serves as reference document on the nuclear fuel cycle, guides the analysis of open source information (including nuclear-related trade data) and provides the basis for conducting an acquisition path analysis (APA). The physical model helps to identify proliferation indicators and to prepare verification activities.

1.2 Use of the Physical Model in Acquisition Path Analysis

The acquisition path analysis is *"a structured method used to analyse the plausible paths by which, from a technical point of view, nuclear material suitable for use in a nuclear weapon or other nuclear explosive device could be acquired. Each path is made up of the steps that would be required to acquire nuclear material and process it into a form suitable for use in a nuclear weapon or other nuclear explosive device. Acquisition path analysis is used to establish technical objectives for a State with a comprehensive safeguards agreement. An acquisition path analysis does not involve judgments about a State's intention to pursue any such path"* [3].

In order to identify technically plausible acquisition paths at the State level, the nuclear fuel cycle and related technical capabilities of the State, as represented by its physical model, need to be considered. Figure 1 shows how the physical model of the State could be used in this context. The physical model, when formulated in such a way that the boxes (or nodes) represent the nuclear installations and activities of the State, and the arrows (or edges) symbolize the nuclear material flows between them, can be used to specify a set of technically plausible pathways, which then would be prioritized in terms of safeguards significance.

Different approaches using the physical model were proposed in the literature (e.g. [4–6]) for a more structured and formalized method for APA. The two latter approaches used a reversed formulation of the physical model in order to apply

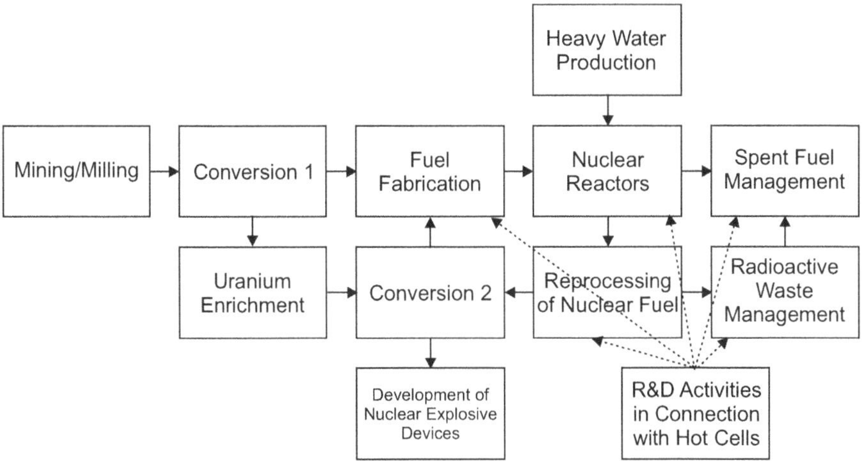

Fig. 1 Acquisition path in the physical model (slevel). Adapted after [1]

a mathematical model (based Directed Graph Methodology), where the nodes representing the nuclear material form and the edges the processes required to turn material from one form to another.

The chapter *Formalizing Acquisition Path Analysis* by Morton Canty and Clemens Listner in this book, details a systems-based approach to acquisition path analysis for nuclear safeguards. The authors present a structured approach for evaluating potential acquisition paths by using graph theory and non-cooperative games. Such an approach uses three basic steps: (1) network modeling, (2) network analysis, and (3) strategic assessment. The network modeling and subsequent evaluations are based on the IAEA's physical model of the nuclear fuel cycle. In order to use the physical model for such a purpose, the material forms (e.g. source material, natural uranium fuel feed, enrichment fuel feed, etc.) are represented in the graphs as nodes and the processes to convert the material forms (e.g. conversion I, enrichment, conversion II, etc.) are represented as edges. The edges and paths are combined to form acquisition paths. For any given state, the model is adapted to the nuclear fuel cycle technology of the state and coupled with the possibility for clandestine facilities along the acquisition path. During the first step of the model, each process or path segment is evaluated based on technical difficulty, proliferation time and proliferation cost to the state to carry out the processes involved in transforming the material form. These three parameters are based criteria proposed by the Generation IV international forum [7]. After the model is created, the second step of the approach is an automated process which analyzes the model and input parameters to determine all possible paths to arrive at the direct use material. Direct use material, as defined by the IAEA is "nuclear material that can be used for the manufacture of nuclear explosive devices without transmutation or further enrichment. It includes plutonium containing less than 80% ^{238}Pu, high enriched uranium and ^{233}U" [2]. The third step uses game theory principles to determine the optimal path for both the state to proliferate and an inspecting entity, such as the IAEA, to verify the state's activities. By using non-cooperative game principles, it is possible to identify the most strategic pathway for the state and inspectorate.

The IAEA's State-level Approach to safeguards demands that each state be considered as an entity whereas a criteria based inspection strategy does not sufficiently consider the state as a whole. The IAEA's model successfully supports evaluation of the state in a reproducible and non-discriminatory way. This chapter will use the same basic model and extend it beyond the point where direct use materials are obtained.

In conclusion, given its rather general definition, different formulations of the IAEA physical model for illustrating the (civilian) nuclear fuel cycle are possible. However, what about the extension of the physical model to the military fuel cycle and the weapons complex?

2 Physical Model and Acquisition Path Analysis in Arms Control Verification

The extension of the model to weaponization is beyond the purview of the IAEA but can assist for the implementation and verification activities related to disarmament treaties or treaties related to weapons of mass destruction. First, this chapter explains how the physical model may be extended to include the weaponization and illustrate the model with an example. The extended model is to be used for analysis and the results will be provided for a fictitious state whose advanced nuclear fuel cycle technologies and military complex would support a nuclear weapons programs. The example state, Trenzalia, has been used at several workshops, meetings and presentations in the past 2 years. The chapter *Investigating Development of Nuclear Arms Control Verification—Requirements at the State Level* by Keir Allen et al. in this book references Trenzalia and provides more details about its use. Afterward, an example of biological weapons modeling will be presented to demonstrate the acquisition path analysis method's application to modeling other weapons of mass destruction. The chapter concludes with an outlook into new possibilities for applying this model for international treaties.

2.1 Adapted Physical Model

The physical model described in the chapter *Formalizing Acquisition Path Analysis* by Morton Canty and Clemens Listner in this book is a variant of the IAEA's physical model and is well-suited for extension to weaponization paths. Figure 2

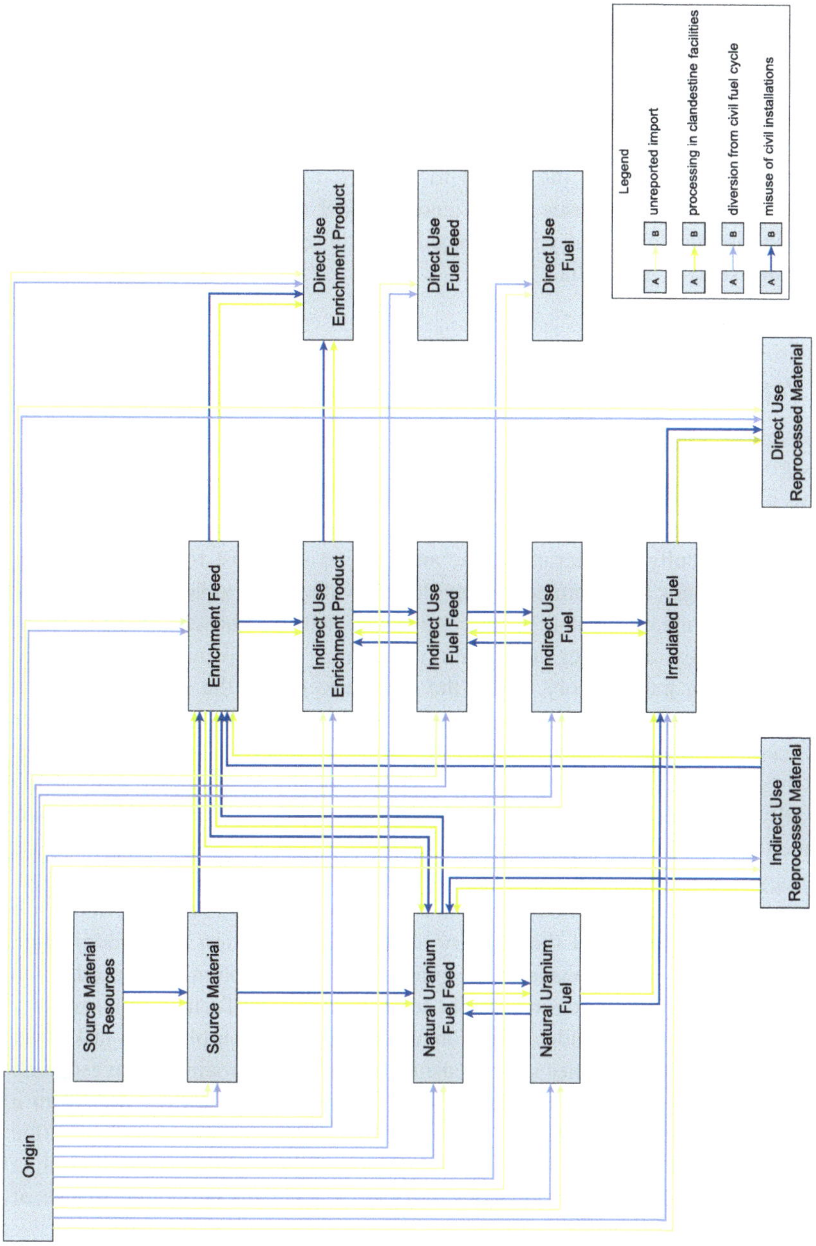

Fig. 2 Physical model with material forms as nodes and processes as edges

shows a template of the graph used to model the nuclear fuel cycle [8]. In this model, all paths end at one of the direct use materials. The adapted physical model in Fig. 3, however, contains edges for military contribution to the acquisition paths. In previous papers, the military dimension was not considered with explicit edges [9, 10]. Materials and processes to reach deployment of a weapon which occur after direct use material is produced are added and thus extend the physical model. This is shown in Fig. 4. This model considers the material form from origin to deployment systems whereas the IAEA physical model stops at direct use material. Such an extension naturally permits the model to be used for disarmament treaties and analysis. The next section uses this extended model to analyze an example state whose civilian nuclear industry is advanced and whose military capabilities mimic the fictitious state of Trenzalia.

2.2 *Pathways Analysis*

For the given scenario of a state with an advanced nuclear fuel cycle and a military complex of Trenzalia, the model was constructed and analyzed. Each path is ranked by its attractiveness to the state. The ranking is based on the evaluation parameters of technical difficulty, proliferation time and proliferation cost for each edge. For each path, the edges are summed and the lowest values are ranked highest. These rankings do not mean the path is the one which will be chosen by the state to follow. Rather they only offer insight into each path's attractiveness. The strategic assessment, using game theory, offers the chance to model the strategic analysis required to estimate the most likely path to be chosen by the state. The chapter *Formalizing Acquisition Path Analysis* by Morton Canty and Clemens Listner in this book explains how the model is mathematically constructed. Over 4000 technically plausible weaponization pathways were output when the model was calculated for Trenzalia. This large number illustrates that a state with the given technology has a wide array options to acquire and deploy a nuclear weapon. The first six paths are exclusively in the military domain (Fig. 5). This result highlights the importance of the military domain's contribution to weaponization paths. Furthermore, this emphasizes the rationale for a focus on military production routes. Future verification regimes ought to include a large focus on the military domain. How such treaties and verification regimes might accomplish this task is yet to be determined but the treaties should ensure that military domains are investigated with priority. Among the ranked paths, the first one to include diversion from a civilian facility is path 31 (Fig. 6) and also highlights the importance of clandestine facilities in the path analysis. The most attractive path including misuse of civilian enrichment is path 64 (Fig. 7) and the most attractive path including misuse of civilian reprocessing is path 91 (Fig. 8).

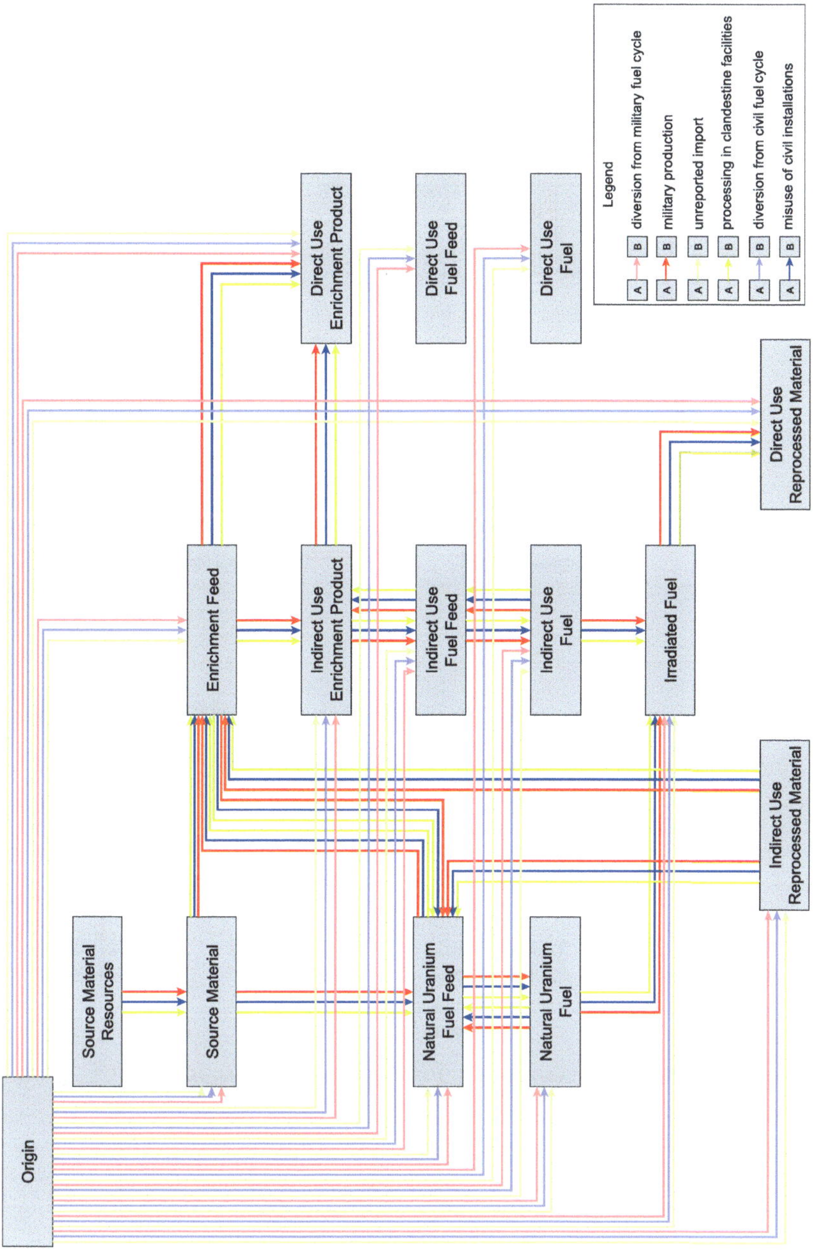

Fig. 3 Physical model with military edges considered as distinct paths

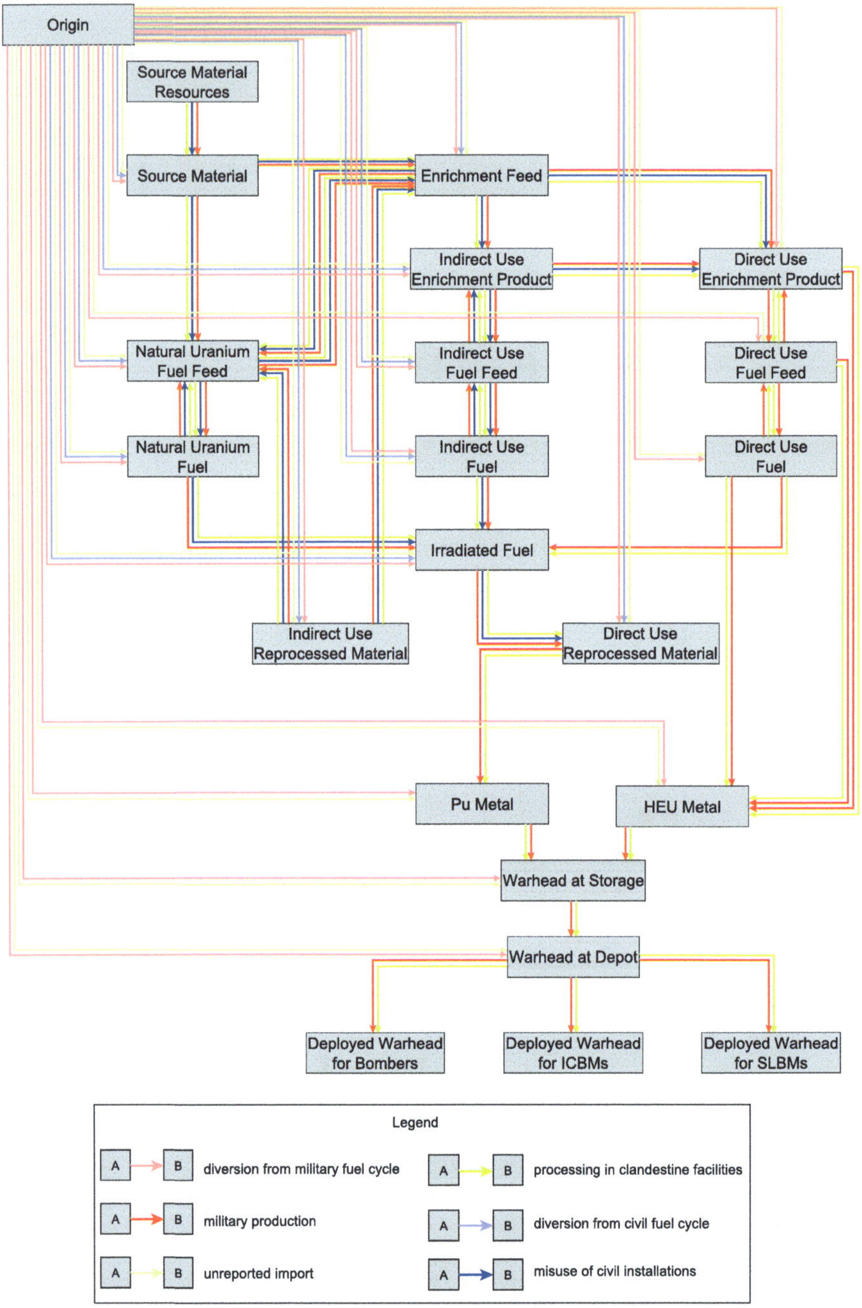

Fig. 4 Physical model extended to include weaponization nodes and edges

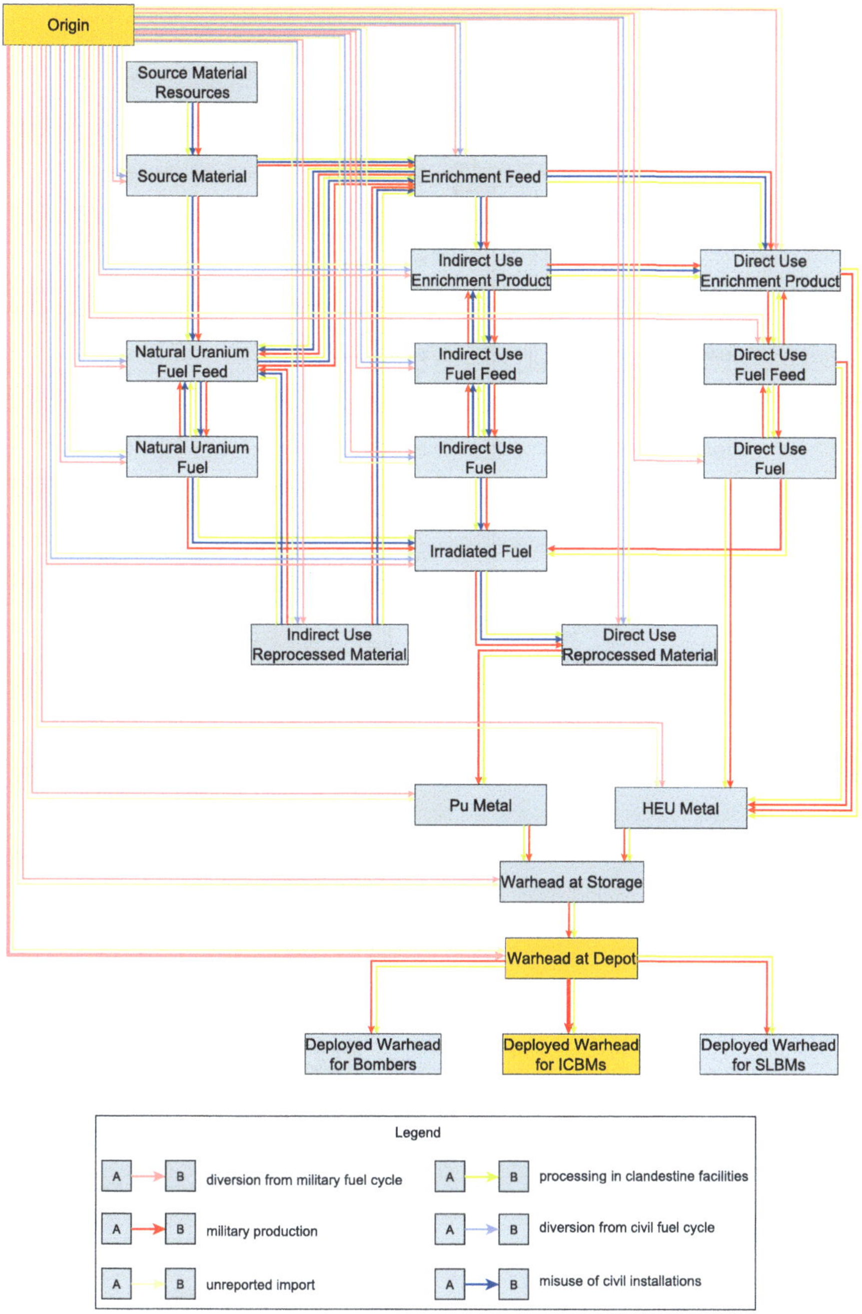

Fig. 5 Top ranked path based attractiveness of state with advanced nuclear fuel cycle and military domain

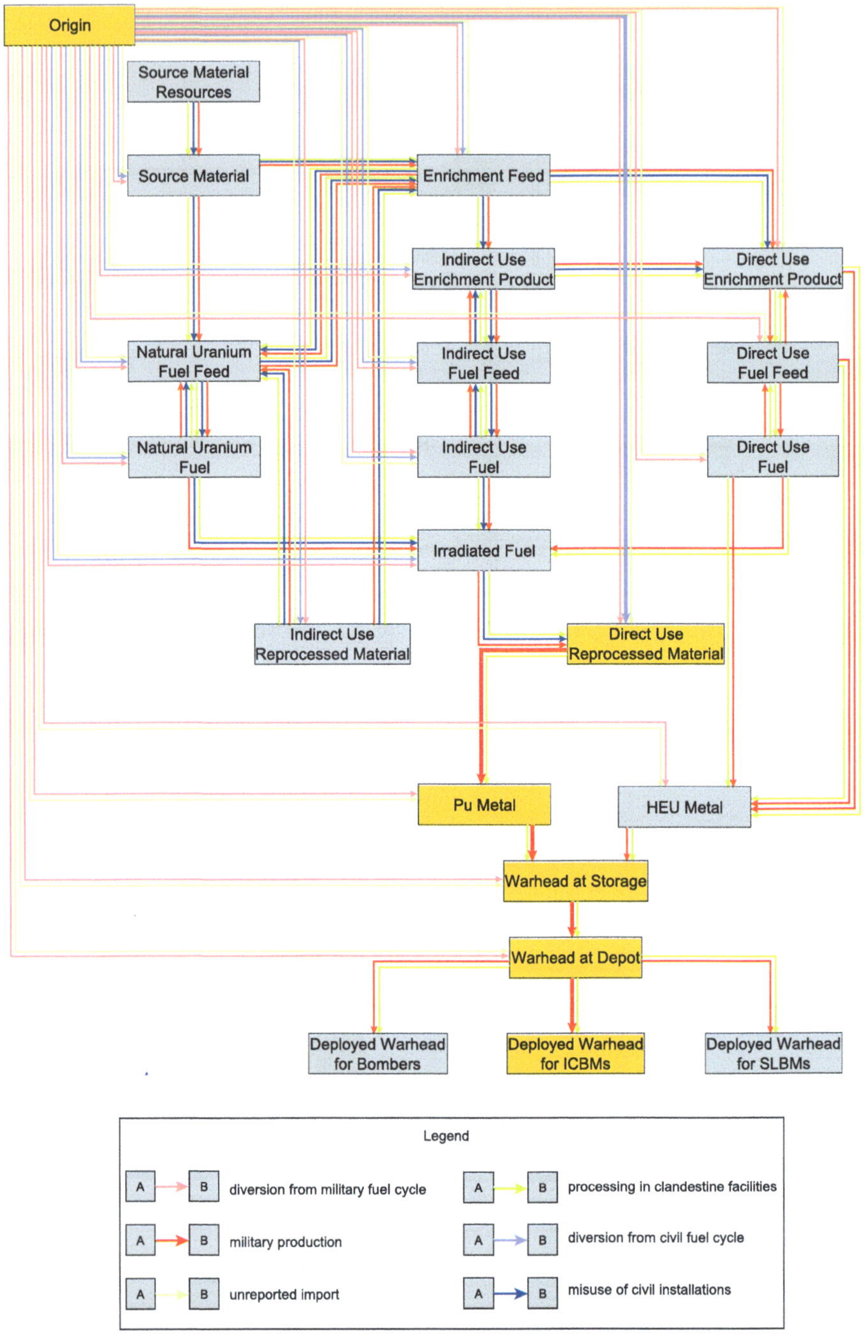

Fig. 6 First path including diversion from civilian facility

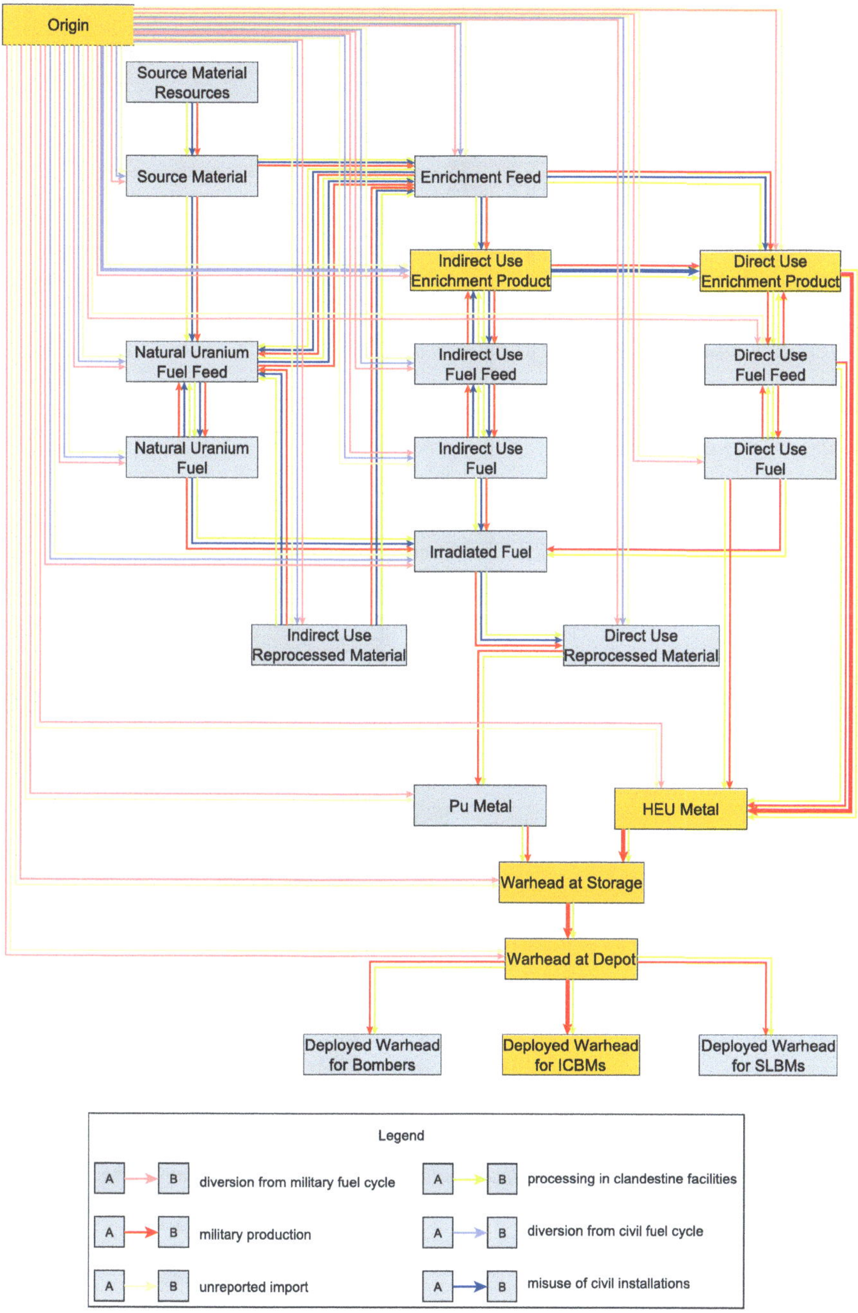

Fig. 7 Most attractive path including misuse of civilian enrichment

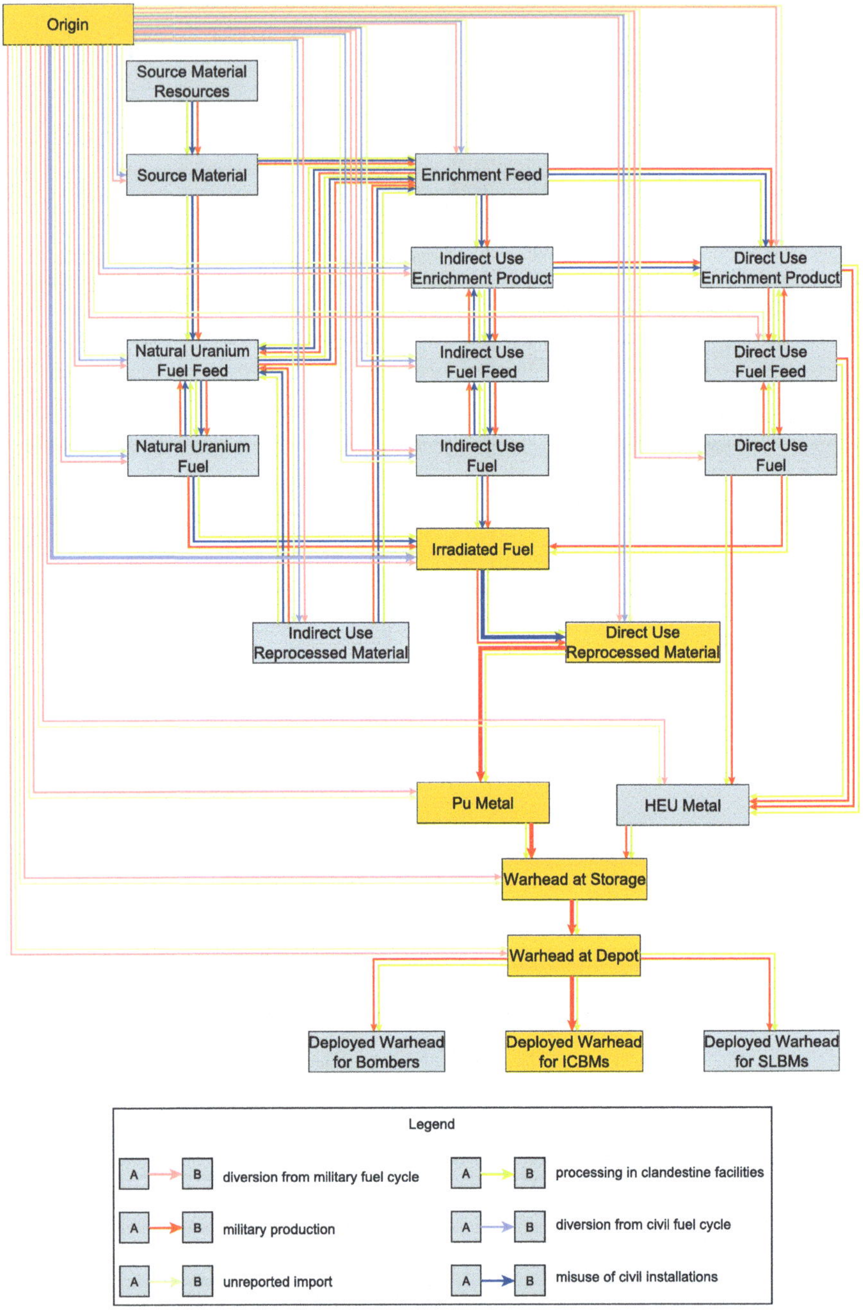

Fig. 8 Most attractive path including civilian reprocessing

3 Physical Model and Acquisition Path Analysis in Arms Control Verification

Based on the lessons learned from using a physical model in IAEA safeguards and the prior case study related to arms control, a systems approach is also expected to provide a generic and structured approach to the complex and sensitive issues in disarmament verification (top-down). A systems approach including a physical model could help design a transparent state-level systems framework to define verification objectives, processes, and timescales for an effective verification regime based on the strategic goals of a given disarmament treaty. Similarly to applying APA in the context of State-specific safeguards approaches, a physical model could provide the basis for identifying and assessing cheating pathways (CP) in the disarmament process, specifying and prioritizing state-specific verification goals, and identifying verification measures to address the verification goals.

In order to further exploiting the use of a physical model and the APA in a systems approach to nuclear disarmament verification, some questions need to be addressed, *inter alia*: How to formalise the model? Do we need a separate 'Disarmament physical model' next to the civilian/military nuclear fuel cycle? What are the treaty accountable items? How to assess path attractiveness and path prioritisation? How to take non-technical factors (strategic considerations) into account?

4 Application for Biological Weapons

Just as the physical model for the nuclear fuel cycle graphs the flow of uranium to direct use product, so would each type of biological weapon require such a graph. The graphs need to consider both the material and the processes (misuse of declared facility, clandestine facility, etc.) just as the extended model in this chapter does for nuclear disarmament. The materials and processes for each type of biological weapon would require a separate model where the process paths for creating the weapon would follow the possible material forms. Figure 8 demonstrates a possible biological weapon acquisition path for a state using a generic material form. The figure emphasizes that each material form may be achieved through either misuse of a declared facility, through a clandestine facility or through undeclared import of the material. Each biological weapon category for bacteria, fungi or virus would require a model which follows the possible processes and materials required to manufacture and deploy a biological weapon (Fig. 9).

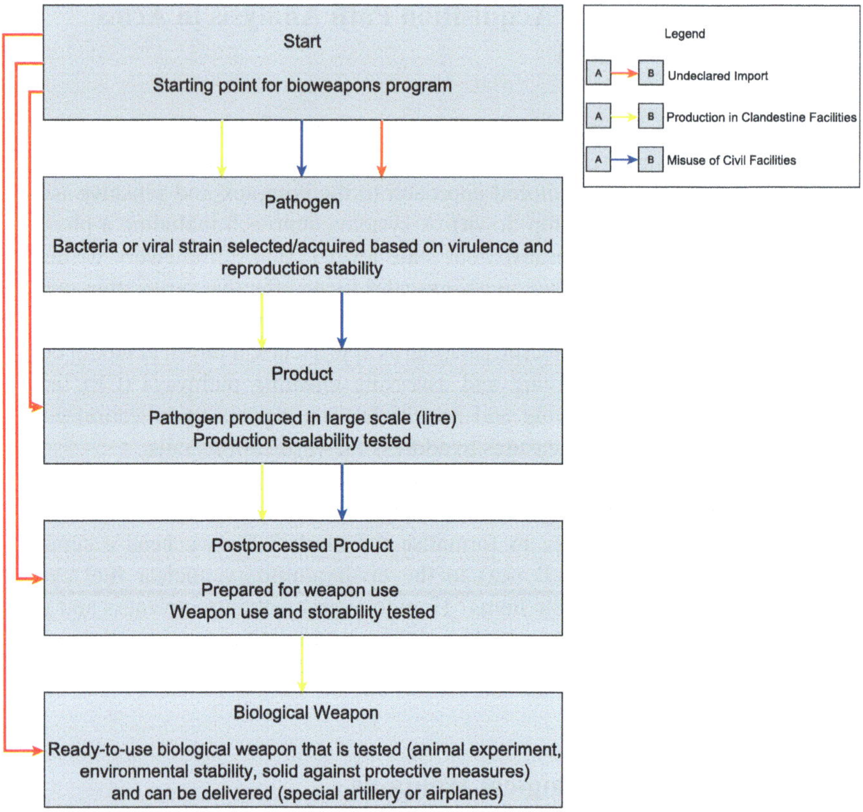

Fig. 9 Biological weapon acquisition path model

5 Conclusion

This chapter has highlighted a systems approach to modeling verification of nuclear and biological arms control treaties based on previous research on non-proliferation as well as other chapters within this book. The chapter provided a structured thinking about the problem (top-down) which sets out clear goals (to understand the possible paths and determine the most strategic path), and is applicable to fields considering international, regional or bi-lateral agreements which mandate a verification regime. Models like the one presented here require essential input from expert judgments, which heavily impact the results. Therefore, proper understanding of the underlying principles used in the approach is necessary to execute the model or extend it to other fields. The model is a decision support tool and should not be thought of as a decision-making tool. The formal methodology requires quantification during the strategic assessment of the paths. Such quantification requires further research, especially in the justification of detection probability values for declared and clandestine paths. This systems based approach to modeling enables would-be treaty authors to focus attention on the most attractive pathways for states in order to ensure the treaty adequately authorizes the verification authority to carry out its mission in a proper manner. The military domain's dominance in the most attractive paths in this chapter shows how such treaties must consider the military domain in future treaties.

References

1. Liu Z, Morsy S (2001) Development of the physical model. In: Proc. IAEA Symposium on International Safeguards, IAEA-SM-367/13/07, Vienna, 2001
2. International Atomic Energy Agency (2002) Safeguards Glossary, 2001 Edition. International Nuclear Verification Series No. 3
3. International Atomic Energy Agency (2014) Supplementary Document to the Report on The Conceptualization and Development of Safeguards Implementation at the State Level (GOV/2013/38), GOV/2014/41
4. Budlong Sylvester KW, Pilat JF, Murphy CL (2011) The Future Use of Pathway Analysis in IAEA Safeguards. In: Proc. INMM/ESARDA Joint Workshop, Aix -en-Provence, France, 2011
5. Vincze A, Nèmeth A (2014) Effect of State-specific Factors on Acquisition Path Ranking. In: Proc. IAEA Symposium on International Safeguards, IAEA-CN-220, 2014
6. Listner C, Niemeyer I, Canty M, Stein G (2016) A strategic model for state compliance verification. Naval Research Logistics 63(3):260–271
7. GEN IV International Forum (2006) Physical Protection Evaluation Methodology Expert Group of the Generation IV International Forum,"Evaluation Methodology for Proliferation Resistance and Physical Protection of Generation IV Nuclear Energy Systems, Revision 5,"GEN IV International Forum
8. Rutkowski J, Canty MJ, Niemeyer I, Stein G, Rezniczek A (2016) Acquisition Path Analysis Modeling Using the Joint Comprehensive Plan of Action as a Case Study. In: Proc. 57th Institute of Nuclear Materials Management Annual Meeting, Atlanta, Georgia, USA, July 2016
9. Listner C, Canty MJ, Rezniczek A, Niemeyer I, Stein G (2012) A Concept for Handling Acquisition Path Analysis in the Framework of IAEA's State-level Approach. In: Proc. 53rd Institute of Nuclear Materials Management Annual Meeting, Orlando, Florida, USA
10. Listner C, Niemeyer I, Canty MJ, Murphy C, Stein G, Rezniczek A (2015) Acquisition Path Analysis Quantified–Shaping the Success of the IAEA's State-level Concept. Journal of Nuclear Materials Management 43(4):49–59

Investigating Development of Nuclear Arms Control Verification: Requirements at the State Level

Keir Allen, Mona Dreicer, Cliff Chen, Irmgard Niemeyer, and Gotthard Stein

Abstract Over the course of 2014 and 2015, a group of international experts held a series of technical meetings to explore how to use a systems approach to promote the technical analysis of structural factors associated with arms control agreements and guide the development of creative verification solutions. The goal was to develop a framework that could be effectively applied to the satisfaction of multiple stakeholders and provide a means for identifying how technical verification approaches could be better aligned to meet strategic concerns, in support of state security priorities and objectives. Participants found that scenario based exercises provided a practical way to further develop the methodology and could be employed to improve communication between international experts with diverse backgrounds, as well as encourage new ways to approach this difficult issue. One aspect explored during the meetings was how to identify solutions that might satisfy states with diverging perceptions regarding their respective security environments. Interesting ideas for future work on how a systems approach could be used to address verification challenges are presented in the final section.

K. Allen (✉)
Kings College London, London, UK
e-mail: keir.allen@kcl.ac.uk

M. Dreicer · C. Chen
Lawrence Livermore National Laboratory, Livermore, CA, USA
e-mail: dreicer1@llnl.gov; chen41@llnl.gov

I. Niemeyer
Forschungszentrum Jülich GmbH, Jülich, Germany
e-mail: i.niemeyer@fz-juelich.de

G. Stein
Consultant, Bonn, Germany

I. Niemeyer et al. (eds.), *Nuclear Non-proliferation and Arms Control Verification*,
https://doi.org/10.1007/978-3-030-29537-0_28

1 Introduction

The development of existing nonproliferation and arms control verification regimes have been directly linked to the specific treaty or agreement to be implemented. As the international community considers the next feasible steps to reduce the threat of nuclear weapons, technical experts in the nonproliferation and arms control community have begun to explore how to apply analytical concepts, that have evolved over the past decade, to arms control verification.

Through the State-level Concept (SLC), the International Atomic Energy Agency (IAEA) implements safeguards in a holistic manner that considers a state's nuclear and nuclear-related activities and capabilities, as described by Jill Cooley in the chapter titled *The Evolution of Safeguards*, in this book. By using integrated processes to support safeguards implementation, the IAEA applies a systematic approach to:

- collect and evaluate safeguards-relevant information;
- develop customized state-level safeguards approaches;
- plan, conduct, and evaluate safeguards activities; and
- draw safeguards conclusions.

The SLC includes the application of analytical tools such as acquisition path analysis (APA), which is conducted to establish and prioritize technical safeguards objectives for a state. Based on the IAEA's physical model of the complete nuclear fuel cycle, the APA can provide a visual tool for identifying all of the processes required to turn naturally-occurring materials into end-use (weapons-useable) fissile materials. Material forms are represented as the nodes in the model, and the processes required to turn material from one form to another are represented as edges that link the nodes, as presented in *Formalizing Acquisition Path Analysis* by Morton Canty and Clemens Listner, in this book. A series of processes link up to form a route or pathway from one form of material to another. The physical model thus maps the relationship between material and processes along all pathways that result in the acquisition of end-use material. Besides visualization, APA also provides information about the attractiveness of pathways for a state, so monitoring and inspection strategies can be developed that sufficiently guard against the exploitation of those pathways in a resource-efficient manner. The verification strategy put in place can draw upon all of the means available to verify compliance, from analysis of open-source information and export control data to material accountancy data collected through safeguards technologies and environmental monitoring.

A method of systematically identifying verification options is attractive for developing future nuclear weapons control agreements since an important aspect of the viability of any such treaty will be the suitability and sufficiency of the verification provisions [1]. Our hypothesis was that the application of a systems approach such as the SLC to arms control agreements could help to identify pathways of concern with regard to noncompliance and the systems of information exchange and verification that could reduce concerns and increase confidence under future treaty conditions. To investigate this, a series of technical meetings were organized over the course of 2014–2015 to explore the utility of a systems framework in the context of nuclear arms control.

2 Applying Systems Level Analysis to Arms Control

International experts from a range of communities including *inter alia*, nuclear safeguards, arms control verification, radiation detection, political science, and defense studies considered how to apply a systems approach to arms control and identify knowledge gaps. To provide context, exercise scenarios involving two fictitious nuclear weapons possessing countries, named Trenzalia and Azaria, were created with relatively simple physical models of their nuclear weapons enterprises. A fictitious bilateral treaty was also proposed. Further details about the scenarios can be found in the Appendix at the end of this chapter.

An initial exercise to analyze potential diversion pathways and verification measures, that could be applied in a nuclear weapons state, took place during a meeting of the European Safeguards Research and Development Association (ESARDA) Verification Technologies and Methodologies Working Group at the Joint Research Centre, Ispra, Italy, in autumn of 2014 [2]. A second short exercise was held by the Institute of Nuclear Materials Management (INMM), hosted by the Center for Global Security Research (CGSR) at the Lawrence Livermore National Laboratory (LLNL) during the summer of 2015, based on the some of the factors discussed during the first exercise [3]. The final discussion of key issues was held at the 8th INMM–ESARDA Joint Workshop at Jackson Hole, Wyoming, in October 2015 [4]. This session was not directed towards any particular scenario, but instead focused on the topics of technology, information security, and application of systems engineering approaches.

As a result of the discussions, the general framework shown in Fig. 1 was developed. The results of the two exercises are presented in the context of these elements.

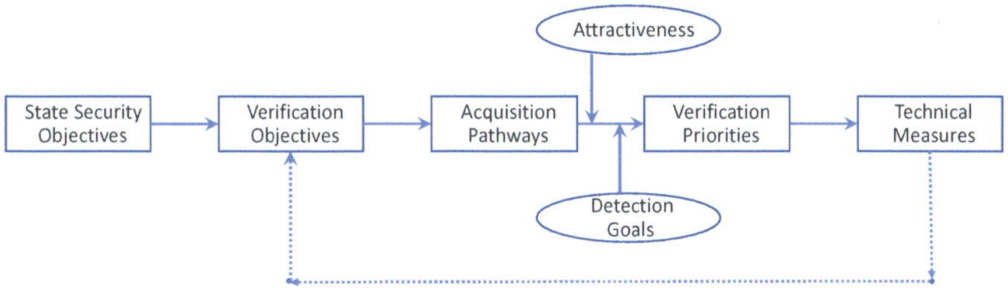

Fig. 1 Framework used to explore the usefulness of a systems approach to development of a treaty verification regime

3 Exercise One

The use of fictitious states and a model treaty during the technical meetings effectively facilitated the discussion on concepts and structural factors that are relevant to the application of a systems approach. This exercise explored systemic factors that might influence the verification of the model bilateral treaty. The treaty required each state to agree to:

- Deploy no more than 500 nuclear warheads, and
- Cap the size of its total stockpile of warheads at its current size for the duration of the treaty.

To facilitate this, each state would make declarations about their warhead stockpile and categorize each weapon as being either deployed or non-deployed.[1]

The focus of the exercise was the verification of Trenzalia's stockpile declaration of a total of 1970 weapons, of which 500 were declared to be deployed on delivery systems. The meeting participants considered Trenzalia to be in breach of the treaty, if undeclared warheads above the initially declared total of 1970 existed anywhere, or if any declared warheads were deployed above the maximum of 500. It was understood that some information detailed in the exercises could be security-sensitive and that it might not be possible for states to share such information under present security conditions, though this did not adversely affect the discussion of possible verification approaches. Key insights are discussed below.

3.1 State Strategic Security Objectives

Arms control agreements are negotiated to support a state's strategic security goals, which are shown as the first step in the framework illustrated in Fig. 1. In considering how to verify a treaty, one must consider why a state has entered the agreement, why it might not comply, and if cheating could be achieved undetected, would there be a significant strategic advantage for the cheater? What level of cheating would be sufficient to counterbalance the risk of being detected? How quickly will cheating build to an unacceptable level? These factors will drive how quickly cheating would need to be detected and would drive the state's requirements for a certain level of confidence in the ability of the verification regime to verify compliance.

The level of cheating at which a significant strategic advantage might be gained by a cheater over the complier, would then drive verification objectives, detection goals and priorities of the verification system that could detect non-compliant behavior before the level could be reached. In the Exercise One scenario, it was

[1] The terms nuclear warhead and nuclear weapon were used interchangeably in the exercise and in the model treaty, and "deployed" was taken to mean that the warhead was physically mated to a delivery system.

useful to explore motivations for one of the states to clandestinely maintain or obtain more weapons, when both states could openly deploy 500 warheads.

The participants discussed the deterrent value of a number of undeclared deployed or stockpiled warheads and concluded that any such value was limited. Deterrence was not considered a motivating factor for cheating. A completely secret weapons capability would not have any deterrence value and any deliberate ambiguity regarding the existence of additional weapons would immediately raise questions regarding adherence to the treaty, potentially causing it to fail. A stronger motivation might be to enable the cheating state to gain a strategic military advantage by overwhelming the other state's ability to respond to a surprise attack. To do this, the cheating state might need to amass more warheads than the other state, in order to ensure a successful first strike. If this were the goal, it seemed that a small number of warheads would not make any material difference in the strategic relationship between Trenzalia and Azaria. Rather, a state following this path might seek a significant numerical advantage. Even though numerical disparity is in reality only one of a number of strategic factors associated with capabilities that might affect strategic stability, it was a useful and illuminating simplification for exercise purposes.

Further, it was postulated that two potential cheating goals for Trenzalia could be the ability to rapidly deploy 150 warheads above the 500 limit or stockpile an additional 250 warheads without being detected. The rapid deployment of an additional 150 warheads was chosen to represent a significant imbalance between the two sides. A stockpile of 250 was postulated in order that some proportion of them might be deployed successfully in a critical situation, given that they may not all be kept in a state of operational availability. The goal of this verification system was therefore to detect non-compliant behavior before either 150 additional warheads could be deployed, or before 250 undeclared warheads could be stockpiled.

It was interesting to note that since states may view comparative advantage differently, some states may require much tighter limits on the detection parameters than others and some may be more concerned about one aspect of its adversaries capability, whilst others may focus of different aspects. This insight was a key driver of the second exercise, discussed in further detail in section four of this chapter.

3.2 *Verification Objectives*

Verification objectives need to be set to ensure that detection of noncompliance would be in line with the state security objectives. In the Exercise One scenario, compliance would be demonstrated by regularly submitting a declaration of the total number of weapons in the inventory and allowing it to be verified. The group recognized that familiar conceptual verification objectives, such as determining the correctness and the completeness of a state's declarations, could be applied in this case, since deviations from the declared inventory would signify non-compliant behavior.

To demonstrate declaration correctness, the verification system would need to provide confidence that no more than the allowed 500 systems were deployed at any point in time. With respect to completeness, the verification system should provide confidence that the overall arsenal is no larger than initially declared—i.e. that no undeclared weapons exist within the country.

The specific verification objectives postulated to meet the goals of verifying the completeness and correctness of a declaration were:

- Ensure the absence of undeclared, deployed warheads
- Ensure declared warheads could not be uploaded without being detected
- Ensure the absence of undeclared reserve warheads
- Detect the accumulation of warheads close to delivery systems
- Detect the undeclared production of warheads
- Detect the undeclared production of components

Though not specifically regulated in the treaty scenario, the accumulation of warheads close to delivery systems could indicate an intention to 'break out' of the agreement by exploiting spare upload capacity. It was recognized that the monitoring of stockpile inventory declarations could provide assurance that this was not about to occur, since it could indicate when large numbers of stockpiled warheads were being amassed in proximity to spare capacity on delivery platforms. The detection of accumulating warheads was therefore included even though it wasn't strictly controlled under the proposed treaty.

3.3 Acquisition Pathways

The next step taken in the exercise was to identify pathways that could be used to acquire nuclear warheads or weapons above the treaty-mandated limits. The group started with the physical model and APA methodology used for IAEA safeguards, but it was necessary to expand the existing APA to include the broader weapons and military complex. As shown in a case study that applied the APA methodology in a hypothetical weapons state, there are thousands of possible paths to consider. See *Physical Model and Acquisition Pathway Analysis—Applicability to Nuclear Disarmament Verification* by Irmgard Niemeyer, Joshua Rutkowski and Gotthard Stein, in this book. However, which pathways might be considered most attractive, and why, must be considered specifically for a cheating state that wishes to benefit from deploying or stockpiling additional weapons.

Three examples of cheating pathways Trenzalia could exploit in order to deploy an additional 150 warheads are shown in Fig. 2.

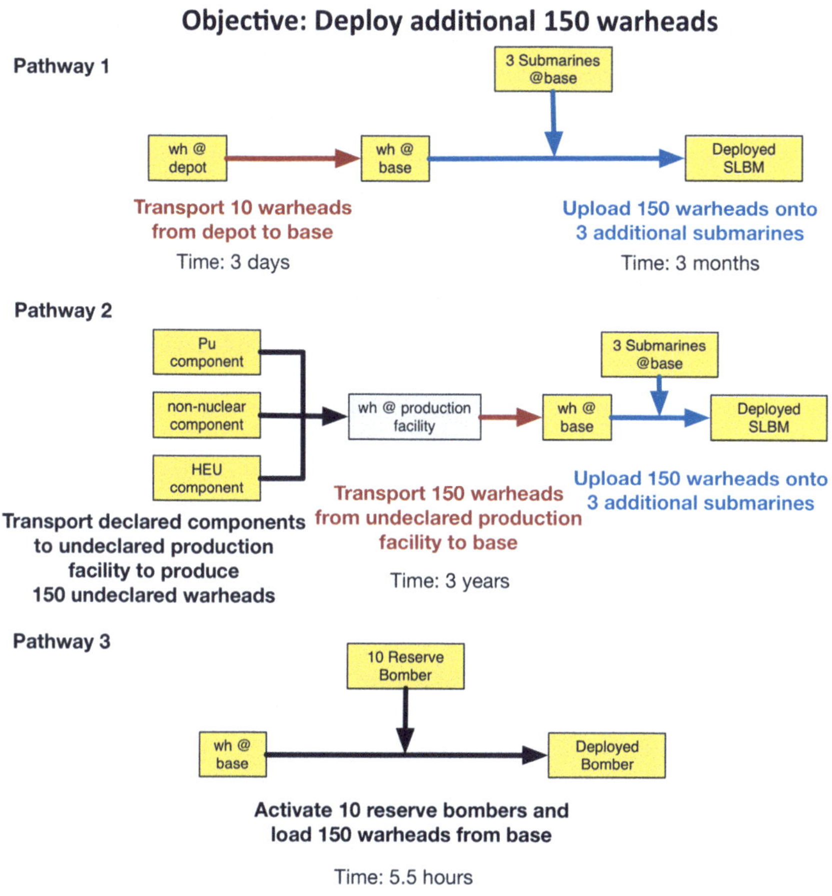

Fig. 2 Examples of cheating pathways that Trenzalia could exploit to deploy an additional 150 warheads above its treaty-mandated limit of 500

3.4 Pathway Attractiveness

Next, one must consider the parameters that might make one potential path for cheating more attractive than another. Together with detection goals, this could be used to set verification priorities and the technical measures needed. Setting priorities allows for the distribution of verification resources across potentially many cheating pathways. Rather than reinvent suitable attractiveness metrics, the six proliferation resistance metrics defined in the 2011 report *Evaluation Methodology for Proliferation Resistance & Physical Protection of Generation IV (Gen IV) Nuclear Energy Systems* [5] were used in Exercise One by the experts group. The Gen IV metrics were found to be a broadly suitable starting point for assessing attractiveness—with some adaptations. A general summary of the experts group discussion is presented below, but it is clear that further work on the definitions and relative values of metrics would be needed to adapt to the aims and objectives for a specific arms control agreement.

Proliferation Technical Difficulty It was assumed that the ability to deploy existing weapons, build additional weapons, or modify stockpiles was already established in Trenzalia, because, in the scenario, it maintained a full weapons enterprise. Therefore no significant technical barrier to exploiting a particular pathway existed within the state. Furthermore, existing non-prohibited processes could be manipulated to try to mask prohibited activities. Nevertheless, some pathways might require mobilization and coordination of greater resources than others, adding to the operational difficulty involved in exploiting certain cheating pathways. For example, the deployment of reserve warheads onto reserve bombers may be accomplished more simply than building a clandestine stockpile of new weapons and secretly loading them onto a submarine (see Fig. 2). A generalized observation may be that for states with existing technical capabilities, an analogy to *Proliferation Technical Difficulty* was mostly relevant in terms of the stealth (or denial and deception) required to successfully exploit the pathway.

Proliferation Cost The unequal cost of employing certain pathways relative to others is evident, as illustrated in the previous point. Maintenance and operational costs for existing capabilities are likely to already be included in national budgets. Using those capabilities to cheat might require relatively fewer resources, so only pathways requiring significant capital investment might be deemed less attractive to a state wishing to cheat. On the other hand, the overall cost may not be a primary factor in the decision to exploit a particular cheating path. Stealth and the overall likelihood of fulfilling the overall cheating goal might be much more significant factors, in an arms control context.

Proliferation Time How long will it take to achieve the cheating goal following a specific path? For the scenario in this exercise the analogy to proliferation time varies between the minimum time required to deploy a strategically significant quantity of additional weapons onto battle-ready delivery systems/platforms and the amount of time taken to build and stockpile an even larger number of weapons in storage.

Fissile Material Type Material types were not discussed in great detail since material stockpiles were not declared under the treaty in the scenario, therefore a possible analogue could be warhead types or their operational readiness.

Detection Probability The Gen IV metrics assume that a standard set of safeguards measures and practices could be implemented at any facility subject to safeguards and would result in a standard detection probability. No analogous standard set of detection measures exist for nuclear arms control purposes. The absence of a standard set of detection measures provides an opportunity to suggest bespoke solutions fit for fulfilling verification objectives. Different combinations of technical verification measures could be developed on an iterative basis to reach the necessary detection probability for different facilities, processes or pathways. A cheating state might find the least detectable pathways attractive so detection probability could be linked to the technical difficulty metric through the concept of stealth. Alternatively, the ability to exploit a pathway quickly might be attractive. From these alternative views of attractiveness it is evident that a high detection probability would be required both for the detection of low-level diversion aimed at building up a clandestine stockpile and for detecting attempts to rapidly deploy large numbers of weapons in excess of agreed deployment limits.

Detection Resource Efficiency The efficient use of staffing, equipment, and funding to apply verification measures across different parts of the weapons enterprise would need to be considered in the development of a regime. Efficiency might be enhanced if there are points in the nuclear weapons enterprise where the ability to verify the completeness and correctness of declarations with high confidence is evident, particularly if the monitoring of such points decreases how 'stealthy' the path is or increases the exploitation time associated with cheating pathways.

3.5 Detection Goals

Nuclear safeguards detection goal is to detect diversion of nuclear materials from within a state before one 'Significant Quantity' (1 S.Q.) of nuclear material can be accumulated. One S.Q. is the amount of material required to build a single nuclear weapon and in the case of safeguards 1 S.Q. is deemed to be the unacceptable limit. The safeguards system must therefore be capable of detecting much lower levels of noncompliant behavior in order to provide timely warning before this limit is reached. But what is a suitable analogy to a significant quantity when considering stockpiles of existing nuclear weapons?

The verification regime to be considered during the exercise scenario was to provide high confidence that noncompliance would be detected before an additional 150 weapons could be deployed, or before an extra 250 warheads could be stockpiled by a cheating state. Monitoring system sensitivity must be considered in determining whether detection of noncompliance is possible. Specifically, how accurate would a verification system need to be to detect non-compliance that might lead to the accumulation of hundreds of weapons? How timely does detection of noncompliance have to be to prevent a cheater from reaching its cheating goals before being detected? Clearly, the determination of how accurate and timely the system needs to be will be a factor in the level of intrusiveness required. Together, the parameters of accuracy and timeliness will influence the verification priorities, and alongside intrusiveness, will help determine which inspection techniques and technologies might be most suitable.

Accuracy For the purpose of exploring detection goals, technical experts considered one nuclear weapon to be the discreet unit of account. While each country is still able to maintain some 500 deployed warheads, under the model treaty in

Exercise One, it was suggested that small fluctuations around this number may not be considered strategically significant, since a cheating state would need to accumulate at least an additional 150 weapons. There was potentially room for tolerance in the accuracy of stockpile inventory accounts. Since states may hold concerns over the provision of some details about their weapons inventory, some tolerance might be attractive. However, it was noted that 'strategic significance' might be different in situations where large numbers of warheads remain deployed, compared to some disarmament scenarios where deployed warhead numbers are considerably reduced. Moreover, accounting accuracy that is tolerant of discrepancies, when permitted weapons inventories are high, may cause uncertainties that propagate through the accounting and verification process, and complicate future verification efforts. Furthermore, perceptions of strategic significance and significant quantities may differ for each for each treaty signatory, and so any agreed tolerance must be acceptable to all. Finally, it was noted that the detection of low-level discrepancies was key because they may indicate larger-scale deception. The group suggested that the verification system would be required to detect the existence of a single undeclared stockpiled weapon and the deployment of a single nuclear weapon above the treaty limit. The inventory would also need to be accurate down to the level of individual weapons.

Timeliness As part of the Exercise One scenario, Fig. 2 illustrates a pathway available for Trenzalia to deploy an additional 150 weapons, in as little as 5.5 h. If fewer weapons were considered strategically significant, then Trenzalia could accomplish the goal in even less time. These types of short timescales might drive the technical requirement for real-time monitoring of this pathway, in order to detect discrepancies in a timely manner. The same pressure for real-time monitoring might not exist at other points in the nuclear enterprise, for example,where it is not possible for a cheating state to simply upload weapons onto delivery platforms. A timeliness goal measured in weeks, months, or years might be more appropriate for ensuring the absence of an undeclared stockpile of weapons or an undeclared capability to produce them.

3.6 Verification Priorities

The verification priorities that drive the allocation of monitoring resources to fulfill the verification objectives will take into account the factors discussed above, such as the potential attractiveness of identified cheating pathways, the potential pathway exploitation time, and the stealthiness required.

In Exercise One, the spare upload capacity on Trenzalia's declared delivery vehicles was of concern because that capacity could rapidly be filled with warheads from the declared stockpile. For treaties focused on the numbers of deployed warheads, frequent monitoring of vehicles and very frequent or even continuous monitoring of non-deployed warheads in proximity to the delivery systems, might be a priority. Because the treaty also applied to the overall stockpile, the existence of a large undeclared stockpile of warheads was considered a risk, though one that might be managed using a different set of monitoring techniques. Stocks of material, excess production capacity, and facility production rates could be monitored and inspected periodically to ensure undeclared weapons were not being hidden amongst legitimate activities.

A state-level assessment could help highlight the most pressing issues and help select the most effective verification strategy given limited resources. State level assessments may even highlight where constraints and verification measures could be applied in a nonreciprocal but mutually satisfactory way, in order to meet overall treaty aims whilst taking into account the unique aspects of the individual signatories.

Intrusiveness and Transparency The degree of accuracy and timeliness required to meet the verification goals can directly impact the level of intrusiveness associated with the verification system. The technologies and methodologies that make up the system must balance intrusiveness with respect for restrictions, based on nationally-sensitive or proliferative information, and on the need to minimize disruption to the operating schedule of any location being monitored.

The accuracy and timeliness goals that emerged from the exercise would require very careful preparation and management if they were to be implemented in reality. In order to verify the correctness of inventory declarations, increased transparency over the location and deployment status of declared warheads might be required. The participants discussed how logistical details, such as maintenance schedules and weapon retirements, could also be useful. Information of this type would enable declared warheads to be monitored accurately through the enterprise and could help ensure that undeclared warheads, if found, could not be explained away as part of the declared stockpile. Conceivably, a formal update could be required every time a warhead changes status or location. Nevertheless, such detail may have national security implications and so the mechanism by which such information was shared and verified would need careful development. Inspectors would potentially need access to military facilities and other industrial support sites frequently enough to verify declared changes to the stockpile. How to achieve this suggested additional transparency, facilitate the associated inspections and manage the intrusion was not discussed in detail.

A mechanisms that would allow inspectors to verify the absence of undeclared warheads, either deployed or stockpiled, was also discussed. The system would need to ensure that undeclared warheads were not being kept either in declared warhead related sites such as production facilities or in unrelated locations throughout the country. Overall, it was evident that an inspections to verify the lack of undeclared activity, under the terms of the model treaty, could be very disruptive and intrusive into the operational aspects of the nuclear weapons enterprise, the supply chain, and other sites that could harbor undeclared warheads.

Although transparency into the nuclear enterprise might be considered necessary, it is reasonable to consider that logistical and operational information such as warhead locations, facility storage capacities and warhead production rates or schedules, would be considered to be sensitive by Trenzalia. This highlights the recurring theme of balancing national security with the desired level of confidence in the verification regime. The balance struck between transparency and maintaining opacity may result in the acceptance of a lower overall level of confidence in compliance.

An interesting aspect of the discussion during the exercise was to note that a focus of the technical arms control research community has been on the protection of weapons design information and concern about the intrusiveness of techniques used to verify weapon characteristics. However, little of the information declared or exchanged under the treaty proposed between Trenzalia and Azaria was concerned with warhead design details. Instead, weapons specific information related to identity, location and deployment status, primarily operational information that would require much less physically intrusive techniques for verification, was desired.

During the third meeting exploring the application of system level concepts, in Jackson Hole, Wyoming, the group considered three types of information related to nuclear weapons that could not be exchanged: design information, vulnerability concerns and use control information [6]. For the right purpose, and with sufficient preparation, other forms of enterprise relevant information might be shared, most likely only between signatories or with an inspection body, rather than publicly.

Correctness and Completeness The goals of verifying both the correctness and the completeness of a declaration may be familiar to those with knowledge of IAEA safeguards. The same concepts could be applied to the exercise scenario, as described earlier.

Declaration correctness verification measures would focus on verifying the deployment status of the declared stockpile—ensuring that no more than 500 systems could be considered deployed at any point in time. Characteristics for identifying deployment status might be a physical distinction that could actually be verified (perhaps simply a distinction by location), but might not be agreeable to all sides. At its most extreme, the participants concluded that a distinction between the status of two warheads might only be administrative. Retirement of a warhead provides a good example of this challenge—retirement could be a purely administrative process that categorizes a particular non-deployed warhead differently from others, but a retired warhead may remain co-located with other non-deployed warheads for some period of time. The retired warhead, in this case, would not be physically distinct. Whilst verifying the status might be challenging if based on administrative differences, it was interesting to note that the model treaty did not stipulate that the items declared to be nuclear weapons must truly be nuclear weapons, so there might be less need to prove that an item declared to be a warhead really was one.

Indeed, it is difficult to determine a suitable method for confirming that an object is a nuclear weapon without the exchange of sensitive information (e.g. design information), which may not be acceptable. A final decision would have to balance the resources needed to verify that an item designated as a warhead truly was one and the limited contribution this verification might provide towards completeness. This form of verification might, however, limit the ability of states to declare vastly inflated stockpiles of weapons when making an initial declaration about its inventory and might be more relevant under differing treaty conditions.

Declaration completeness verification measures would need to provide confidence that the overall arsenal is no larger than initially declared—i.e. that no undeclared weapons exist within the country. Any other item exhibiting characteristics that might indicate that it was a nuclear weapon would need to be investigated and suitably discounted from the treaty or accounted for in the inventory. The evaluation of completeness would benefit from the clear separation of treaty accountable items from other potentially suspect objects. Objects declared not to be warheads could then be measured and discounted as nuclear weapons. This could be complex; for instance, the verification system might need to include a means of accounting for the components of nuclear weapons (to provide assurance that they are not in fact weapons themselves and as a means of deterring low-level clandestine production of undeclared warheads) whilst allowing for the production of new weapons using those components so long as the new weapons were correctly declared and entered onto the inventory.

3.7 *Technical Verification Measures*

The experts participating in the exercise postulated possible verification measures that could be employed as part of the system to verify compliance, detect diversion within a suitable detection time and potentially reduce the attractiveness of specific cheating pathways. Possible verification system elements were divided into declarations and monitoring activities according to whether the site being monitored was declared to hold nuclear weapons or not. During treaty negotiation, a down selection process to determine the most suitable verification measures (e.g. declarations, information exchanges, verification technologies, inspector access) would require an iterative process of testing and enhancing options, taking into account all constraints related to intrusiveness and potential disruptiveness. This process is illustrated by the dotted line leading back to the verification objectives stage of the framework.

The experts proposed that Trenzalia be split into three zones (Fig. 3): one encompassing sites and facilities associated with the inventory of weapons, another encompassing sites and facilities associated with military fissile materials, and the third capturing the rest of the country. Each zone could be split further based into appropriately sized 'mass balance' areas to ensure the physical inventory of deployed or stockpiled weapons was equal to the declared accounts for each area. The zones were suggested to indicate where an emphasis on certain verification methods and measures might be more appropriate for fulfilling the verification objectives associated with the completeness and correctness of the inventory items given the verification parameters outlined. The ideas reflecting the discussion of

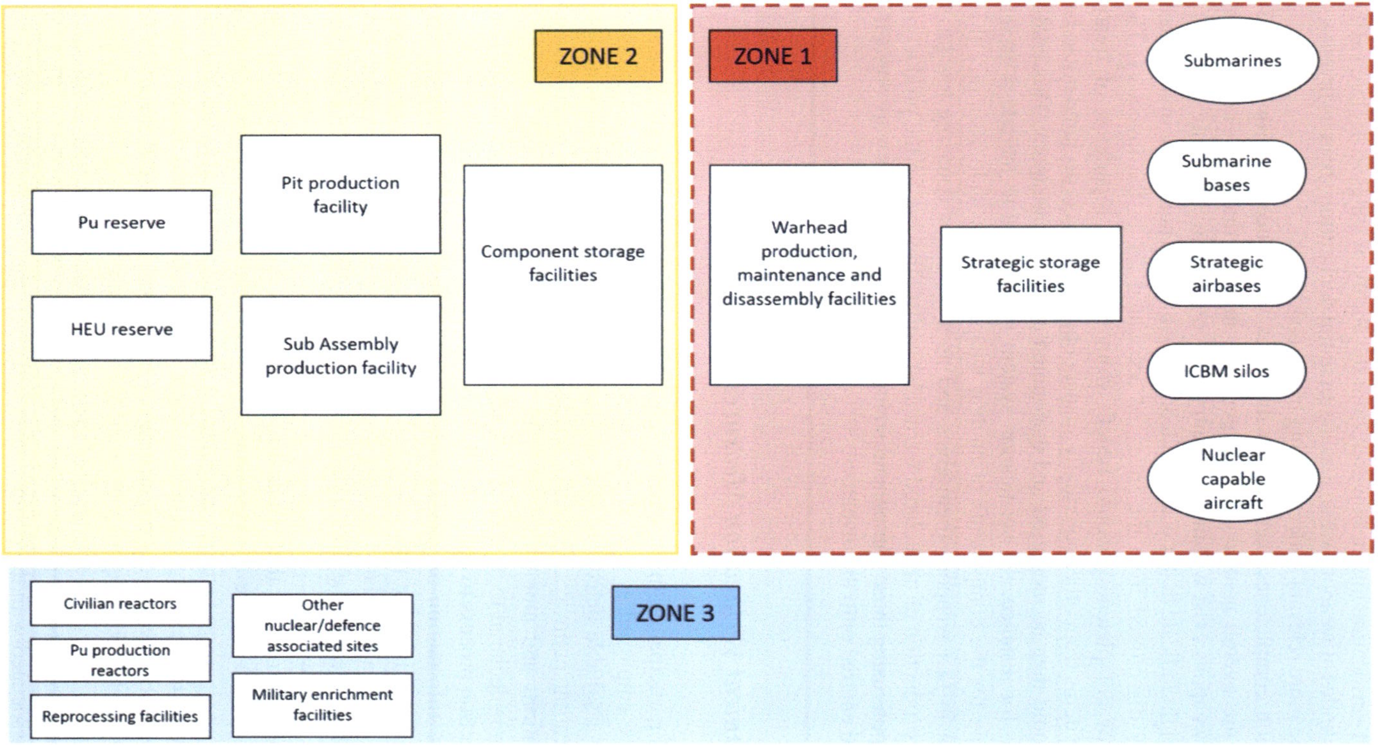

Fig. 3 Zoning off Trenzalia for implementing a verification regime. Zone 1, contained within the dashed red border, encapsulates all declared facilities where weapons could be stored or deployed. Zone 2, in the solid yellow border, captures all declared facilities that handle fissile materials for the weapons program. Zone 3 contains the rest of the country

technical experts are presented in Tables 1, 2, and 3. They should not be considered definitive or complete.

Zone 1: Military Nuclear Sites The focus of the verification system for Zone 1 was to ensure that declarations made about the inventory of weapons was correct and complete and to detect breakout in a timely manner, given the potentially short time scales involved.[2] The military nuclear sites in Zone 1 encompass all locations where nuclear weapons could be expected to be stored or deployed, and where there was a relatively straightforward route to mating weapons with delivery vehicles. The military locations included in this zone were for warhead production, storage, and deployment.

Robust monitoring would be needed to ensure that warhead locations and status remain correct to a high level of confidence and ensure that they are not deployed in breach of treaty commitments. The group believed that high confidence would be required to ensure that only declared items could interact with declared delivery systems and that a status change could not occur rapidly and stealthily in large numbers. i.e. those weapons declared not to be deployed could not be deployed rapidly and stealthily. Detection probabilities for undeclared items at declared deployment sites would need to be sufficiently high, and the detection time for undeclared items would be rapid. Table 1 presents some possible verification measures, along with the frequency of monitoring and types of technologies that could be employed in military nuclear sites.

Zone2: Military Industrial Capacity Zone 2 encompassed the industrial capacity of the state to produce fissile materials and the fissile components of nuclear weapons. Zone 2 facilities would directly feed into warhead production. The focus of the verification system here would be to ensure that declared facilities and materials were not being used for the low-level build up and diversion of an undeclared stockpile of warheads. The monitoring system would need to be capable of detecting long-term trends of overproduction or other deviations from scheduled activities. The proposed measures are shown in Table 2.

Zone 3: Rest of Country Zone 3 comprised the rest of the country. Zones 1 and 2 include locations where nuclear weapons-related materials might reside, so the purpose of monitoring in Zone 3 was predominantly to ensure the absence of undeclared stockpiles or warheads deployed on undeclared delivery vehicles. Possible verification measures are shown in Table 3.

[2]Breakout describes the situation where warheads could rapidly be deployed onto delivery systems in significant numbers to rapidly produce a strategic imbalance.

Table 1 Possible verification measures for sites where warheads could be stored (Zone 1) in the Exercise One scenario

Verification regime elements	Purpose	Information/verification measures	Frequency	Potential techniques and technologies
Inventory declarations	Set expectation: Deviation from declaration may indicate non-compliant behavior	Warhead unique ID, status declaration, and location (at appropriate level of accuracy)	Every time there is a change	Unique reference identity generation and recording
Deployed weapon verification	Verify declared inventory of deployed weapons	Select delivery platforms for inspection—verify number of weapons deployed against inventory	Frequent: multiple inspections per year	Observation; Radiological means of discounting any item declared not to be warheads
Stockpile verification	To ensure accuracy of declared inventory of stored weapons and prevent rapid, undetected upload/status change	Monitor location of all non-deployed weapons; Monitor status of all weapons; Physical inspection to confirm location and status against periodic declaration	Continuous monitoring and regular physical inspection	Physical inspection; Tagging technologies; Containment and surveillance technologies; Portal perimeter monitors; Satellite surveillance
Completeness verification	To verify absence of items with warhead characteristics from locations where non have been declared	Challenge inspection of chosen locations. Activities at chosen location frozen and physical access enabled within a defined timeframe	Frequent at locations in close proximity to delivery platforms/systems Reasonably frequent at production and storage facilities elsewhere in country	Observation; Radiation detection to demonstrate absence of nuclear weapon characteristics; Satellite surveillance/ monitoring under open skies

Table 2 Possible verification measures for military industrial sites for material production (Zone 2) in the Exercise One scenario

Verification regime elements	Purpose	Information/verification measures	Frequency	Potential techniques and technologies
Inventory declarations	Set expectation: Deviation from declaration may indicate non-compliant behavior	Location of fissile components and materials; ID of containers containing fissile material or components; Production schedules and outputs;	Every time there is a change; Prior to production campaigns or other changes in form	
Stockpile verification	To ensure accuracy of declared inventory of stored material and components	Physical inspection to confirm location and ID against periodic declaration	Infrequent	Physical inspection/ observation; Unique reference identity generation and recording
Completeness verification	To verify absence of undeclared items with fissile material characteristics	Challenge inspection of chosen locations. Activities at chosen location frozen and physical access enabled within a defined timeframe	Infrequent	Inspector random selection of objects for testing; Observation; Radiation detection to demonstrate absence of fissile material from items not on inventory; satellite surveillance/ monitoring under open skies

Table 3 Possible verification measures for the rest of the country that are not designated military or military industrial sites in the Exercise One scenario

Verification regime elements	Purpose	Information/verification measures	Frequency	Potential techniques and technologies
Inventory declarations	Set expectation: Deviation from declaration may indicate non-compliant behavior	Register of fissile materials including location and quantity	Update register periodically	
Completeness verification	To verify absence of items with warhead characteristics from locations where non have been declared	Challenge inspection of chosen locations. Activities at chosen location frozen and physical access enabled within a defined timeframe	Infrequent but routine (e.g. one per year)	Inspector random selection of objects for testing; Observation; Radiation detection to demonstrate absence of nuclear weapon characteristics; Satellite surveillance/monitoring under open skies

4 Exercise Two

A second short exercise was held by the Institute of Nuclear Materials Management (INMM), hosted by the Center for Global Security Research (CGSR) at the Lawrence Livermore National Laboratory (LLNL) during the summer of 2015. It focused on states' security and verification objectives and how these influence arms control priorities. A scenario was designed to observe how each state might focus on different acquisition pathways, depending on their objectives and security requirements to limit the capabilities of their adversaries [3]. To facilitate the discussion and focus on the security objectives and priorities of each individual state, two groups were asked to consider scenario in the context of the framework presented in Fig. 1.

The Exercise One scenario was simplified so that Trenzalia and Azaria more clearly represented different levels of development, capability and population size. The simplified physical model of the states' weapons complexes and their stockpile details are presented in the Appendix at the end of this chapter. For Exercise Two, the model treaty limited the total nuclear forces at the existing levels for a period of 10 years, designating strategic/tactical weapons and deployed/non-deployed status. The number of each type of delivery system, deployed warheads, and total number of warheads was capped. The development, testing, and deployment of new types of warheads and delivery systems was prohibited.

As was found during Exercise One, strategic security considerations defined the states treaty verification goals. The group representing Azaria gave priority to assessment of national political objectives and developed a strategic doctrine and a national security strategy. This was then used to analyze Azaria's own strategic and tactical weaknesses, which directly linked their verification objectives. The Trenzalia group took a more systematic approach to develop technical verification objectives, accepting the scenario-defined security objectives. This group worked to identify key acquisition pathways and verification priorities, presumably because they were satisfied with the status quo.

After considering likely acquisition pathways and possible verification measures, the groups developed the following set of verification priorities for their respective fictitious countries:

- Azaria, as the weaker state, was concerned about detecting signs of preemptive attack and identifying the development of new capabilities that could impact the effectiveness or survivability of their nuclear arsenal. Its potential verification priorities focused on understanding and detecting changes to Trenzalia's deployment patterns and identifying the development of new delivery systems.
- Trenzalia's security concerns focused on deterring Azarian aggression in the future. Its potential verification objectives centered on detecting new capabilities or postures that threatened the existing strategic balance. Specific objectives included detecting the development of new delivery systems and the enhanced research and production capacities for future arsenal growth.

A key lesson derived from Exercise Two was that the results were heavily influenced by the perspectives of the experts/decision-makers on each team, being that the short exercise time did not allow the teams sufficient time to explore all the security implications. Azaria team members concentrated on strategic security, leading to the identification of a specific set of acquisition pathways to be monitored in accordance with its national security strategy. Azaria's strategy was to accede to the treaty to cap development of Trenzalian capabilities while it developed its own economy, conventional forces, and weapons complex (in compliance with the treaty) to improve its long-term strategic position.

Trenzalia's more technical approach followed the prescribed framework and developed lists of verification objectives and pathways to monitor in order to address all of the potential cheating risks. This team focused on the ability to detect further development or modernization of Azaria's capabilities, being that Trenzalia was confident in its present-day military superiority. Since the nuclear arsenals and security concerns of the two fictitious states were not directly matched, the verification goals and approaches varied significantly. It is not clear if this result would have been maintained had there been more time to work through the exercise.

The differences in arms control objectives highlighted the challenge of identifying mutually acceptable constraints among unequal partners. During the second exercise, consideration of reciprocal verification measures was deferred to stimulate discussion and present opportunities for creative solutions. Suggestions included formalising asymmetries in the application of verification measures in each state; i.e. that treaty constraints and verification measures did not necessarily have to be the same for each state or even govern the same types of capabilities so long as the agreement enhanced mutual security by meeting an overarching goal.

Reciprocity has traditionally featured strongly in arms control agreements and would likely be strongly advocated in future. Consequently, it would be unusual to implement asymmetries in this way but there may be scope for future research efforts to consider the possibility of creative verification solutions tailored to the concerns of each state that would not necessarily result in direct reciprocity of verification measures.

5 Lessons Learned

The exercises and targeted meetings, which involved a wide range of international technical and policy experts provided valuable insight into how a systematic approach could help identify nonproliferation and arms control verification requirements. By applying a transparent state-level systems framework it was possible to make an initial effort to define the verification goals, objectives, processes, and parameters needed to design an effective verification regime based on the strategic goals of a specific treaty whilst simultaneously factoring in facility, location and other potential restrictions.

Authorities could benefit from using a systems approach to assess potential verification preferences and drawbacks associated with potential future treaties. Moreover, the discussions illustrated the potential for such an approach to advance the dialogue between potential treaty partners—particularly in a broader international context—such as the ongoing International Partnership for Nuclear Disarmament Verification [7]. The different security environments, degrees of weapons knowledge and verification experience coupled with the requirement of upholding Nuclear Nonproliferation Treaty Article VI to protect the spread of nuclear weapons information, makes substantive dialogue a challenge in a broad international environment. With a rational, transparent and systematic framework, greater insights could be gained in the complexities of achieving deeper disarmament goals. Nonetheless, caution should also be taken since end-users of systems analysis should be aware of how the perspectives of the experts or decision makers providing input into the system model can affect the results.

Although the experts group agreed this was a very promising approach, there was only a limited amount of effort possible during three ESARDA and INMM meetings. Rather than presenting definitive verification approaches and measures, the group generated more questions and challenges. Possible future activities could include:

- A concerted effort to link the material and weapons sectors of the nuclear weapons complex to achieve broader confidence in a state's activities and intentions. A spin-off of such an activity could help create a new generation of experts that take a broader view and do not focus on just nonproliferation or arms control
- Satisfying the competing needs for effective verification and protection of national security information. If clear verification objectives can be defined, the design of declarations, inspections, and managed access procedures could be further explored in a diverse community.
- Defining the treaty-controlled items in such a way that declarations can be verified effectively. This may be a challenge when establishing a suitable means of verifying the validity of an administrative designation in the absence of a physically verifiable change but it also opens up new possibilities.
- Use of a systems approach to analyze the pros and cons of possible verification regimes that would allow for the conduct of a form of sensitivity analysis that could provide feedback and a better understanding of achievable confidence levels. This analysis could then be used to set priorities and help determine needed R&D to fulfill technology requirements.

Bringing together a diverse set of experts to focus on different aspects of the nuclear weapons life cycle has opened up avenues for discussion and possible new ways to apply emerging technologies. The efforts of a relatively small group of technical experts has shown that with a systems approach and transparent framework, it is possible to bridge communities, forge a common lexicon and benefit from advances that have already been achieved in other areas relevant to effective verification and international security.

This work was partly authored by Lawrence Livermore National Security, LLC under Contract No. DE-AC52-07NA27344 with the U.S. Department of Energy. The views and opinions of authors expressed herein do not necessarily state or reflect those of the United States government or Lawrence Livermore National Security. LLNL-BOOK-714538

Appendix: Fictitious States of Trenzalia and Azaria and Model Treaties

To investigate the use of a systems framework like the IAEA State Level Approach in the context of nuclear arms control, a series of technical meetings were organized over the course of 2014–2015. Exercise One took place during a meeting of the European Safeguards Research and Development Association (ESARDA) Verification Technologies and Methodologies Working Group, held at the Joint Research Centre in Ispra, Italy, in 2014. Exercise Two was held at an Institute of Nuclear Materials Management (INMM) meeting hosted by the Center for Global Security Research (CGSR) at the Lawrence Livermore National Laboratory (LLNL) in 2015. The scenarios included the two bordering fictitious states of Trenzalia and Azaria. The details of their nuclear enterprises are presented here. Not all the details were used during the exercises and expert discussions but are presented here for possible future utility.

Exercise One

During this exercise, the group analyzed the diversion pathways and possible verification measures that could be employed for only Trenzalia, so the details regarding Azaria were not relevant and are not included here.

The model treaty was a bilateral agreement between the two nuclear weapons states (Trenzalia and Azaria) to limit the total number of warheads deployed and stockpiled. Each state agreed to limit the number of nuclear warheads it deployed to no more than 500 and to declare its total stockpile of warheads, both deployed and nondeployed. That number was set as the upper limit to its overall future stockpile. No further details were specified. During Exercise One the meeting participants considered the existence of undeclared warheads above the initially declared total of 1970, and any warheads deployed above the maximum of 500 would constitute a form of cheating.

In order to visualize the different paths Trenzalia might take to cheat, the group considered the flow of fissile materials through the weapons complex starting from a material stockpile (of either military or civilian stocks) and ending up in a weapons system. Stages included use of material in the production of subcomponents, the assembly of full warhead systems, the storage of warheads the maintenance of warheads, the mating of warheads with delivery systems and the deployment of systems onto delivery platforms. It also included the potential for the eventual retirement of warheads and the reclamation of the fissile material. Consideration was given to both a declared nuclear enterprise and an undeclared nuclear enterprise, which could duplicate any or all of the declared facilities, materials, warheads, delivery systems and delivery platforms. A conceptual view of Trenzalia's nuclear weapons enterprise is shown in Fig. 4.

One can imagine a simple unidirectional flow diagram of material through a weapons enterprise, illustrating both declared and undeclared routes. For example an undeclared stockpile of plutonium could be introduced into declared facilities to produce undeclared components and warheads, which were then stockpiled in an undeclared storage location.

Details of the hypothetical distribution and configuration of warheads in Trenzalia are presented in Tables 4 and 5. In Table 6, notional times that are operationally relevant to nuclear warhead production and deployment are presented.

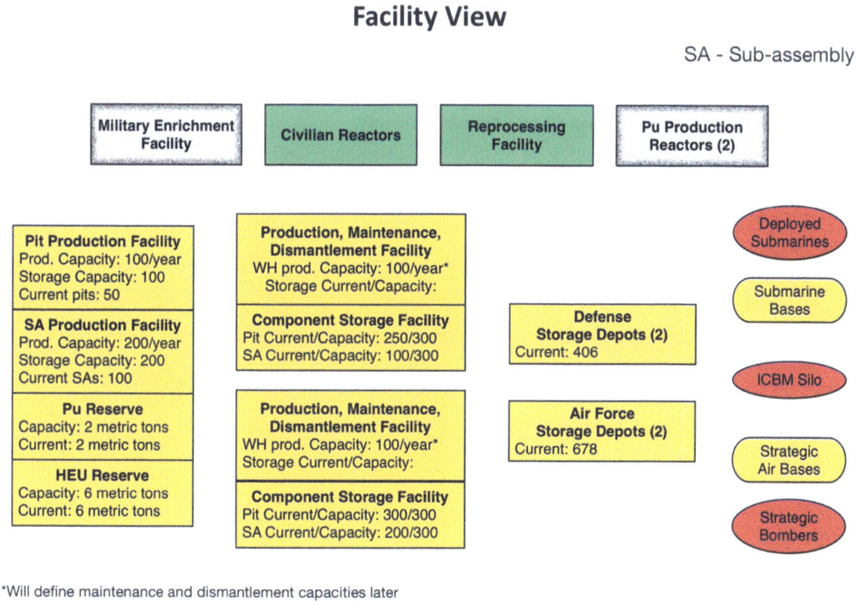

Fig. 4 A conceptual view of the nuclear weapons enterprise of one of the fictitious states, Trenzalia, used in Exercise One

Table 4 The distribution of warheads in Trenzalia for Exercise One

Delivery system	Warhead type	Current deployed	Current forward deployed storage	Current defense storage	Current delivery capability
4 boats with 8 SLBMs with 8 warheads	I	28	60	44	4 boots, 1 always at sea
	II	36	80	66	
	III				
100 ICBM with 4 warheads	I	14	6	38	51 deployed ICBMS in silos, 49 reserved
	II	48	14	74	
	III	96	24	184	
60 bombers with 16 warheads	IV	152	34	130	20 deployed bombers, 20 active reserve, 20 other
	V	78	120	322	
	VI	48	48	226	
Totals		500	386	1084	
Total in Storage = 1470					
Total Number = 1970					

SLBM submarine-launched ballistic missile, *ICBM* intercontinental ballistic missile

Table 5 Configurations of the number of warheads loaded onto Trenzalia's deployed ICBM and aircraft operational delivery platforms (Exercise One)

Deployed bomber configuration (20 deployed bombers)			
Warhead type	Configuration of 6 bombers	Configuration of 14 bombers	Total warheads
IV	2	10	152
V	6	3	78
VI	8	0	48
Sum			278

Deployed ICBMs (15 deployed ICBMs)				
Warhead type	# of missiles	# of warheads	Totals	Spare upload capacity
I	32	4	128	0
II	11	2	22	22
III	8	1	8	24
Sum			158	46

Table 6 Notional times operationally relevant to nuclear warhead production and deployment for Trenzalia in Exercise One

Times	Component production	Production	Strategic depot	Base storage	Deployment
Bomber	1 week	4 weeks	3 days	3 days	90 min
ICBM	1 week	4 weeks	3 days	2 days	2 days
SLBM	1 week	4 weeks	3 days	3 days	4 days

Exercise Two

The lessons learned in the first exercise prompted the revision of the hypothetical states of Azaria and Trenzalia. They were reconfigured to represent different levels of development, capabilities and populations. Both states were considered to have first-generation nuclear arsenals and weapons complexes but with different types of capabilities to meet their strategic needs.

Trenzalia, the larger power, is a moderately advanced industrialized state with regional military and economic dominance and growing global influence. Its population is 200 million, it has a moderately advanced industrialized country, and it is a dominant regional military and economic power with ambitions for broader global influence. It has a sophisticated nuclear weapons enterprise consisting of civilian and military nuclear fuel cycles (indigenous), with a total of 322 nuclear warheads. The details are presented in a simplified physical model is presented in Fig. 5 and Table 7.

Azaria is a newly industrialized, ascending power with a recently developed nuclear capability and a population of 100 million. It has a modest conventional force and is reliant on a primitive nuclear deterrent. Azaria's nuclear enterprise consists of both civilian and military nuclear fuel cycles, and it possesses a total of 110 warheads. The details are presented in a simplified physical model is presented in Fig. 6 and Table 8.

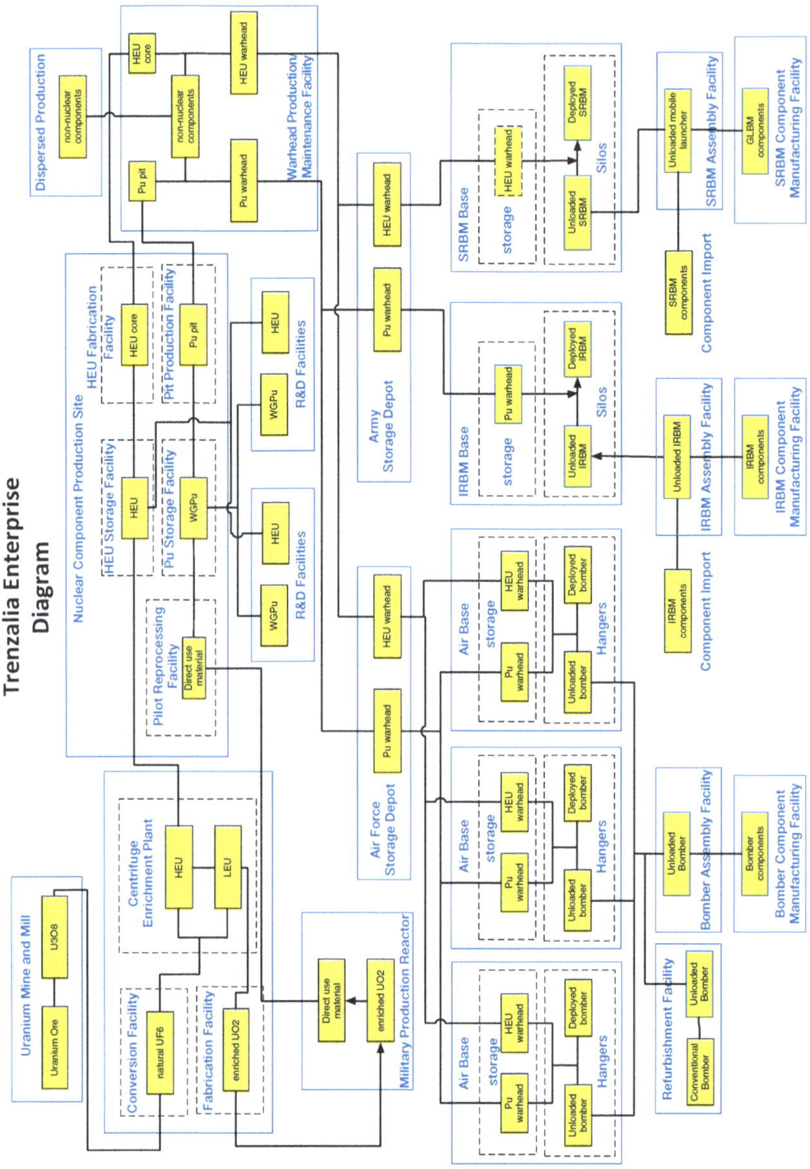

Fig. 5 A simplified physical model of the nuclear weapons enterprise of the fictitious state, Trenzalia, used in Exercise Two

Table 7 Nuclear weapons, deployment systems, and nuclear material possessed by Trenzalia in Exercise Two

Country of Trenzalia				
Delivery system	Available weapons platforms	Warhead type	Deployed (includes on-base storage)	Storage depot
SRBM	46	HEU implosion	64	30
IRBM	28	Boosted Plutonium	36	18
28 Bombers × 4 warheads	112	HEU implosion	82	52
		Boosted Plutonium	34	6
Total	186		216	106
Material	kg	Location		
Pu	800	Component production facility		
HEU	1600	Component production facility		

SRBM short-range ballistic missile, *IRBM* intermediate range ballistic missile

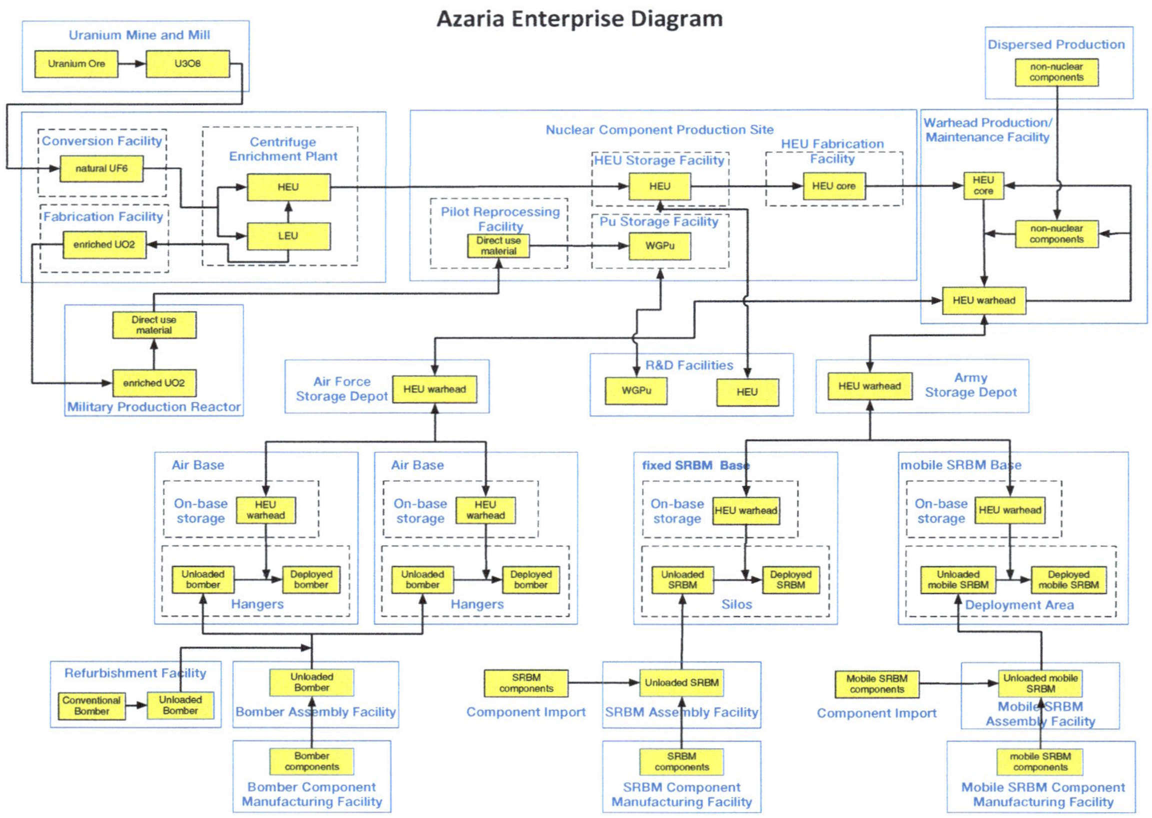

Fig. 6 A simplified physical model of the nuclear weapons enterprise of the fictitious state, Azaria, used in Exercise Two

Table 8 Nuclear weapons, deployment systems, and nuclear material possessed by Azaria in Exercise Two

Country of Azaria				
Delivery system	Available weapons platforms	Warhead type	Deployed (includes on-base storage)	Storage depot
12 Tactical mobile launchers × 1 warhead	12	HEU implosion	18	8
SRBM	12	HEU implosion	18	10
16 Bombers × 4 warheads	32	HEU implosion	42	14
Total	56		78	32
Material	kg	Location		
Pu	40	Component production facility		
HEU	500	Component production facility		

References

1. Dreicer M, Stein G (2013) Applicability of nonproliferation tools and concepts to future arms control. ESARDA Bulletin No.53:95–99
2. Allen K, Dreicer M, Chen C et al (2015) Systems approach to arms control Verification. ESARDA Bulletin No. 53:83–91
3. Chen CD, Dreicer M, Proznitz, P et al (2015) Systems Concept for Arms Control Verification. CGSR Workshop Report. https://cgsr.llnl.gov/content/assets/docs/arms_control_verification_report.pdf. Accessed 28 Feb 2019
4. INMM/ESARDA Working Group 2 Report—Arms Control. 8th INMM/ESARDA Workshop "Building International Capacity", Jackson Hole, Wyoming, USA, October 4–7, 2015. https://www.inmm.org/INMM/media/Documents/Presenations/8thINMM-ESARDAJointWorkshop2015/INMM_ESARDAWorkingGroup2ArmsControlSummary.pdf. Accessed 28 Feb 2019
5. Proliferation Resistance and Physical Protection Evaluation Methodology Working Group of the Generation IV International Forum (2011) Evaluation Methodology for Proliferation Resistance and Physical Protection of Generation IV Nuclear Energy Systems, Revision 6, GIF/PRPPWG/2011/003
6. Feary BL (2015) Challenges to Arms Control—Protection Sensitive Information while providing transparency. 8th INMM/ESARDA Workshop "Building International Capacity", Jackson Hole, Wyoming, USA, October 4–7, 2015. https://www.inmm.org/INMM/media/Documents/Presenations/8thINMM-ESARDAJointWorkshop2015/5T-B-Fearey_(LANL)_Challenges_to_Arms_Control.pdf. Accessed 28 Feb 2019
7. International Partnership for Nuclear Disarmament Verification (IPNDV). https://www.ipndv.org. Accessed 28 Feb 2019

The manufacturer's authorised representative in the EU is Springer
Nature Customer Service Centre GmbH, Europaplatz 3, 69115 Heidelberg,
Germany. If you have any concerns regarding our products, please
contact ProductSafety@springernature.com

Printed and bound by CPI Group (UK) Ltd, Croydon, CR0 4YY
30/09/2026
02225456-0020